Henry Dresser and Victorian ornithology

MANCHESTER
1824

Manchester University Press

Dedicated to my mother, for all her encouragement

Henry Dresser and Victorian ornithology

Birds, books and business

Henry A. McGhie

Manchester University Press

Published by Manchester University Press
Altrincham Street, Manchester M1 7JA

www.manchesteruniversitypress.co.uk

British Library Cataloguing-in-Publication Data
A catalogue record for this book is available from the British Library

ISBN 97 8 1784 99413 6   hardback

First published 2017

Typeset in Bembo
by R. J. Footring Ltd, Derby

Contents

List of illustrations and tables

Plates

Figures   *page*

Appendix images 268–78
All images in appendix 1, 'Birds mentioned in the text', are from
Dresser, H. E. (initially Sharpe, R. B. and H. E. Dresser) (1871–82),
A History of the Birds of Europe (from the Biodiversity Heritage Library,
digitised by the Smithsonian Institution Libraries,
www.biodiversitylibrary.org)

Tables

Preface

I have been fascinated by birds for as long as I can remember. I became interested in ornithologists of earlier generations, realising that the collections, diaries and letters they left behind in museums were a fantastically rich source for exploring both the distribution of birds in the past and changes in human relationships with birds over time.

On the day I started work at Manchester Museum (part of the University of Manchester), back in 2000, I opened a large cardboard box to discover diaries, letters and photographs that had belonged to Henry Dresser, one of the most influential ornithologists of the late nineteenth century. I was completely hooked. As I read through the diaries and letters, and came to know Dresser's collections of bird skins and eggs better, the idea of writing this book came to me. As the years went by, I found more and more of Dresser's correspondence in other museums and archives, which helped to fill in the blanks in his story and build up a picture of his life and relationships with other ornithologists.

Several serendipitous events brought new opportunities and made the project even more interesting and rewarding. A Russian colleague at Manchester Museum, Dr Dmitri Logunov, used his links to discover letters from Dresser (written in English) to a Russian ornithologist, Sergei Buturlin, preserved in the Museum of Local Lore, History and Economy in Ulyanovsk, Russia. The curators there provided us with copies of the letters, which I transcribed and Dr Logunov translated, allowing the curators in Ulyanovsk to understand their contents for the first time. We were fortunate to be able to participate in the Second Readings Dedicated to the Memory of Sergei Buturlin, held in Ulyanovsk in 2005. Dr Kirsten Greer contacted me when she was still a PhD student at Queen's University, Ottawa, in 2009. She was interested in ornithologists and the military, and the naturalists of New Brunswick. We developed a collaboration investigating the activities of Henry and his youngest brother Arthur, both of whom spent time in that province.

My work on Henry Dresser and on understanding birds (and ornithologists) was presented at seminars in Oslo as part of the project 'Animals as Things, Animals as Signs' (2008–12), generously funded by the Norwegian Research Council. This project resulted in the book *Animals on Display* (Thorsen *et al.*, 2013) and an exhibition, 'Animal Matters', in 2012 at the University of Oslo Library, curated by Bryndis Snæbjörnsdóttir and Mark Wilson. I was also extremely lucky to be invited to Spitzbergen in 2015 as part of a workshop on interdisciplinarity and the Arctic, again supported by the Norwegian Research Council. All of these projects have been enormously enriching, bringing together great collaborators. They have helped me develop my thoughts on the roles and potential roles of collections for exploring the past and shaping the future.

One last incident worth mentioning concerns an album that Dresser maintained, consisting of photographic portraits and letters from his scientific correspondents. I knew he had such an album from his correspondence and wondered what had become of it until it was sold at auction in 2006. I was fortunate to be able to acquire it on behalf of Manchester Museum and John Rylands Library, with the aid of a substantial grant from the PRISM fund. The album had been in private hands for almost a century so I was very lucky that it became available when I was writing this biography. Many of the photographs reproduced in this book come from the album.

When I started this project, I did not know just how interesting and satisfying it would be to write. I have been extremely lucky to have had many first-rate original sources to work with, including Dresser's unpublished diaries and photographs in Manchester Museum, free access to his bird and egg collections, and a great deal of information found in other museums, libraries and archives, not to mention a great deal of wisdom, help, encouragement and patience from colleagues in museums and libraries. This book has been an absolute delight to write and has led me down many paths that I never anticipated. It has occupied me for many evenings, weekends, and my commute to and from work. It is intended for anyone interested in birds and ornithology, travel and exploration, and the history of natural history. I hope that the reader finds something of enjoyment and interest in the twists and turns of the book.

Acknowledgements

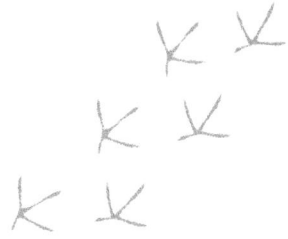

This book would not have been possible without the help of many, many people and I take this opportunity to thank them all for providing me with information, ideas, advice and support. I thank Manchester University Press for making the book possible and the Centre for Heritage Imaging and Collection Care, University of Manchester Library, for assistance with illustrations. I am grateful to Ralph Footring and Martin Hargreaves for their help in producing the book.

My colleagues at Manchester Museum have been very encouraging, especially Dr Nick Merriman and Tristram Besterman. Dr Dmitri Logunov has collaborated with me on Russian ornithologists and Paddy Moss translated a Finnish passage for me. Other colleagues throughout the University of Manchester, particularly librarians of the University Library and of John Rylands Library, have been very supportive (and patient).

Dr Kirsten Greer of Nipissing University has collaborated with me on the Dressers and I am very grateful to her for her help and encouragement. I am also grateful to the late Prof. John Pickstone, as well as Prof. Matthew Cobb (both University of Manchester), Prof. Patience Schell (University of Aberdeen), Dr Sam Alberti (National Museums Scotland) and Prof. Tim Birkhead (University of Sheffield) for encouragement and useful discussion. Andrew Cole was very helpful with information relating to his books on egg collecting (Cole, 2006; Cole and Trobe, 2000). Stan da Casto and Prof. Joel Weintraub (California State University) helped me understand Henry Dresser's time in Texas and his association with Adolphus Heermann. Prof. Nigel Collar (University of Cambridge) and Philip Hall helped with information on Robert Swinhoe and Dr Nicholas Keegan helped me with information on the British consular service. Prof. Zbigniew Wszolek (Mayo Clinic, Jacksonville, Florida) helped me with information on Benedykt Dybowski. Dr Karl Schulze-Hagen helped me with information on Joseph Wolf. I thank Paul Hobson for very useful discussion on Henry Seebohm. I thank Alexander Buturlin of Moscow for inviting Dr Dmitri Logunov and me to participate in the Second Readings Dedicated to the Memory of Sergei Buturlin – his father – held in Ulyanovsk in 2005, which made for a particularly memorable experience.

I thank my collaborators from the project 'Animals as Things, Animals as Signs', particularly Prof. Liv Emma Thorsen and Prof. Brita Brenna (University of Oslo), Prof. Karen Rader (Victoria Commonwealth University, Richmond, Virginia), Dr Nigel Rothfels (University of Wisconsin–Milwaukee), Prof. Brian Ogilvie (University of Massachusetts) and Dr Adam Dodd (formerly University of Oslo, now University of Queensland, Australia). Our collaborations helped me get my thoughts in order for the present book, and have been enormously stimulating. Prof. Thorsen also read an entire draft of the book, and her comments were extremely useful. I thank Prof. Marit Hauan (Tromsø University Museum) and Prof. Lena Aarekol (Polar Museum, Tromsø) for giving me the opportunity to participate in workshops and seminars in Spitzbergen and Tromsø exploring Arctic narratives, exploration and collecting, including some of my work on Henry Dresser.

I am grateful to Jim Dresser of the Dresser Family Worldwide Genealogy Center, USA, for assistance of various kinds, to Richard Tetley for assistance in finding information on Joseph Dresser Tetley and to the late Richard Walmisley for assistance with information on Eleanor Walmisley Dresser. I thank John Hickman, Richard Lines (of the Norwood Society) and Stephen Oxford for assistance with information on Norwood, and Nicholas Reed for very useful discussion on the time spent in Norwood by the artist Camille Pissarro.

I am most grateful to Mike Whittam for his help in developing this book – discussing ideas, reviewing chapters – and for general tolerance and encouragement.

Museums

I thank colleagues in many museums, both in the UK and around the world, who have provided me with information on collections and collectors. I especially thank Bob McGowan (National Museums Scotland) for help in many, many ways and for lots of stimulating discussion about Victorian collectors, and Stephen Moran (formerly of Inverness Museum) for general encouragement. I thank Dr Clem Fisher (World Museum Liverpool) for helping in many ways with this project, and Tony Parker for information on specimens in the collection of World Museum Liverpool. I thank the curators of the bird collections at the Natural History Museum (NHM, London and Tring), Dr Robert Prŷs-Jones, Mark Adams, Dr Jo Cooper, Douglas Russell and Hein van Grouw, for help in many ways. I thank Robert in particular for sharing with me his unpublished notes on the relationship between Richard Bowdler Sharpe and Henry Dresser. Jeremy Adams (formerly of the Booth Museum of Natural History, Brighton) helped me with information on Edward Booth and on Thomas Parkin's collection. Dr Michael Brooke and Matt Lowe (Cambridge University Museum of Zoology) provided information on collection holdings. Igor Fadeev (Moscow State Darwin Museum), Tony Irwin (Norwich Castle Museum), Maggie Reilly (Hunterian Museum, University of Glasgow) and Dr Tony Walentowicz (formerly Chelmsford

Museum) have all been very helpful in providing information on collections in their museums. James Hamill was very helpful regarding Arthur Dresser's manuscripts in the British Museum.

I thank all those curators who have helped me in my searches for specimens of Ross's Gull and other birds collected by Sergei Buturlin (see McGhie and Logunov, 2005, for a full list).

Libraries and archives

I thank the following individuals and institutions for assistance in locating archives and providing copies, in no particular order: Louise Clarke, Daisy Cunynghame, Hellen Pethers and John Rose (NHM Library and Archives); Dana Fisher (Ernst Mayr Library, MCZ, Harvard University); Patsy Hale (Harriet Irving Library, University of New Brunswick, Fredericton); Lucy Jardine (Provincial Archives of New Brunswick, Fredericton); Daryl Johnson and Peter Larocque (New Brunswick Museum, St John); Eleanor MacLean and Dr Richard Virr (Rare Books and Special Collections, McGill University Library, Montreal); Sandra Taylor (Lilly Library, Indiana University); the late Peter Meadows and Adam Perkins (University of Cambridge Archives); staff at the Canadian National Archives (Ottawa); Lindsay Ould (Croydon Borough Archivist) and staff of the Museum of Croydon; Heljä Strömberg (Central Archives for Finnish Business Records (ELKA), Mikkeli, Finland); Andrew McDougall (National Museums Scotland Library); Tad Bennicoff (Smithsonian Institution Archives, Washington, DC); Laura Outterside (Royal Society, London); Gina Douglas (Linnean Society, London); and staff of the University of Glasgow Archives.

I am especially grateful to Olga Borodina, Curator of the Museum of Local Lore, History and Economy, Ulyanovsk, Russia, for providing Dr Dmitri Logunov and me with copies of letters from Henry Dresser and Otto Ottosson to Sergei Buturlin.

Excerpts from correspondence from Henry Dresser to Alfred Newton are reproduced by permission of the Syndics of Cambridge University Library. Excerpts from correspondence from Henry Dresser to John Harvie-Brown are reproduced by kind permission of National Museums Scotland. Excerpts from letters in the Blacker-Wood Autograph Letter Collection are reproduced courtesy of McGill University Library (Rare Books and Special Collections). All material from the Natural History Museum Library and Archives (London) is used here by permission of the Trustees of the Natural History Museum.

Conventions

Common names for birds that have been recorded in Britain are taken from Harrop *et al.* (2013). For other birds, names are taken from the IOC World Bird List, Master IOC list v5.4, www.worldbirdnames.org/ioc-lists/crossref (accessed 8 April 2017). Common names together with scientific names are given in appendix 1. Species names are given initial capitals; group names (e.g. thrushes, falcons) are not.

Place-names follow those used by Henry Dresser in his correspondence and publications, followed by their modern equivalent where different.

Introduction

This is the story of the life and activities of Henry Dresser (1838–1915), one of the most productive English ornithologists of the mid–late nineteenth and early twentieth centuries, but it is not just his story. It is an exploration of ornithology in Britain during a period when the subject changed dramatically in many ways. Dresser came from a wealthy Yorkshire family and had a very early adventurous life, travelling widely on business in Europe, New Brunswick and even to Texas during the American Civil War. Following on from this adventure he settled into business in London in the timber and iron trades. He had a lifelong interest in birds and built enormous collections of bird skins and eggs, which were – and still are – among the finest of their kind (see plate 1). These collections formed the basis of over 100 publications on birds, notably the enormous and very beautiful *A History of the Birds of Europe*, issued in eighty-four parts during 1871–82 and the standard work on the subject for many years (see plate 2).

Writing in 1959, Philip Manson-Bahr, an English ornithologist (and expert on tropical medicine), recalled meeting Dresser many years before, when Dresser was nearing the end of his long ornithological career:

> Dresser was possessed of demoniacal energy, boundless enthusiasm and immense application. When he took up a subject, he saw it through regardless of any difficulties or obstructions. He had a striking appearance which was arresting in any company. His countenance was pale, clean-shaven with sharp nose and striking intense features. The oddity of his appearance was heightened by a rather ill-fitting wig, because he was completely bald. He had a rapid method of elocution and the words poured from his mouth and almost stunned his audience. He was in fact quite a character in an age of individualism. He possessed vast ambitions and was visibly proud of his achievements in ornithology.
>
> A born collector, he would converse on birds for hours to the exclusion of all other topics and he was most ambitious in acquiring valuable eggs, [bird] skins, and other rare specimens. (Manson-Bahr, 1959: 59)

Dresser was one of the prime movers in ornithology; he witnessed and played a part in many of the transformations that took place in the discipline. Who was he? What did he do? Why? How? These are the questions that this book is concerned with. To answer them requires an understanding of his working life as a businessman, ornithologist and publisher as well as the relationship between these different activities during what Barber (1980) calls 'the heyday of natural history'. His success in ornithology stemmed from his position within a web of related activities, including field collecting, cabinet collecting (where specimens were bought and exchanged), in scientific societies and society more generally, in publishing and with his readership. These were underpinned by his success in business – which provided the capital to support his ornithological activities – and his position in society, which enabled him to mix with the most aristocratic of naturalists. Ornithology was a contact sport, as its devotees collaborated and competed with one another. An understanding of Dresser's relationships with his contemporaries is crucial to understanding his own story and activities. In answering these questions, this book is intended to explore the motivations and aspirations of someone who was, on the one hand, a most singular and successful individual and, on the other, a representative of a group of men with 'serious hobbies'. More generally, it is an exploration of the process by which baseline ornithological knowledge was created through the activities of individuals in networks, notably the private gentlemen-naturalists who dominated nineteenth-century ornithology.

Dresser was familiar with the leading naturalists of the day and he was a prominent figure in two of the most notable societies, the British Ornithologists' Union (BOU) and the Zoological Society of London. His life spanned a period that saw the development of scientific ornithological societies, scientific journals devoted to birds, the bird conservation movement and institutionalised museum collections in Britain, as well as the emergence of professional ornithologists and international standards for the scientific naming of species. An investigation into his life provides an insight into these changes and how scientific society was transformed from a pursuit of private dilettantes into a landscape of institutions, professionals and practices. During the mid-nineteenth century, private individuals such as Dresser formed the backbone of ornithology in Britain but as the century wore on the British Museum (Natural History) – usually known as the BM(NH) – rose to take control of the ornithological scene. Dresser was sometimes in open conflict with his peers and with some of those most closely associated with the BM(NH). These subjects are explored in order to fully understand the events that took place and the contribution that Dresser and others like him made to ornithology.

This book is based on a large body of previously unpublished archival material, including ten years' worth of Dresser's unpublished diaries (including diaries from his time in New Brunswick and during the American Civil War), letters and photographs in Manchester Museum (see figure I.1). There are 299 letters from Dresser to Alfred Newton, the leading ornithologist in late nineteenth-century Britain, in Cambridge University Library; over seventy letters to George Boardman in the Smithsonian Institution Archives (Washington, DC); twenty-two letters to Richard Sharpe in the Blacker-Wood Autograph Letter Collection at McGill University Library, Montreal; sixty-two letters to John Harvie-Brown in National Museums Scotland; and

I.1 Papers and photographs that belonged to Henry Dresser.

almost 100 letters to Sergei Buturlin in the Ulyanovsk Museum of Local Lore, History and Economy (Ulyanovsk, Russia). One notable source of information is the album of photographs of his scientific correspondents and their letters that Dresser collected, now in John Rylands Library at the University of Manchester. Dresser's published writings on birds form another invaluable source of information; they include large folio books and many scientific papers in journals. His collections of 7,200 bird skins and 6,000 eggs in Manchester Museum are another major source of information, which has been extracted from labels and Dresser's own catalogues of his collections.

Histories of ornithology

The history of ornithology and ornithologists has been the subject of a number of books, really beginning with Erwin Stresemann's classic *Ornithology from Aristotle to the Present* (1951, translated into English in 1975). More recent histories by Michael Walters (2003) and Peter Bircham (2007) deal with worldwide ornithology and ornithology in Britain respectively. David Allen's classic *The Naturalist in Britain* (1976) is remarkably broad in scope but as a consequence gives little detail about the activities of ornithologists. Paul

Farber (1982) studied the emergence of ornithology as a scientific discipline, notably the development of institutional structures and professionals. Mark Barrow's *A Passion for Birds* (1998) examines many of the same themes as the present book, from an American perspective. Professionalised ornithology in America is explored in Daniel Lewis's biography of Robert Ridgway, *The Feathery Tribe* (2012). Barbara and Richard Mearns have produced biographies of many ornithologists (1988, 1992, 1998, 2007). Some of the more prominent nineteenth-century ornithologists have been the subject of biographies, including John Gould (Sauer, 1982; Tree, 1991, 2004; Russell, 2011), Alfred Newton (Wollaston, 1921), Walter Rothschild (Rothschild, 1983, 2008) and Lord Lilford (Drewitt, 1900; Trevor-Battye, 1903). There is an industry based around John James Audubon and his books, including a number of biographies (e.g. Streshinsky, 1998; Hart-Davis, 2004; Rhodes, 2004). Birkhead *et al.* (2014) explored the development of modern ornithology and those who can be linked with current scientific practices, but this approach excludes many of those who were significant figures in their own time, including Dresser.

Thus far there has not been a detailed, critical biography of any of the 'industrialist bird collectors' such as Dresser. The present book has a lot in common with Endersby's biography of Joseph Hooker (2008), although there are some clear differences: Dresser was not in a paid position as a naturalist, and he was an ornithologist rather than a botanist.

Some published histories of ornithology have covered long periods of time, giving an illusion of continuity in the subject (such continuity is contested by Farber, 1982). Notably, hunting and collecting, science, conservation, and concepts of 'amateurs' and 'professionals' have sometimes been written about in ways that would not have been appropriate or even understood in the nineteenth century. Collecting activities have sometimes been sanitised and uncritically justified on the one hand or criticised on the other, in order to produce a single canon of ornithology that links people from the present with those from the past, an approach that is more palatable to modern conservation-minded sensibilities. This book explores the motivations and activities of one particular ornithologist who was both collector and conservationist, and who was active at a time when these now seemingly mutually exclusive activities first started to be disentangled. It also follows the story of an ornithologist who was effectively left behind as ornithology changed beyond recognition.

A cautionary word relating to original sources

This book relies heavily on personal diaries and private letters. I have accepted what is written in these at face value. Dresser's diaries and letters contain personal information about Dresser himself and about those around him. There is no reason to suspect that they contain anything but Dresser's impressions of his own experiences, albeit from a subjective viewpoint (he does not seem to have intended to have them published).

Victorian hobbies and natural history

Dresser's interest in natural history and his ability to follow it reflect a number of widespread movements in British society during the nineteenth century (see Allen, 1976; Barber, 1980; Lloyd, 1985; Armstrong, 2000). In the previous century, hobbies had largely been the preserve of the aristocracy as most people had little free time from work, travel was costly and books were expensive. By the mid-nineteenth century, the middle classes had expanded due to industrialisation, urbanisation, empire and colonialism. Many more individuals had the time and the money to indulge their particular interests, but what was their motivation? Idleness may have been the greatest enemy of any upright Victorian, but the Evangelical Revival dictated that, in order to be permissible, any leisure activity also had to be morally uplifting. This presented a particular dilemma for the upper and middle classes, who, being waited upon by servants, had large amounts of spare time to fill with socially acceptable activities. Natural history provided a suitable hobby, as natural theologians such as William Paley had argued that the study of nature was in effect the study of God's work (see Barber, 1980: 22). Barber writes that natural theology

> gave thousands of amateur naturalists an excuse to kill butterflies and uproot rare plants to their hearts' content.... Motives such as killing boredom might supply the real reason why so many people took up natural history in that period, but natural theology provided, as it were, the excuse, and since the Victorians never allowed themselves to do anything merely for fun, an excuse was essential. (Barber, 1980: 24–5)

Samuel Smiles' *Self-Help: With Illustrations of Character and Conduct*, published in 1859, set out many ways by which to lead a purposeful life, drawing on the life experiences of a variety of successful businessmen, scientists, artists and politicians. Smiles advocated application, taking advantage of opportunities, and developing one's character and broad-mindedness. The book became a best-seller and was reprinted many times through the late nineteenth century.

Natural history became a national obsession in Britain, as it did in Europe and America and, farther afield, in colonies and empires. Innumerable books on natural history encouraged people to read about the subject, from the cheapest pulp to luxurious folio volumes that only the wealthiest could afford. In Britain and America in particular, natural history was an activity enjoyed by all social classes (Allen, 1976; Barrow, 1998). If, as the maxim went, the middle classes talked about things while the working classes talked about people, then natural history was a suitable topic for any drawing room.

Part of the craze for hobbies was the development of a 'culture of collecting'. People collected all manner of things and natural history objects were particularly popular. These need not be collections of dead things and many houses had their Wardian case (a kind of miniature greenhouse that gave protection from air pollution), ferneries, aquaria and hothouse plants.

Natural history could be enjoyed as an indoor activity, occupying evenings at the microscope, or in preparing and arranging collections. Nature could be brought back home, bought in markets, swapped with other enthusiasts, grown in the garden, kept in aquaria and studied indoors. Plants were pressed and dried on herbarium sheets; insects were impaled on pins and kept in glass-topped drawers or in picture frames; tiny objects were diligently arranged on microscope slides. Shells, fossils, minerals, bird eggs and skins were arranged – often impeccably – in drawers in cabinets. Display cases of mounted birds decorated many houses. Thus natural history formed an excellent all-encompassing hobby. Weekend forays into the countryside were frequently accompanied by the butterfly net and killing jar, the vasculum (a tin case for keeping picked plants in a fresh condition), the hand-lens and other collecting equipment, including the gun for those in a position to own and use one. Back at home, spare time could be spent in studying, preserving, labelling and arranging new acquisitions into collections. An industry of suppliers of specimens, cabinets, equipment and literature developed to satisfy public demand (Allen, 1976: 141–57; Jardine *et al.*, 1996).

Society life

There were hundreds of clubs and societies throughout nineteenth-century Britain and many of these were concerned with natural history. For example, the Zoological Society of London met fortnightly in Hanover Square (with a summer recess) from 1830 onwards. Natural history societies offered people an opportunity to mix with their peers and superiors, and even to associate with prospective marriage material: women were eligible for full membership of the Zoological Society of London from 1827, for example, 'on the same terms and with the same privileges as Gentlemen Subscribers' (see Scherren, 1905: 25), and women also played a part in the Botanical Society of London (established in 1836) and the Entomological Society of London (established in 1833). However, for the most part, women played a small part (or were permitted to play only a small part) in natural history societies (Allen, 1976: 150–2; Allen, 1980). Natural history therefore offered an opportunity for social advancement, which was of course high on the minds of the socially ambitious middle classes whose activities so precisely mimicked those of their aristocratic 'superiors' (Allen, 1976; Barber, 1980: 37). Some religious sects, notably Quakers, found natural history particularly attractive. Members of leading Quaker families often intermarried, generating great dynasties that encompassed political ideology, religion and natural history interests. The Barclays, Gurneys and Backhouses were particularly prominent in this respect (Allen, 1976: 89–92) (see figure I.2). However, it can be difficult to disentangle cause from effect: shared membership of societies came as a consequence of existing relationships as well as being responsible for establishing new ones.

I.2 John Henry Gurney junior, from Henry Dresser's album of correspondents.

Little enough was known of natural history for anyone with the time and the inclination to make significant contributions to the subject. Many people had 'serious hobbies' for which they became well known in society, going far beyond what would now be regarded as leisure activities. Businessmen would pursue their subject of choice in the evenings and weekends – as well as from their offices, to judge from the frequency of correspondence that was sent from business addresses. Life in learned societies was an important part of the career of many business figures and great value was placed upon intellectual might. 'Amateurs' – not a particularly useful term in the absence of institutions and 'professionals' – were frequently prominent in the larger societies, sometimes devoting more time to their intellectual pursuits than to business. To take one example, Sir John Lubbock (1834–1913) was the senior partner in a London bank as well as a Liberal MP (responsible for the Bank Holiday Act of 1871). During his 'spare' time, living as Charles Darwin's neighbour (and protégé), he was an expert in insect physiology and behaviour, and President at one time or another of each of the Anthropological, Entomological, Ethnological, Linnean and Statistical Societies. Lubbock (figure I.3) has been described as 'the very exemplar of the high-minded, broad-minded City banker' (Kynaston,

82 PUNCH, OR THE LONDON CHARIVARI. [August 19, 1882.

HOME AND FOREIGN PIGEON-SHOOTING.

PERHAPS the notice of the HOME SECRETARY, in reading the foreign news, may have been attracted by a telegram from Amsterdam, stating that the Minister of Justice has issued an order prohibiting the pigeon-shooting matches which were to have been held the other day on the Rustenburg estate. Why cannot Sir WILLIAM HARCOURT likewise issue an order prohibiting the pigeon-shooting matches which are held at Hurlingham? This is no fool's question; for the HOME SECRETARY has power to withhold permission to perform experiments on living animals from scientific men, and it has been stated, as yet without contradiction, that he has actually refused vivisection certificates to several eminent physiologists. How inscrutable is the wisdom of the law which empowers him to hinder investigators from wounding rabbits or guinea-pigs, even for the advancement of medicine and surgery, but not to forbid idlers from shooting, crippling, maiming, and mangling doves, of which the pigeon-shooter stands charged at the bar of public opinion with causing the eyes to be gouged out previously, for fun! This inconsistency must be conspicuous to everybody outside of Earlswood, but qualified by quantity of reflective faculties to be an object at least as eligible for admission to that asylum as anyone in it.

MUDDLE LODGING-HOUSES. —The Peabody Buildings.

PUNCH'S FANCY PORTRAITS.—No. 97.

SIR JOHN LUBBOCK, M.P., F.R.S.

How DOTH THE BANKING BUSY BEE
IMPROVE HIS SHINING HOURS
BY STUDYING ON BANK HOLIDAYS
STRANGE INSECTS AND WILD FLOWERS!

SHAKSPEARE AND SHOP.

MR. PUNCH,

ARE you quite sure, Sir, that Mr. DUTTON COOK is exact in saying that "in SHAKSPEARE's time the Actors knew nothing of Benefits?" How goes the song in *As You Like It?*—

"Freeze, freeze, thou bitter sky,
Thou dost not bite so nigh
As Benefits forgot."

From the above showing, would it not rather seem that the Actors, whom SHAKSPEARE represented in the way of business, did indeed know something of Benefits, but something too little, and much less than they wished to know. Perhaps SHAKSPEARE, who speaks so touchingly of "benefits forgot," wished to signify that he would like to have them remembered by the patrons of the Drama; thus delicately inviting them to "remember the poor Player." Doesn't this conjecture suggest an association of ideas rather opportune just now à propos of Egyptian and Turkish requirements—Benefits and Backsheesh?

Yours truly,
COMMENTATOR.

FROM THE WELSH HARP.

THE Grand Old Minstrel Boy will not (it is feared) preside with harp and voice at the Eisteddfod. But if the Harpists want an extra Lyre—and a good big 'un too—here's a chance for the ex-War Correspondent of a certain, or recently uncertain Daily Paper!

I.3 Cartoon of John Lubbock, the 'banking busy bee', from *Punch*.

1994: 296). He was only one – albeit a particularly successful one – of many businessmen, including Dresser, who effectively had two 'careers', or a career made up of business and social elements. Notions of 'work' and 'leisure' do not apply in the same sense as they do today.

Social aspects of hunting and collecting birds

While natural history was a mainstream pastime, specialist subjects were associated with particular groups of people. Some, such as botany and shell-collecting, were enjoyed by both men and women. Ornithology was associated with the wealthiest of collectors, and was entirely dominated by men, as it often involved shooting and hunting. Dresser's photograph album of his scientific correspondents includes only two women among 266 photographs, both of whom were the wives of famous ornithologists (so explaining their inclusion in the album). Lucy Audubon (see figure I.4) was the wife of John James Audubon, who was famous for producing *The Birds of America*. Claudia

I.4 Lucy Audubon (wife of John James Audubon), from Henry Dresser's album of correspondents.

Hartert features in a photograph alongside her husband Ernst, a prominent museum curator (see figure 11.2, p. 189).

Ornithology had a moral stamp of approval as it took young men outdoors into the fresh air and promoted manly characteristics of vigour and hunting prowess. Charles Darwin wrote how, as a young man:

> How I did enjoy shooting, but I think that I must have been half-consciously ashamed of my zeal, for I tried to persuade myself that shooting was almost an intellectual employment; it required so much skill to judge where to find most game and to hunt the dogs well. (Darwin, 1958: 55)

Henry Dresser's younger brother Arthur, who did a lot of collecting in New Brunswick, wrote in a similar vein:

> The study of ornithology has many attractions, and will prove a great source of pleasure to all who have inclination and time to pursue it. This branch of Natural History proves deeply interesting to young men on account of having to use the gun so much and also the pleasure of being out in the fresh air.[1]

Many of the photographs in Dresser's album of correspondents show their subjects posed with guns, often in photographers' indoor studio settings, reflecting the image that they wished to present of themselves to others (see figure I.5).

Nineteenth-century ornithology in Britain was dominated by an elite group of (male) collectors. While many other European countries had long-established bird collections in museums, Britain lagged behind its Continental rivals (see Mearns and Mearns, 1998: 71–103). The natural history collections of the British Museum languished until the middle of the century when they were split off to form the British Museum (Natural History) – usually known as the BM(NH) – which finally opened in 1881 (Sharpe, 1906: 82–3; Knox and Walters, 1992). Richard Sharpe of the BM(NH) considered that the lack of a strong natural history museum had been part of the reason for the development of immense private collections of birds and eggs that was typical – and distinctive – of British ornithology and ornithologists. Reflecting on the state of ornithology in Britain in the early–mid-nineteenth century, he wrote:

> It was undoubtedly the want of management on the part of the Museum Curators that led to the formation of the great private collections of the nineteenth century. It was on these that all the sound ornithological work of this country was based, and no one cared to visit the British Museum, unless he were forced to do so for the purpose of examining some special type or historical specimen. (Sharpe, 1906: 84)

On a more local level, many natural history societies, as well as literary and philosophical societies, held their collections in museums that were more or less restricted to their members. Civic museums as we know them today were very much in their infancy, mostly dating after the Museums Act of 1845, which gave town councils the powers to establish them (Black, 2000; Yanni, 2005; Burton, 2010).

The practicalities of bird collecting

Mid-nineteenth-century ornithologists lacked the camera, the telescope (excepting fairly basic 'eyeglasses') or binoculars. Their most popular tool was the gun, for shooting birds; the American Elliott Coues once advised 'the double-barrelled shot gun is your main reliance … *get the best one you can afford to buy*' (Coues, 1874: 5, original emphasis). Others used their wallets and their wits to acquire dead or preserved birds from other collectors, in auctions and markets, and from anyone who could obtain good specimens. Books were, of course, an important tool, the most popular being the ubiquitous 'Bewick', 'Yarrell' and 'MacGillivray': Bewick's *A History of British Birds* (1797–1804) (which even gets a mention in Austen's *Jane Eyre*), Yarrell's *A History of British Birds* (1843) and MacGillivray's *A Manual of British Ornithology* (1840–42). Books were, for the most part, illustrated by black and white engravings that were of limited value for identifying birds. A collection of preserved birds was

I.5 Horatio Wheelwright (the 'Old Bushman'), from Henry Dresser's album of correspondents.

an indispensible tool for an ornithologist, something that could be referred to as an identification aid, even if it did take up a bit of houseroom. Being attractive, collections of mounted birds doubled up as household ornaments, as trophies of successful hunting trips and as specimens for natural history study.

Ornithology was a specimen-based discipline: bird specimens were studied closely to distinguish new species and to understand how the plumage of each species varied and developed over time. Specimens also provided solid evidence of the occurrence of particular birds in particular places. The maxim 'what's hit is history and what is missed is mystery', used by Horatio Wheelwright (the 'Old Bushman' – figure I.5) and many others, was frequently cited as a justification for killing rare birds, even into the twentieth century. In the days before photography, the bird itself was the only possible material evidence that could be obtained to prove the occurrence of rare species. For anyone who aspired to become an ornithologist, a collection of bird skins or their eggs was *de rigeur* if they wished to be taken seriously.

Birds required comparatively complicated preservation methods, involving the removal of the skin from the body, cleaning tissue and fat from the skin and making an artificial body, whether posed or preserved as a flattish study skin (that lay on its back, to fit easily within a drawer in a collections cabinet, with data labels attached to the legs and safe from the ravages of daylight and insect pests). Mounted specimens were common enough in the homes of 'ordinary' people, but ornithologists mainly collected study skins: specimens

11

that were remarkably uniform in appearance as a result of conventions in their preparation. This uniformity enabled birds from different places and different times to be readily compared with one another, and exchanged between collectors. Study skins were also easier to transport than ungainly mounted birds.

Bird collecting had one particular drawback, shared with insect and plant collecting, in that bird skins and feathers were susceptible to being eaten by the larvae of various beetles and clothes moths. Bird collectors used 'arsenical soap', consisting of soap mixed with powdered white arsenic, salt of tartar, camphor and powdered lime. This was worked into a lather that was applied to the inner surface of bird skins, producing an effective deterrent against pests (Farber, 1977; Morris, 1993; Mearns and Mearns, 1998: 43). Arsenical soap had been invented by a French apothecary, Jean-Baptiste Bécoeur (1718–77), and popularised in Britain during the 1820s and '30s by Thomas Bowdich (Anon., 1820) and Captain Thomas Brown (1820) in manuals on taxidermy. Brown's *The Taxidermist's Manual* ran to twenty-seven editions, the last appearing as late as 1876 (Rookmaaker *et al.*, 2006). The spectacular rise in popularity in bird collecting that took place through the first half of the nineteenth century was attributed to the popularisation of arsenical soap and changes in gun technologies (Farber, 1977, 1980, 1982; Morris, 1993; see also Schulze-Hagen *et al.*, 2003; Steinheimer, 2005), although the step-by-step instructions for preserving bird skins given in collecting manuals may have been equally important. The invention of new gun technologies, such as special 'dust shot' (fine lead shot that reduced the level of damage to specimens) for use in shotguns, certainly contributed to the increase in popularity of bird collecting, as did the growth of travel (Mearns and Mearns, 1998: 51–3; Morris, 2010).

By 1870 or so, there were numerous manuals available (in addition to reprints of older books) that covered all of the features of collecting: sourcing specimens, preserving them, arranging them, labelling them and so on. To give one example, Edmund Harting's *Hints on Shore Shooting* (1871) described methods for keeping bird corpses from getting soiled by blood and other bodily fluids, skinning and preparing study skins, and poisoning bird skins with the ubiquitous arsenical soap, giving useful step-by-step instructions. In America, the Smithsonian Institution published notes on how to collect and preserve animals to ensure that budding collectors had the necessary skills to provide the Institution with good-quality specimens (see chapter 6). Elliott Coues issued *Field Ornithology* in 1874 with detailed instructions on shooting birds and cleaning guns, selecting a good dog to help with collecting, killing injured birds, transporting dead birds, taxidermy, labelling and making observations; the book was reprinted many times.

Serious collectors maintained a catalogue of their collection to manage their specimen-related information. Edmund Harting issued a blank catalogue in 1868 that collectors could use to list their own collections, entitled a '*Catalogue of ... in the Collection of ...*' (Harting, 1868). This led one reviewer to comment that 'a collection without a catalogue is reduced to half its real value' (Anon., 1868). A number of other publications were intended to be cut into labels with which to label collections in cabinets. Edward Newman published such a species list in 1845 for British vertebrates (Anon., 1845).

Natural history periodicals were an important means of keeping up to date with what was happening and two in particular were popular among bird enthusiasts: the *Zoologist*, founded in 1843 (issued monthly; it was amalgamated into the journal *British Birds* in 1916), and *The Field*, which began circulation in 1853 as a weekly newspaper (now a monthly magazine) on field sports and country matters (Bourne, 1988). Many natural history clubs and societies published their own proceedings and journals.

How many species, and what to call them?

The discovery, categorisation and description of nature were fundamentally important during the nineteenth century, as the world was opened up as a result of travel, exploration, imperialism and colonialism. It would have been thought entirely appropriate that the natural productions of the Empire should be sent back to the seat of Empire, just as if they were commercial commodities. The collecting and transporting of natural productions were important elements of 'scientific imperialism', a movement that combined scientific enquiry, often in the form of natural history, with economic growth and political relations. For many aspiring scientific travellers (cum collectors and writers), the formation of a collection was of the utmost importance. Collecting specimens was a necessity as they would need to be compared with others in Western museums in order to establish what species they belonged to – or, more excitingly, whether any belonged to new species.

If a naturalist came across ('discovered') a species that they thought was unknown to science, the accepted scientific practice was that it had to be described in a book or scientific journal, given a unique two-part scientific name (for example *Turdus merula* for the Blackbird, *Turdus* being the genus, to which a number of similar species belong, and *merula* referring only to this particular species) and a description that set out the characteristics that can be used to separate it from similar species. Towards the end of the nineteenth century, some naturalists began to give a third part to scientific names, referred to as a trinomial, to recognise local variations (subspecies) within particular species. The specimens that were used as the basis of such descriptions, called type specimens, were considered to be particularly valuable and were deposited in private collections and museum collections for indefinite preservation (Farber, 1976, 1982; Johnson, 2005). In earlier times, when preservation techniques were less well developed, illustrations of new species were especially important (as specimens often deteriorated), and artists accompanied many of the great voyages of discovery (Farber, 1977, 1982). Most type specimens were kept in European and North American museums or private collections during the nineteenth century (with the notable exception of the Indian Museum, which had been established in Calcutta as early as 1814). Consequently, any

13

supposedly 'new' species needed to be assessed in the West, by comparing them with type specimens.

Almost as important as identifying new species was the delineation of the distribution of each species. Every specimen could potentially add empirical knowledge on the geographical range of species, so collections need not be confined to rare things and it was better to build up a representative selection of all the species found in a particular area (especially when travelling in remote regions). As it became easier to acquire specimens, 'serial collecting' (acquiring series of each species) became increasingly common, so that serious collectors were no longer content to possess one or two specimens of each species. Taken together, these developments saw scientific travelling reach new heights during the mid- and late nineteenth century. Vast amounts of information and specimens were accumulated as a consequence. One of the less recognised tasks of late nineteenth-century ornithologists was the synthesis of great amounts of information into a coherent whole. For instance, scientists, travellers and collectors named species that they thought were new over and over again, so that a single species could have many different names, causing great confusion. While less exciting than discovering new species, this synthesising of information was just as important a task.

Note

1   National Archives of Canada, Ottawa, Microfilm reel A-1536, *Birds of Canada* by A. R. Dresser, Preface.

1 Family background and early life

Henry Dresser was born in 1838, but our story really begins three generations earlier, when his great-grandfather, Joseph Dresser II (1737–1809),
took on the management of a corn mill on the River Swale, near Topcliffe
in North Yorkshire. The mill, five storeys high and twenty metres long (see
figure 1.1), was the source of great wealth for the Dressers and Henry's grand-
father, Joseph Dresser III (1770–1846), owned much of Topcliffe (Graham,
2000: 88–90).[1] Sometime around 1820 Joseph opened a private country
bank – one of a thousand in England at the time – in the centre of Thirsk,
eight kilometres from Topcliffe. He used his depositors' funds to run Topcliffe
Mill and controlled the sale of the produce in local markets (Phillips, 1894:
263; Sykes, 1926: 99).

Henry Dresser's father, also named Henry (1803–81), was a second son.
He set out in business on his own, working in Hull with a shipping merchant,

1.1 Topcliffe, the waterfall and mill, c.1955.

Robert Garbutt, who mainly dealt in timber from the Baltic.[2] Henry married
Garbutt's daughter Eliza Ann (1806–89) in 1824 and the young couple had
their first child, Ann Eliza, the following year. They moved back to Thirsk
so Henry could work in the family bank. The couple lived in the bank and
had five more children there. The first four of these children were all girls:
Rebecca (born 1828), Emily (born 1830), Sophia (born 1834) and Caroline
(born 1835). It is their next child who forms the subject of this book: Henry
Eeles Dresser was born on 9 May 1838, and was named after his father and
his grandmother's (Rebecca Eeles) side of the family; he was christened on
28 June 1838.

Henry senior's career developed rapidly at the bank of Dresser and Co.: he
was a partner by 1830, along with his father and his brother-in-law, William
Tetley (Twigg, 1830). The bank was taken over by the newly established Yorkshire District Banking Company in 1835, and Henry became the manager. He
moved his family to Leeds in 1840 when he became the general manager of
the Yorkshire District Banking Company head office.[3] In 1843 he became the
first manager of the Yorkshire Banking Company, which was formed from the
older company (White, 1840: 533; Phillips, 1894: 263, 414–17; Sykes, 1926;
Pemberton, 1963). During their five or so years in Leeds, Henry and Eliza
Ann had three more children: Joseph (born 1842), Frederick (born 1843) and
Arthur (born 1845).

The move south

Like many wealthy families of the mid-nineteenth century, Henry senior and
his family moved to London in search of fortune in the rapidly expanding
city, arriving there in 1846. Other family members went further afield: Joseph
Dresser Tetley (1825–78), Henry Eeles Dresser's ambitious cousin (son of
William Tetley of the Thirsk Bank), emigrated from North Yorkshire to New
Zealand in 1857 (Loftus, 1997; Hunt, 2001). Henry senior moved his family
to a property twenty-one kilometres south of London in Lock's Bottom (now
Locksbottom) near Farnborough, bought for 'the life interest of a gentleman',
in other words for the duration of his life. The property, Chalk Farm, was a
seventeenth-century 'genteel family residence' with 158 acres of farmland,
woodland and labourers' cottages,[4] together with 'finely timbered and taste-
fully displayed pleasure grounds, kitchen garden and stabling'; Henry senior
renamed the house Farnborough Lodge.[5] Previous occupants included the
banker Sir John Lubbock (1774–1840) and the father-in-law of Sir Robert
Peel (Bavington Jones and Pike, 1904: 36). Soon after moving into their new
home, Henry senior and Eliza Ann had another child, Clara, in 1847. The
Dressers lived a comfortable life at Farnborough Lodge, looked after by a team
of servants and farmhands.[6] They remained there until 1858, when Henry
senior moved the family to 107 Westbourne Terrace in the West End of
London, close to Hyde Park (he leased out Farnborough Lodge and the farm).

The West End had become a fashionable district for the rich and the Dressers lived among merchants, MPs and lesser ranks of the aristocracy (Kynaston, 1994: 140; *Kelly's* Post Office directories). The Dressers' household included a butler and a sixteen-year-old page (as recorded in the 1861 Census).

Henry senior established himself as a commission merchant in 1848, based at 14 Great St Helen's in the heart of the City, trading as Henry Dresser and Co. He ran a major operation (his offices had their own money vault) and worked in association with his father-in-law and former partner, Robert Garbutt of Hull.[7] Dresser and Co. received cargoes of timber from traders in the Baltic and New Brunswick; they sold these in London in return for a commission on the price of the sale. They also arranged for cargoes of goods to make the return trip, ensuring that the ships always went with full holds, to help generate money. To give some idea of the scale of Henry senior's business, one timber deal alone would have made him the equivalent of £600,000 in modern terms. One of his main clients was Hackman and Co., a leading firm of merchants based at Vyborg (Viipuri in Finnish, then in southern Finland but now in Russia).[8] Dresser and Garbutt had very close relations with the Hackman family and hosted their sons a number of times during the 1850s as apprentices, in order to prepare them for business (Tigerstedt, 1940, 1952; Mead, 1968).

In the same year that Henry senior established himself in London, he purchased a large lumber and sawmill business near St John in New Brunswick (Thompson, 1980: 6). The Baltic might have been closer, but there was a trade tariff to promote trade with the British colonies in North America and to ensure that Britain had a safe supply of timber in the event of war in Europe. The values of Baltic and colonial timber varied greatly from year to year, meaning that their relative values were highly volatile (Wynn, 1981: 31).

Dresser's education

Henry senior's business and aspirations had a great influence on the course of his eldest son's life. Young Henry was first sent to a private school in Lewisham in 1847, when he was nine years of age.[9] However, if he was to take over the reins of business in due course, he would require a quite different education. Young Henry was sent abroad, alone, at the age of fourteen, to be schooled in the same manner as the sons of the Hackman family. This was a common enough practice among merchant families at the time and *The Times* carried many adverts offering 'commercial education' abroad. Henry was schooled in Ahrensburg, near Hamburg, from 1852 to 1854, to learn German. He spent the years 1854–55 being educated in Sweden (and Swedish) at Gefle (now Gävle) and Uppsala. As a result of this 'commercial education', Henry became fluent in German, Swedish, Finnish, Danish, Norwegian and French; he could also speak a smattering of Spanish and Italian but never learnt Russian. This form of schooling singled Henry out from his brothers and sisters in a number of ways: he was physically isolated from them for a number of years, although he was sometimes home for Christmas and holidays. His outlook would have been heavily influenced by those with whom he lived and studied. His brothers had more conventional educations in England, although still with an eye to business. Joseph and Frederick both attended Merchant Taylors' School (in the City of London) between the ages of around fourteen and fifteen (around 1855–58); Frederick was also educated at Cheltenham College from 1859 (Hunter, 1890: 179; Anon., 1904c; Hart, 1936). Less is known of Henry's sisters' schooling, suggesting that their education was down to their governess; his youngest sister, Clara, was educated at a small boarding school in Blackheath.

By the time he went to Ahrensburg, Henry was already a devoted naturalist and collector: he had a keen interest in field sports and could already prepare bird skins reasonably well. His first specimens were trophies of his own shooting and collecting adventures, but even by the age of sixteen he had begun to build up a comprehensive collection of European bird skins and eggs (according to a biographical note that Henry probably wrote himself in 1909). After shooting birds (or otherwise acquiring corpses), he would skin them and fill the 'empty' skin with soft material to produce a natural-looking shape. To prevent his precious specimens from being eaten by insect pests, he painted the inside of the skins with lather from a bar of arsenical soap, quite a serious undertaking for a teenage naturalist. Henry sometimes mounted his specimens by inserting wires into the wings and legs to pose them in lifelike positions and gave them expensive glass eyes. He was evidently quite skilled in this: he helped to mount birds at the Gothenburg Museum, which he visited when he was on his way back to England. More commonly, he prepared skins into 'study skins', with the bird laying on its back, without glass eyes, and with data labels tied to the legs. Henry probably kept his study skins in some kind of wooden cabinet or case, where they would be protected from light (which caused colours to fade), dust and pests, and for ease of transport.

As well as collecting bird skins, Henry was an avid egg collector. He used a small handheld conical drill, called a rose drill, to grind a neat circular hole in the side of an egg. Once the hole was drilled, a small pipe (usually made of brass and narrower than the hole in diameter) was inserted into the hole and the contents were forced out by blowing into the pipe. Once the egg shells were clean and dry, Dresser would have laid his treasures out in a cabinet, nestling in cotton wool. From these childhood exploits, Henry developed to become one of the leading ornithologists of the nineteenth century.

Work abroad, 1856–62

During 1856–62, Henry Dresser was sent three times to Finland and twice to New Brunswick, to learn the ropes of business (see figure 1.2). In 1856, he worked at Hackman and Co. in Vyborg; his duties involved a lot of travelling and he took a two-month tour of all the towns from Vyborg to Uleåborg

1.2 Photograph of Henry Dresser aged around twenty.

(Oulu in Finnish) in northern Finland. In March 1857 he travelled all round the Baltic coast; he travelled to Marseilles and Italy the same year, where he contracted 'malaria' (as he described it), which laid him up until March 1858. Once he was well again, he returned to the Baltic and travelled round the Finnish and Swedish coasts assessing timber.

Whenever work allowed, Dresser would squeeze in some time for ornithology and collecting. A small pocket diary that he kept during 1856–58 – beautifully written in his characteristic low, looping writing – gives details of his day-to-day collecting activities, his meetings with local naturalists and brief notes on the birds he saw.[10] He had a high social standing so, although only a young man (of nineteen), he was able to befriend the most notable ornithologists and he was certainly confident enough to approach them. These included Dr Edvin Nylander (1831–90) of Uleåborg; Magnus von Wright (1805–68) (a famous wildlife artist and author) and Professor Evert Julius Bonsdorff (1810–98) of the Institute of Anatomy in Helsingfors (now Helsinki). These were the perfect companions from whom to learn about the birds of the country and with whom to exchange eggs. Dresser also encouraged students, peasants and forest workers to bring him dead birds and eggs, paying them for their troubles. It is clear from his diary that wild birds of all kinds were systematically hunted down, trapped and killed by local people, for amusement and presumably also for the pot.

During his second trip to Finland, in July 1858, Dresser made his first major scientific discovery. He and two friends had sailed to Sandön, a small island close to Uleabörg, to enjoy a couple of days' bird collecting. There they discovered a breeding pair of Waxwings, beautiful and unusual birds that visited Britain

1.3 Mounted Waxwing chicks and nest, collected by Henry Dresser in Finland in 1858.

in winter. Dresser found the birds' nest, which contained five chicks and an infertile egg (see figure 1.3). This chance discovery would have a great impact on the development of his scientific career (discussed below and in chapter 4).[11]

During his 'apprentice years', Dresser made two trips to New Brunswick. The first was in June 1859, when he took a shipment of goods for the store at his father's sawmill. Dresser was also to oversee tree-felling operations and the shipment of timber back to Britain – quite a responsibility for a young man of twenty-one years of age. After sailing to St John (the largest town in New Brunswick), Dresser travelled the twenty-two kilometres to his father's property, the Inglewood Manor at Musquash (see figure 1.4). The Inglewood Manor consisted of 32,000 acres – 129 square kilometres – of spruce, larch,

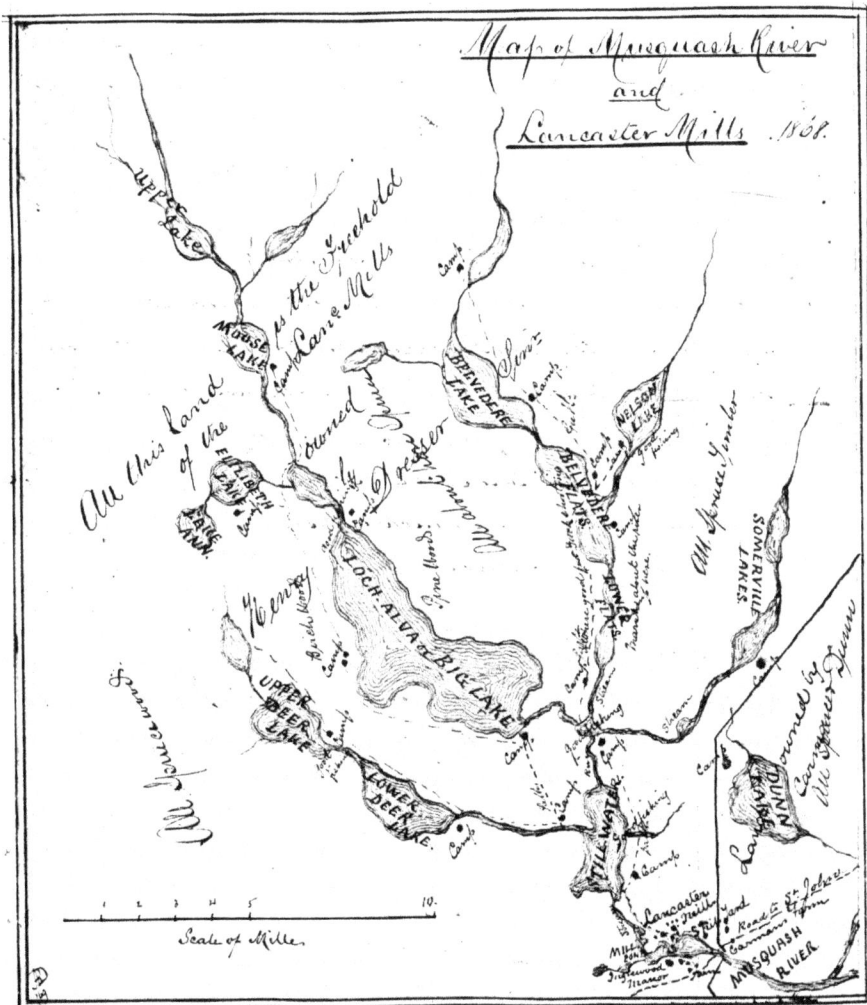

1.4 'Map of Musquash River and Lancaster Mills, 1868' by Arthur Dresser.

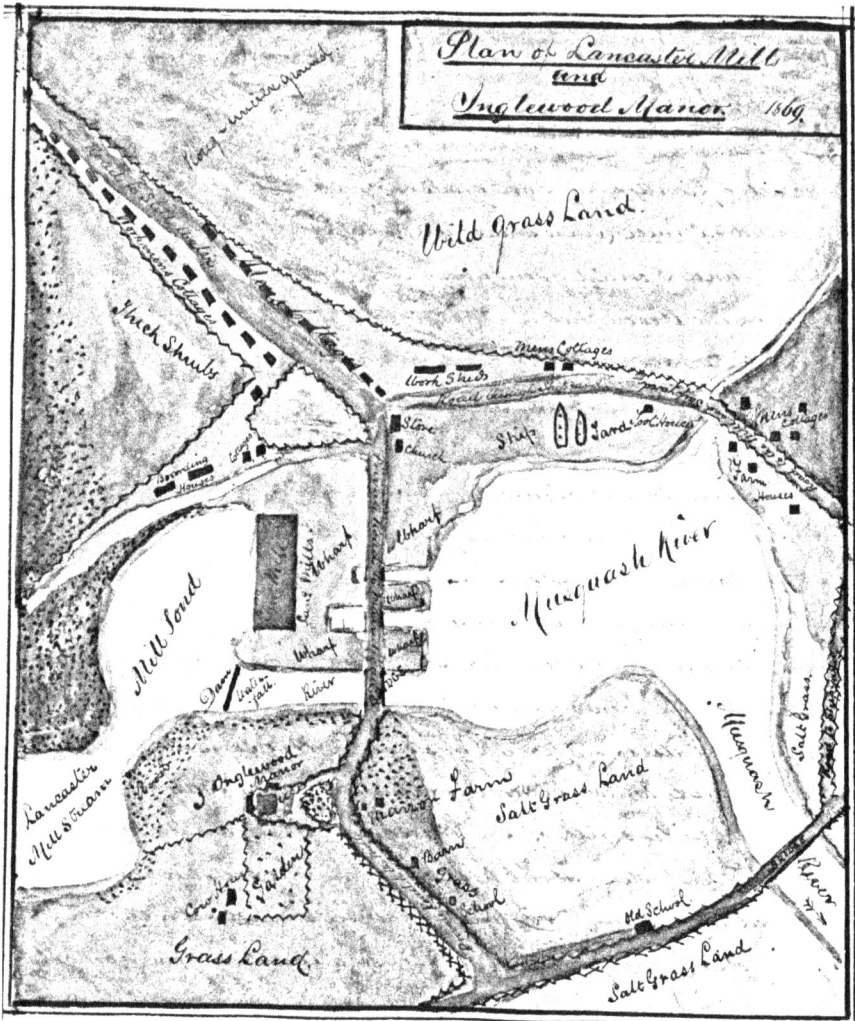

1.5 'Plan of Lancaster Mills and Inglewood Manor 1869' by Arthur Dresser.

cedar and birch forest. Twenty-one lakes were scattered over the property; streams from the lakes flowed eastwards to join the Musquash River where the Lancaster Mills were situated. There were a number of other sawmills in the area, but the Lancaster Mills were among the largest (Thompson, 1980: 6; Acheson, 1985: 14) (see figure 1.5).

In addition to the great sawmill, the estate included a mansion house, workers' houses, a church, a schoolhouse, a store and office, a farm and farmland, and a shipyard; the buildings alone had cost over £25,000 to develop, equivalent to £2 million in modern-day terms.[12] When the estate was first established, the developer had renamed many of the geographical

Lancaster Mills, from the Manor House). winter 1870.

1.6 'Lancaster Mills from the Manor House winter 1870' by Arthur Dresser.

features after places and characters in Walter Scott novels, possibly because many of the local people were Scottish and Irish immigrants. The estate was named Inglewood Manor; the lakes included Lochs Robin Hood, Friar Tuck, Sherwood, Little John and Alva. Dresser lived with his uncle Henry Garbutt (1817–95) and his family in the two-storey wooden mansion house. Dresser's father had set Garbutt up as the manager of the estate in 1848, with full power of attorney (Allen, 1861). He lived in the style of a colonialist country squire along with his wife and two children.

Work on the estate consisted of logging through the winter and moving the logs onto the frozen lakes. When the spring thaw set in, the logs floated downriver to the mill (see figure 1.6). The water from the streams was dammed at the mill to drive two enormous water wheels, eight metres in diameter, that were linked to a variety of belts and saws that cut the wood. The mill could cut about 274,000 metres of timber per month into various sizes, the bulk of which were 'deals' (long planks) and battens (Wynn, 1981: 55). Two or three ships were built each year at the shipyard with timber from the forest; these in turn would have brought some of the timber back to Britain. Work in the office consisted of overseeing logging contracts, paying and arranging the workers' wages and maintaining their credit accounts.

Dresser mainly worked in the store that sold goods to the workers. They lived on a system of credit, running up tabs at the store that were deducted from their wages when they were paid for cutting and moving timber. As the son of the owner, Dresser was practically an aristocrat and he kept aloof from the workers. When not working, he spent most of his time hunting and collecting birds and he had plenty of opportunities to do so. The lakes and rivers were full of trout; salmon and lobster could be caught on the Musquash estuary. Grouse were abundant in the forest and the shore was alive with ducks, geese and waders in winter. Deer were plentiful and Moose were found oc-casionally. Smaller predatory mammals such as Otter, Mink, Skunk, Puma and Lynx were common and there were bears in the forest. Wild birds and animals were an important part of people's diet, and Dresser's diary records that he ate American Robins (recording that they were tasty after they had been eating blueberries) and American Bitterns, as well as the more usual ducks, waders, grouse and other gamebirds. He found the Passenger Pigeon (now extinct) to be very common.

Over the course of a year, Dresser encountered just over 100 species of birds and kept notes on them in his pocket diary.[13] There were beautiful Great Northern Divers on the lakes, Bald Eagles along the shore and many species of small birds in the forests. Dresser's hunting focused on ducks (particularly Buffleheads, which he called 'Dippers') on the rivers, and Ruffed Grouse ('Birch Partridges') in the forest. He consulted everyone he encountered on the birds of the district, including George Thomas, the lighthouse keeper at nearby Point Lepreaux (now Point Lepreau), and Robert McCawley, who worked in the mill store alongside Dresser; both men provided him with many dead birds for his collection. Dresser had a young live Bald Eagle for a while, which was looked after by a fisherman until his wife dispatched it with a broomstick for killing her chickens.

Early ventures in ornithological society

The second half of 1860 and early months of 1861 were spent working in London, excepting September and October, which consisted of a two-month family holiday with relatives in St Leonards-on-Sea in Sussex. While he was there, Dresser enjoyed shooting trips to the shore with Edward Booth (1840–90), a young man near Dresser's age who became famous (infamous) as a collector and an author (see figure 1.7). Dresser first became acquainted with Alfred Newton (1829–1907), a prominent naturalist from the University of Cambridge and a leading figure in ornithological society, in August 1860 (see figure 1.8). Newton had a fanatical interest in birds and especially their eggs; he was no doubt tempted to correspond with Dresser because of his discovery of the breeding Waxwings on Sandön (Newton had seen these at the naturalist Leadbeater's in Piccadilly in 1859) (Newton, 1861: 104). This relationship was pivotal in Dresser's scientific career as Newton publicised his discovery of breeding Waxwings in an article in the ornithological journal *Ibis* in 1861. Dresser and Newton became firm friends and remained so for over forty years; Newton was also Dresser's mentor, and he consulted the great Newton on all matters concerning birds.[14] Following their first meeting, Dresser worked hard at assisting Newton, delivering eggs for him to a curator in St Petersburg in 1861 while travelling on business and offering to exchange specimens on his behalf when on his travels.[15]

1.7 Edward Booth, from Henry
Dresser's album of correspondents.

1.8 Alfred Newton, from Henry
Dresser's album of correspondents.

It was during this period that Dresser started to attend the fortnightly
scientific meetings of the Zoological Society of London, held at the Society's
offices at 11 Hanover Square (discussed further in chapter 4).

Dresser returned to Finland on timber business in early 1861, again
availing himself of every opportunity to meet with collectors and to develop
his collection. He also arranged for collectors in Stettin and Danzig to collect
for him. In Finland, Professor Bonsdorff gave him an egg of the Smew, a
scarce northern-breeding duck, which was almost as much of a treasure as
the Waxwing egg to English ornithologists (see plate 3). Dresser recorded
notes on the birds he encountered in his diary: in March he saw a dead
Eurasian Eagle-Owl nailed to a barn door; in St Petersburg he saw many Pine
Grosbeaks – large colourful finches – on sale (probably for eating) in a market.
While in Finland he continued to enjoy the outdoor life in between work,
collecting in boy's-own style, cutting woodpecker eggs out of a tree branch

with his Bowie knife and sailing to offshore islands with friends to hunt for eggs. He also began to keep notes on the habits of birds: on one occasion he went to watch (not shoot) Black Grouse on their lek, a place where the male birds display together to attract mates. Two weeks later he wrote how he 'shot a Starling as I had my gun out and in my hand'.[16]

Dresser Bros. and Buckland

Following his return to London in July 1861, Dresser went into business with his brother Joseph and John Buckland, an associate of his father, trading as Dresser Bros. and Buckland.[17] Henry Dresser was twenty-three and Joseph not even twenty years of age. The partners took premises at 7 New Broad Street, close to Henry Dresser senior's offices. They worked as commission merchants, being listed in the *Export Merchant Shippers of London* for 1873 as shipping to the Baltic and dealing in 'colonial produce'. They traded in anything from which they could make a profit, from fishing flies to tinned reindeer tongues. Dresser also worked as his father's translator for business correspondence. Most of his working day was spent buying supplies to ship to New Brunswick or to Hackman and Co. in Finland. Each day, he visited the Commercial Sale Rooms at 37 Mincing Lane, where coffee, tea, sugar, spices and other colonial products were auctioned; business was conducted in a gentlemanly fashion, in top hats (Kynaston, 1994: 143). Dresser also spent a lot of time at the London docks – the beating heart of the Empire – seeing in consignments of timber and sending goods to Finland. He dealt with a wide variety of people, including Scandinavian sea captains and a Venezuelan with the fine name of Harniodia Montezuma. One sea captain gave him some live monkeys from Cochin (Kochi, India) to give to his brother Joseph.[18]

Although the life of a merchant had a fair amount of romance attached to it, with London 'a kind of emporium for the whole earth' (Kynaston, 1994: 10), the reality was far from glamorous and much of a merchant's life consisted of humdrum repetition:

> For ultimately what abided – day after day, week after week, year after year – was not so much the larger environment as the actual, grinding routine: the volumi-nous ledgers, the salient account books, the endless, pernickety correspondence. There lies the truer, inner, inscrutable history of the City. (Kynaston, 1994: 247, original emphasis)

The monotony of Dresser's work was relieved by bird collecting: while at the docks arranging cargoes of timber and other goods he would also be seeing about shipments of bird skins or eggs to and from his ornithological associates in Finland, Stettin, Danzig and New Brunswick. In between visiting merchants, Dresser would spend some time at the naturalists' shops: Thomas Cooke's at Oxford Street, Leadbeater's Emporium in Piccadilly and the premises of John Gould (the 'Bird Man') on Charlotte Street; Gould (1804–81) was one of the most prominent ornithologists in Britain, famous for producing some of the very finest bird books (see figure 4.1, p. 65). The poulterers' shops in Leadenhall Market (a convenient short walk from the Stock Exchange) were also a favourite calling place for Dresser and other City men interested in natural history, as great numbers of wild birds were on sale there. Dresser's business letters were mixed among ornithological letters and his days consisted of buying and selling, wheeling and dealing, whether in coffee or in bird skins, all dealt with in exactly the same manner and using the language and tactics of business.

Dresser's family lived away, in Dover, at this time; he and his brother Joseph each worked alternate weeks in London, spending the evenings alone at Westbourne Terrace or with friends. Dresser often occupied himself with his ornithology in the evenings, writing to arrange exchanges and packing up specimens to send off, and restuffing bird skins to improve their appearance. Some evenings were spent with his circle of middle- and upper-class friends (including Edward Booth from Brighton), some of whom had their own bird and fossil collections. Alfred Newton visited Dresser at home on 9 August 1861, spending two hours talking about birds and inspecting his egg collection; Dresser gave him some eggs.[19]

Return to New Brunswick

Soon after Dresser returned from his second trip to Finland, in October 1861, he received a letter from the storekeeper at Lancaster Mills accusing Dresser's uncle of swindling his father. Henry Dresser senior dispatched him to Lancaster Mills almost immediately, leaving on 7 November 1861, so he had only just enough time to pick up the essential (to a collecting ornithologist) arsenical soap. On the voyage from Portland to St John, Dresser met an American naturalist, George Boardman (see figure 1.9). This was the beginning of a long and fruitful collaboration: Boardman (1818–1901) had been a partner in a large lumber company in Maine, before giving up work to devote his time to natural history and collecting (Boardman, 1903; Barrow, 1998: 26–7).

Almost as soon as he arrived at Musquash, Dresser confronted his uncle with the allegations. His uncle denied the charges but handed over the running of the entire business to Dresser.[20] Awkwardly, Dresser had to live with his uncle and his family in the mansion house. He worked in the store, investigating various allegations that had been made against his uncle, all of which wore heavily on his spirits. He also had to organise and finance the season's logging. This involved some adventure as he had to travel deep into the snow-laden forest to check on the temporary roads, piles of logs and frozen streams, sometimes with teams of men and sometimes alone. Dresser stayed at campsites in the forest on several occasions, when the men would sing and dance to fiddle music around campfires (see figure 1.10). Once the logs were in the water, in March, Dresser had more time to relax and he spent some time

George. A. Boardman of Calais US. 1.9 George Boardman, from Henry
Dresser's album of correspondents.

shooting at Point Lepreaux with the lighthouse keeper. He also had to arrange
for the building of a ship in St John.

Dresser maintained a pocket diary throughout his stay in New Brunswick,
mixing his business, ornithological and personal matters.[21] His diary provides
many insights into his relationships with the local people. He encountered the
indigenous Micmac (Mi'kmaq) tribe several times; on one occasion Dresser
allowed a group of them to settle near the mill on the condition that they
erected 'a respectable wigwam'.[22] The dangers of forest working are also made
clear: on one occasion Dresser met a sleigh party that was transporting a man
who had been knocked unconscious by a tree during felling.

As the leading figure at Musquash, Dresser had to deal with drunken
workmen several times, usually when they disputed payments and debts, and
he had clearly learnt how to handle himself.[23] A workman threatened to shoot
him if he ever went near his home: 'I hauled the little beggar well over the
coals before I finished with him and intend to go up at the first appointment to
see if he dare do as he says.'[24] On another occasion a neighbouring landowner
came out to verbally abuse him, and Dresser wrote how 'the old boy had not
the advantage of me in gab so he beat a retreat'.[25] The most challenging person
to deal with was Dresser's own uncle. Dresser had the power of attorney from
his father and had to be very firm with his uncle when he tried to remove
various articles from the manor house and store. Dresser tried arbitration but
to no avail and the case of *Garbutt* v. *Dresser* rumbled on for two further years.[26]

My camp and hovel at Deer Brook, Lancaster Mills, Musquash. N.B. while logging during the Winter of 1870 & 1871

1.10 'My camp and hovel at Deer Brook, Lancaster Mills, Musquash N.B. while logging during the winter of 1870 & 1871' by Arthur Dresser.

Throughout his time at Musquash, Dresser spent as much time shooting as work allowed, for ducks and waders on the stormy shore and rivers, grouse and small birds in the forests and deer driven out on the frozen lakes. One especially fine example from his diary concerns some impromptu collecting on the way home from church on a Sunday:

> Had a poor congregation but a pretty good sermon … on passing up to the house saw a small flock of Shorelarks and could not resist the temptation of a shot so I ran up for my gun and returned just as the rest of the congregation came up and shot at the birds just over them bringing down a couple.[27]

The locals brought him many animals, mostly dead but occasionally alive, for example a Flying Squirrel, which he kept until June 1862. He showed local lads and George Thomas at the lighthouse how to prepare bird skins and eggs so that they could continue to send him specimens after he returned to England, making up 'a whole lot of soap of arsenic' for one young helper.[28]

During his six-month stay, Dresser collected ninety-five bird skins, including those of a Snowy Owl and a Bald Eagle.[29] To judge from his diary, bird collecting gave him welcome relief from endless arguments with his uncle and various other tribulations.

In among the various goings-on relating to his uncle and the responsibility of overseeing the forest work, Dresser spent New Year's Eve visiting a family, the Boyles, at Chance Harbour near Point Lepreaux:

> Thus I spend the last of this year in a small cottage on the sea coast of New Brunswick over a comfortable fire with good honest people round me, hearkening to

the waves which almost wash up to the door and the wind which is freshening up for a storm, and little did I think that I should spend New Year here. This year has however been an eventful one for me, for I have in it commenced business for myself, travelled through the whole of North Germany, Russia and Finland up to Tornea [i.e. Torneå] and down through Sweden home and am now here in America to watch an uncle who is swindling my father who has assisted him by giving him an agency and has been repaid by being swindled to an extent which will I fear never transpire. As to my own business I cannot tell how we have done until I receive the balance sheet but do not think my share will come to more than £100 to £150. And now I will return thanks to the Almighty who has preserved me during the present year from many dangers which might have befallen me and has dealt more graciously with me than ever I could deserve. And I pray to Him that He will still watch over me and bless me, for I feel and know that without his aid, however much I may strive I cannot do any good thing. To Him who watches over and will bless and preserve those that put their trust in Him be all Honour, Power, Might, Majesty and Dominion henceforth and for ever.[30]

This was clearly a time for reflection and we get a rare glimpse of Dresser's religious feelings, as well as his heartfelt compassion for his good-hearted hosts, regardless of whether or not they were as socially elevated as his more usual peers.

Return to England

In the end, Dresser could not reach a resolution with his uncle so he handed the running of the store to John Robson, the store manager. He left Musquash in May 1862, visiting George Boardman at Milltown on the St Croix River. On the voyage home, Henry travelled with a coffee planter from Ceylon, a naval lieutenant in charge of the observatory at Quebec and a French Canadian critically ill with tuberculosis.[31] On deck, he saw huge icebergs standing ninety metres out of the water;[32] a Yankee (Dresser's term, 'not so bad as most of them') showed off his shooting talents, shooting seabirds at 100 metres distance with a Smith and Wesson revolver.[33] Dresser arrived in Liverpool, laden with tobacco (with a kilogram of it hidden in his waistcoat to avoid paying duty).[34]

The summer was spent enjoying the high life, with summer balls and dinner parties (three in one week alone), and Dresser was regularly out dancing till 3 a.m. before working in the City the following day. His diary gives glimpses of his character, revealing a sharp intellect and wit. A neighbour once 'fainted in church or at least did her best to do so'.[35] A few days later he was at the theatre: 'I had a glorious time of it for I sat next stall to a very jolly girl from the country with her father and I did the agreeable awfully and had the old man awfully oiled and her equally pleased'.[36]

During September–October 1862 Dresser took a walking holiday to Scotland with a friend and a servant. They walked across Perthshire moors, Skye and Argyllshire, covering thirty kilometres in one day alone (Dresser

1.11 Roualeyn Gordon-Cumming, from Henry Dresser's album of correspondents.

had to get his boots resoled three times). They visited the main tourist stops: cruising down Loch Ness, and visiting the Hermitage at Dunkeld, Fingal's Cave on Staffa and various castles. They met with Roualeyn Gordon-Cumming (1820–66) – a big-game hunter who styled himself as the 'Lion Hunter' – at his private museum in Fort Augustus: 'He was very fantastically dressed in tartan and whole Highland dress and had long hair and beard' (see figure 1.11).[37]

Conclusion

By the time he was in his early twenties Dresser had already had many unusual experiences, between his education for business abroad, his apprentice years in Finland and his business dealings in New Brunswick. His diaries reveal him to be fiercely independent, entrepreneurial and self-confident, fulfilling his father's ambitions that he should be equipped to take over the family business in due course. This confidence, together with a high social standing, facilitated his early bird hunting and collecting; the nature of business, with long periods of relative inactivity between frenetic bursts of action, also provided opportunities for his beloved collecting. One feature that comes across from his writings is his very shrewd character, unsurprising given his business education. For example, when visiting Skye, Dresser and his two companions once travelled in company with some French and German tourists:

> Saw J. McKenzie the landlord about a carriage and a guide for today and then put all our baggage with the others to have sent to Sligachan. We were arranging about the payment we having 3 pieces the Frenchmen 6 and the others 6 and they wished us to pay one third of what the landlord asked viz 11/- [i.e. 11 shillings] but not quite finding that, I suddenly winked at the landlord and proposed to pay 1/- per package paying down on the nail and the others paid before they could see it, thus the landlord had the advantage and we also and he of course looked so much the better for [i.e. after] us.[38]

Here we see the development of Dresser's singular character, independent and sharp-witted. He would need these on his next, biggest adventure, the subject of the following chapter.

Notes

1   See Graham (2000) for the history of Topcliffe and Topcliffe Mill, and for information on Joseph Dresser II and III. The mill is now a block of luxury apartments.
2   Information on the partnership between Henry Dresser senior and Robert Garbutt comes from *The Observer* newspaper for 24 September 1827, p. 1, col. a, announcing the partnership's dissolution.
3   As listed in *White's 1840 Directory of Leeds*.
4   'Sales by auction: Farnborough, Kent – Life interest and policy of assurance', *The Times*, 19 March 1846, p. 12, col. c.
5   As described in the sale of the property, 'Farnborough, Kent – Life Interest in a Residential Estate', *The Times*, 22 August 1866, p. 16, col. c.
6   As recorded in the 1851 Census.
7   London Metropolitan Archives, MS11316, *London Land Tax Records 1692–1932*.
8   Information on the business relations between the two firms can be gleaned from roughly 300 letters from Hackman's to Dresser and Co. dating from 1855–60, preserved in the Central Archives for Finnish Business Records (ELKA) in Mikkeli, Finland.
9   He was still there at the time of the 1851 Census, as was his brother Joseph.
10 Manchester Museum (MM hereafter), ZDH/7/1.
11 One of the adult birds, the chicks, nest and single egg are in Manchester Museum.
12 The description of the Lancaster Mills estate, its buildings and machinery are taken from *Particulars and Conditions of Sale of an Important and Valuable Freehold Property known as The Lancaster Mill Estate*, a copy of which is in the New Brunswick Museum Archives. The sale was advertised as 'Sales by auction: In Chancery – New Brunswick – Important and Valuable Freehold Property, distinguished as the Lancaster Mill Estate', in *The Times*, 21 May 1868, p. 16, col. c. Details of the building of the manor house come from Thompson (1980: 6). Details of the work on the estate come from Henry Dresser's diaries in Manchester Museum and *Travels in New Brunswick, Canada and Manitoba*, by A. R. Dresser, in the National Archives of Canada, Microfilm A-1536. The latter manuscript (of which there is a copy in the British Museum) includes hand-drawn illustrations and photographs

by Arthur Dresser, some of which are shown in this chapter and in chapter 3. A notice of the work of the mill company can be found in the St John newspaper *Morning Freeman*, 7 May 1861, p. 2, col. b. The relative value of the investment was calculated as the purchasing power, using www.measuringworth.com, accessed 11 March 2017.

13 MM, ZDH/7/1.

14 Henry Dresser (HED hereafter) wrote constantly to Alfred Newton (AN hereafter) after he returned to England from Texas, often on the smallest matters; 299 letters from HED to AN are preserved in Cambridge University Library (CUL hereafter), MS.Add.9839.1D.

15 CUL, MS.Add.9839.1D.211, B376, letter from HED to AN, 1 February 1861; CUL, MS.Add.9839.1D.212, B389, letter from HED to AN, 9 February 1861.

16 MM, ZDH/7/1, 24 April 1861.

17 The partnership of Dresser Bros. and Buckland was dissolved in January 1867, as announced in the *London Gazette*, issue 23,213, 29 January 1867, p. 514.

18 MM, ZDH/7/2, 27 September 1861.

19 MM, ZDH/7/2, 9 August 1861.

20 MM, ZDH/7/2, 22 November 1861.

21 MM, ZDH/7/2.

22 MM, ZDH/7/2, 11 March 1862.

23 MM, ZDH/7/2, 2 December 1861.

24 MM, ZDH/7/2, 28 April 1862.

25 MM, ZDH/7/2, 9 May 1862.

26 Correspondence between HED and Robert Hazen, a leading political figure and lawyer, 1862–64, is preserved in the Archives and Special Collections, Harriet Irving Library, University of New Brunswick, MG H14 (R. L. Hazen Papers), MS2: 2.95, 'Garbutt vs. Dresser'.

27 MM, ZDH/7/2, 6 April 1862.

28 MM, ZDH/7/2, 26 March 1862.

29 MM, ZDH/7/1, 'List of bird skins collected at Musquash, New Brunswick'.

30 MM, ZDH/7/2, 31 December 1861.

31 MM, ZDH/7/2, 18 May 1862.

32 MM, ZDH/7/2, 20 May 1862.

33 MM, ZDH/7/2, 18 and 22 May 1862.

34 MM, ZDH/7/2, 29 May 1862.

35 MM, ZDH/7/2, 8 June 1862.

36 MM, ZDH/7/2, 19 June 1862.

37 MM, ZDH/7/3, 2 September 1862.

38 MM, ZDH/7/3, 6 September 1862.

2 Texas: the big adventure

If Henry Dresser's early life was eventful, this was eclipsed by his next 'adventure', when he spent time in Texas and Mexico during the American Civil War in 1863–64. Information on this period comes from his pocket-sized diary; this has evidently lived through some tough times and there are marks that look like they were caused by an animal's claws across the front.[1] Copies of many of Dresser's business letters and letters to his ornithological network are also still in existence, and he published a detailed account of his trip in the *Ibis* (see figure 2.1, and plate 5 for a map of Dresser's travels in Texas).[2]

The American Civil War broke out in 1861 as a result of tension between a broadly anti-slavery North and pro-slavery South (although the full causes were complex). With little industry of its own, the Confederacy depended on imports from the Union and elsewhere. The Union tried to stage a war of attrition by setting up blockades by both land and sea. Europe, especially Britain, was heavily reliant on Confederate cotton to serve the cotton mills of

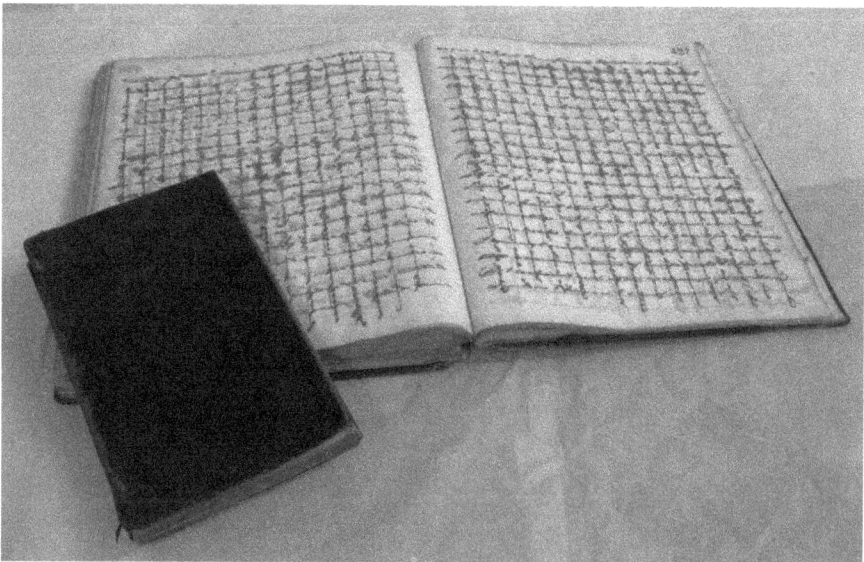

2.1 Henry Dresser's diary of his time in Texas, and his copybook of letters.

the Industrial Revolution. Liverpool was the destination for three-quarters of the cotton from the southern states, to be distributed throughout Lancashire and further afield. Initially, the Confederacy thought that Britain and France would support its cause and the war would be short, so-called 'King Cotton diplomacy'. This did not transpire, so a system of bonds for purchasing Confederate cotton was established in Europe.[3]

The demand for cotton in Europe and goods in the Confederacy meant that great profits could be made by anyone who could get past the blockade. It was broken an estimated 8,000 times during the war; most blockade-runners originated in Britain, especially Liverpool. The blockade was imperfect as the Union could not interfere with the Rio Grande separating Mexico and Texas. This was an international boundary and to interfere with it would lead to widespread political outrage. This 'backdoor' into the Confederacy developed into a major trade route: ships sailed to Matamoros on the Mexican side of the border and sent their wares to Brownsville on the opposite side of the river in Texas. Any goods except military supplies (arms and ammunition, uniforms, saddles, cloth, ship timber and sailcloth) could be legally transported into the Confederacy via Mexico. Paper, blankets, shoes, drugs and medicines were especially in demand in the Confederacy and were traded at greatly inflated prices in exchange for cotton (Bernath, 1970; Kearney and Knopp, 1991). Many arms were smuggled into the Confederacy, especially British Enfield rifles (Huse, 1904: 26; Thompson, 1935: 17–18).

Early in 1863, Dresser was commissioned by some businessmen to take a shipment of goods to the Confederacy in return for cotton. He wrote to George Boardman of the invitation:

> I suppose there will be some little danger as they wish me to go into Texas and that country is unsafe but I am of so strictly neutral a turn that I shall keep pretty clear of it and think when one is determined not to get into a scrape one can do it. Of course I will collect [birds and eggs] as much as I can.[4]

Dresser's employers were very influential figures: the firms Leech, Harrison and Forwood of Liverpool, and Sichel, Alexander and Co. of Manchester. The Liverpool firm was a large commission merchants run by two brothers of the Forwood family.[5] They had especially strong Confederate leanings: one of the brothers had been arrested, in New York in 1861, on suspicion of spying for the South (Forwood, 1910: 70–1). Their firm ran a number of blockade-running operations involving at least eight ships (Wise, 1988). Some ships made a number of successful runs, while others were captured on their first attempts, leading Arthur Forwood to complain to the British government.[6] The most notable, the *Stettin*, was seized on its first run in 1862 with a 600-tonne cargo worth an estimated £100,000 at the time, including forty tonnes of cognac, saltpetre and half a tonne of quinine.[7] Sichel, Alexander and Co. were also commission merchants, mainly dealing in textiles. The firm was run by Julius Sichel, from a German-Jewish family. Sichel's father, Augustus, had been arrested by the Austrians on the border with Italy in 1853 and imprisoned for three weeks on suspicion of being a spy, having come from the 'radical nest' Manchester; he was appointed as Honorary Vice-Consul as a reparation.[8] Julius served as Honorary Vice-Consul after his father's death in 1859, continuing in the role until 1874.

Voyage to Texas

Dresser sailed from Liverpool on 30 April 1863 on a ship named the *Orizava*:

> Found her [the *Orizava*] only a small iron built brigantine 196 tons regular and a cabin one could not whip a kitten round. My cabin about 5 feet long by 3 wide above and not that below and destitute of everything, paint included but with a fair sprinkling of dirt.[9]
>
> So soon as we were fairly off I took to putting all in trim and found all pretty bad, scarcely any crockery aboard and altogether the vessel as shabbily found as any old collier. Had to start with salt meat and sea food the first day and found that not too good and the table furniture worse, only 3 knives and forks between us and scarcely any plates.[10]

As well as Dresser there were a captain, two mates, a steward, two sailors and two apprentices. There were also some chickens and an old dog that had wandered on board in Liverpool; the dog was thrown into the sea by one of the sailors when they passed the Azores, as it drank too much of the fresh water. The journey took three monotonous weeks; to pass the time, Dresser turned his hand at casting bullets, making wooden items for the decks and cutting the crewmembers' hair. He harpooned fish by hanging over the front of the ship and once harpooned a porpoise that was boiled down for oil. The only accident of the voyage was when Dresser fell on deck during a rough sea and the ship's harpoon went in his leg up to the barb, but the injury was not serious. By the time they reached Cuba, the heat was so great that Dresser was forced to spend the night sleeping on deck to keep cool, with his mattress on top of the hencoop.

The *Orizava* arrived at the Mexican coast on 22 June: 'We came to an anchor and certainly the promised land does not look very promising as we can only see a line of sand and ships, the former looking woefully barren'.[11] The Rio Grande was blocked by a dangerous sand-bar lying in shallow water with strong currents. After four days the party finally managed to cross the bar in a flat-bottomed pilot boat, in water just over a metre deep. Many ships lay wrecked in the shallows and sharks were abundant; Dresser heard how half a dozen people drowned there practically every week. The party first visited Bagdad (Boca del Rio Grande), a Civil War boomtown:

> The town or village here (Boca del Rio Grande) is merely a collection of huts of the roughest sort, some merely mud wattled and some of unplaned boards so badly put together that one can pass one's hand through every crack. The people are of

the worst sort: the Mexicans half naked and copper coloured and the whites of the most sinister looks, all armed with knife and revolver.… At 8pm we started [for Matamoros] and now I am sorry I did not bring my revolver as there are queer chaps about and indeed I only have a carpet bag with a single change of linen.… We were stopped by a very queer looking Yankee who asked one for some powder to load up as there were robbers about and he had valuable horses. I thought it a try on so I told him I had quite enough for my own revolver and none to spare on which he departed in peace. All along the road are carcasses of horses and oxen that have fallen by the way and, being in various stages of putrefaction, the perfume is not pleasant. Lots of parties transporting goods and camping by the roadside we also passed. At about 12 our driver pulled up at a rancho to rest the horses and I seeing a bull hide hanging over a fence examined it to see it was clean [of maggots] and, finding it so, spread it on the ground and throwing myself on it was quickly asleep, the other two [Captain Burton and a New Orlean refugee] however being unaccustomed to roughing it remained in the carriage and got no sleep at all.[12]

Unwittingly, Dresser had arrived in Mexico just one month after French forces had ousted the Mexican government (Kearney and Knopp, 1991: 124–5).

Arrival in Matamoros

Dresser and Captain Burton arrived in Matamoros on 27 June, another boomtown on a bend of the Rio Grande with some fine colonial buildings; directly across the river, only a couple of hundred metres away, they could see Fort Brown, which was guarded by a garrison of 1,500 soldiers (Thompson and Jones, 2004). Many other merchants were in town, disease was rife and accommodation was hard to come by: Dresser spent his first few days living in an open-ended shed. The man who was to be his business partner, Samuel Simpson, was away in Houston on business (he did not return to Matamoros until 11 August). Simpson, an experienced blockade-runner, had been in charge of the cargo of the *Stettin* when it was seized and had narrowly escaped being imprisoned.[13] Dresser did find Samuel's brother, John, who was to work with him in the town. Dresser's first days in Matamoros were spent meeting the local merchants and the army. He spent some of his spare time with other British and Canadian merchants, but he kept clear of the locals, criticising the men for spending too much of their time watching fighting cockerels at the cockpit, and the women, who (he thought) spent a lot of time on their balconies combing their hair to rid them of headlice.[14]

John Simpson was sent to Bagdad to collect Dresser's luggage and samples of merchandise, but disaster had struck: the whole lot had been lost when the small boat carrying them capsized at the Boca del Rio Grande:

I am now left in what I stand and still worse cannot replace the clothes here at any price.… I am at a nonplus as I only possess one change [of clothes] (and that dirty).… On the whole I do not bother myself about it as that is certainly no use.[15]

In spite of offering rewards for his luggage, Dresser managed to retrieve only his revolver, gun and boots. This was a serious situation, as the wartime shortages meant it was very difficult to replace anything. Luckily, Dresser's merchant friends rallied to his help and gave him some clothes.[16] Fortunately, his cargo of goods was safe, but he lost many of his papers and his ornithological books.

Dresser rented a store in Matamoros in July and within a month he and John Simpson were selling paper, drugs, hosiery, coffee and castor oil to government purchasing agents in return for cotton bonds. Various quartermasters and purchasing officers from different parts of the army and government were in competition with one another for goods, helping drive up the price of imports (Nichols, 1964: 4–5), as he described to George Boardman:

> A great deal of business is doing here both with the interior of Mexico and the Confederate States and strange to say that the chief part of the supplies for the latter place come from New York or are shipped by New York men, not only provisions and clothing but also munitions of war headed up in casks like flour and hidden in sundry ways.... I do not think that I shall have to go into Texas or the Confederate States much but probably up into the interior of Mexico, so if you do not hear from me for some time do not be surprised. As for sickness, I have seen plenty of Dango [dengue fever] but no yellow fever as yet but have not been at all unwell myself and trust to pull through safely as I have a strong constitution and don't abuse it. If yellow fever comes bad I shall certainly move up somewhere to the Sierra Madre, probably Monterrey or St. Louis de Potosi for I don't see the fun of braving Yellow Jack, particularly in a place like this, and besides I can get away as I have another person working with me here.[17]

Dresser's work was far from easy: his business partner was unreliable as he was a chronic alcoholic and often unfit for work, so Dresser frequently had to work fourteen-hour days to sell his goods. He lived in a store close to his own so that he could keep an eye on his merchandise, with some unusual pets: a tame Greater Roadrunner, a Plain Chachalaca (a wild, chicken-like bird) and an armadillo. What leisure time he had was mostly spent shooting birds; he occasionally went swimming in the shark-infested waters at Bagdad, where he once shot a large shark.

The political situation was extremely volatile, with reports of approaching Union forces to the north of the river and rival bands of Mexicans to the south. Matamoros existed in a state of widespread lawlessness: on one occasion a man walking next to Dresser in the town plaza was fatally stabbed through the heart. Another incident happened when Dresser was visiting the judge in the town hall; a gunshot was fired into the building, the bullet missing Dresser's head by centimetres. Another gunshot hit one of Dresser's companions on the side of his head.

In early August, the cotton-purchasing system was replaced by a new Texas Cotton Bureau, responsible for all cotton purchase, collection and disposition in the district. This reduced the opportunity for speculators such as Dresser, as the Bureau removed the element of competition among buyers (Oldham and Jewett, 2006).

Collecting at Matamoros

Clearly, Dresser had great plans for bird collecting and ornithology in Mexico and Texas even before he left England. He had been reading up on American birds and took a copy of Wilson's *American Ornithology* (published 1808–14) with him from England. He wrote after the trip:

> The loss of my papers, books, and particularly of my '*American Ornithology*,' [at the Boca del Rio Grande] was very annoying, as I did not know much about the birds of the Southern States and Mexico, and I was fully aware that ornithology would be the only amusement I should find during my stay. (Dresser, 1865: 312)

He shot birds on most days from sunrise till work and then again in the evening, on either side of his long working days. Within a couple of days of arriving he had staked out two lagoons near the river as the best collecting spots, and he usually had the shooting to himself. The lagoons were home to a wide variety of birds, including Roseate Spoonbills, Black Terns, Black Skimmers, American Avocets, Black-necked Stilts, as well as various species of herons, egrets and ibises; many wading birds also passed through on migration. Dresser shot some birds for eating (he considered White Ibises to be particularly tasty), while others were killed for pure sport and his collection; he once shot twenty-three Snowy Egrets in two days (and thirteen with one shot, they were so abundant),[18] and on another occasion shot twenty-five Short-billed Dowitchers (a snipe-like bird) before breakfast (Dresser, 1866a: 36). Dresser had to find time before and after work to skin and preserve birds, which was made all the more difficult as birds went off quickly in the heat.[19] In his pocket notebook he kept detailed notes on the birds he saw and collected. As well as his own specimens, Colonel McCormick (a solider at Fort Brown) gave Dresser twenty-one bird skins that had been collected by Patrick Duffy, a hospital steward who had been stationed at Fort Stockton in remote south-west Texas during 1860–61 (Williams, 1982; Casto, 1995a).[20] Dresser's collecting was not without danger: at the Matamoros lagoons he would wade through the slimy mud to retrieve a 'good' bird if he shot it, even if he was worried about poisonous snakes;[21] on another evening he swam twice across the Rio Grande after birds. Events such as these reveal Dresser's great passion for hunting and collecting birds.

San Antonio

Groups of armed Mexicans (with various allegiances) entered Matamoros, making trade increasingly difficult. Dresser made plans to move his goods inland to the town of San Antonio, the base of the government's main cotton purchasing officer in Texas. He hoped that his trip would be a short one and that he would be back at Matamoros within a couple of months. He packed

2.2 Buff-breasted Sandpiper collected by Henry Dresser in Matamoros.

up his bird skin collection in two large boxes and stored it at Matamoros.[22] Dresser, Samuel Simpson, Captain French (from Fort Brown) and a lad called Dennis left on 4 September in a horse-drawn buggy with two spare horses. The journey was arduous: after crossing hot scrubby chaparral they struggled for two days to cross a desert, with the buggy sunk up to the axles in sand. A rattlesnake appeared in camp once and on another occasion Dresser nearly lay on top of a large scorpion.[23] All along the way, Dresser shot many waders, hares and Greater Prairie Chickens (a species of grouse) to feed the company. One species, the Upland Sandpiper (known as the 'Field Plover'), was the mainstay and they often ate them three times a day. The migrating birds were sometimes so plump that they burst when they hit the ground and they could be cooked by frying them in their own fat. The Buff-breasted Sandpiper was another favourite dish (Dresser, 1866a: 38–9) (see figure 2.2 and plate 4).

Dresser and Simpson travelled to the town of Victoria, where the shops were closed up and wounded soldiers loitered around; Dresser saw cotton growing for the first time.[24] They travelled onwards to San Antonio, arriving on 17 September. They found the town to be as dreadful as Matamoros: on the first day Dresser was there the army hanged a man outside the town and shot another man dead.[25] The town, with around 15,000 inhabitants, was arranged around a central plaza. Vultures and other scavenging birds soared around the slaughterhouse in search of scraps. The following day the soldiers showed Dresser and Samuel Simpson the 300 bales of cotton that had been allocated to them. The town was a hotbed of Confederate sentiments; Dresser attended a rally in support of the Confederacy, being among a crowd of 1,000 people gathered outside the town. Twelve cattle were roasted whole on the embers of a fire to feed the crowd.[26] Dresser also donated goods to a Confederate Aid Committee to curry favour with the officials.[27]

Dresser did have one stroke of luck, as there was another ornithologist in the town: Adolphus Heermann (1821–65), a veteran surgeon-naturalist (his biography is explored in chapter 6). Dresser had been keen to meet Heermann, whose name was well known, even before he left England, but Heermann was rumoured to be dead.[28] Adolphus and his brother Theodore owned two adjacent houses in San Antonio; Dresser arranged to live in one of these. The brothers also owned a fine ranch house with over 2,000 acres of land some way from the town. Many years later, Dresser wrote to Heermann's biographer (Witmer Stone) that he recalled Heermann as 'a strong, broad, sturdy man of about fifty (perhaps a year or so older), tinged with gray, and he must have been a very strong man, but was then rather lame, and stumbled now and then, and it afterwards proved that his lameness was *locomotor ataxia*' (Stone, 1907: 4; Casto, 1995b, 1997). Dresser was mistaken on one count, as Heermann was ten years younger than he supposed. Locomotor ataxia is a neurological condition and a symptom of syphilis. Dresser wrote to Heermann's biographer many years afterwards that Heermann 'never had any idea of "committing matrimony"' (getting married) (Stone, 1907: 5). While this would appear to be an unusual statement (especially to appear in print), it was probably made to defend Heermann's honour.

Dresser entered into business with merchants allied with the Confederate officials, exchanging his goods for cotton (Nichols, 1964: 20; Vance Gillespie, 2010). He arranged for ten wagons to transport his cotton back to Brownsville on 5 October but there was a dispute with Colonel Dickinson, the officer in charge of the cotton:

> After much headracking on my part I managed to hatch up a proposal that met all points and now D [Colonel Dickinson] will let me have it. Went out and dined with him and spent the afternoon there pleasantly enough now that the cart matter is off my mind for it kept me quite miserable, $30,000 & liabilities hanging over one.[29]

Following this dispute, Dresser managed to send off the first 100 bales of cotton on 9 October 1863.

In between frenetic bursts of work, Dresser continued to expand his bird collection. He went collecting on horseback with Adolphus Heermann on a number of occasions, strapping Heermann's legs into the saddle so that he could not fall from his horse (Stone, 1907: 4). On one occasion, near Heermann's ranch, the river level was high so Henry carried Heermann across the river on his back.[30] Heermann was in the habit of making nest boxes for hole-nesting birds from old cigar boxes; Dresser and Heermann collected some of these, with the nest and eggs ready packed within (Dresser, 1865: 484; Ridgway, 1873: 605). Heermann's incapacitated state forced him to stay at home; this worked to Dresser's advantage, as Heermann prepared bird skins for him.[31] Dresser also employed some young lads to collect and prepare birds for him, notably Duncan Ogden, aged seventeen, who was particularly industrious.

Dresser made several longer hunting expeditions when business was slack. On one occasion he found a big Wild Turkey roost and shot a big male gobbler weighing nearly eight kilograms.[32] A few days later, at Mitchell's Lake (now

Mitchell Lake), he and his companions shot a variety of wading birds and geese; they also saw some Snow Geese and Whooping Cranes, the latter now very rare birds. A local governor organised a hunting trip across the mountains to Tynan's Ranch. The expedition was an eventful one and Dresser and his companions once had to flee from a group of Comanches on a marauding trip. Over the next week, Dresser's party discovered some rare Montezuma Quail, shot ten deer and hunted after bears, Dresser and a companion creeping into a bear den, revolver in hand. The expedition had to be cut short, as they received news that the Union forces had seized Brownsville.[33]

Returning to San Antonio, Dresser heard how General Bee, in charge of Fort Brown, had had to abandon the post. He had burned 200 bales of cotton to keep them from the Union forces but the fire got out of control, destroying buildings and igniting three and a half tonnes of gunpowder. Bee ordered wagon teams, including Dresser's ten wagons, to turn back. Dresser sent the teams to Camarga (now Camargo, opposite Rio Grande City) via Laredo, an alternative route into Mexico. Chaos reigned in San Antonio and there were many revolver fights as people panicked; Dresser had great difficulty in raising money orders to pay for his cotton. He finally managed to leave the town on 9 December, travelling to Eagle Pass 240 kilometres distant with two com-panions and a black servant; the cotton teamsters travelled separately. As usual, Dresser provided food for the company by shooting ducks. They travelled rapidly across the dry prairieland, covering the distance in less than four days.

Eagle Pass lay opposite the settlement of Piedras Negras on the Mexican side of the Rio Grande. Dresser contracted with a merchant to send his goods to Matamoros. He left for San Antonio on 16 December; on the journey, Dresser's companions were scared they would be attacked as they slept round their campfire: 'Up early and spent a poor night as the wolves disturbed us so much, coming close up to where we were laying and howling like demons and it was too dark to see to shoot them.'[34]

On reaching San Antonio on 19 December, Dresser was gratified to hear that most of his 100 cotton bales had made it to Laredo. He settled his accounts with Colonel Dickinson and organised another consignment of cotton. Although the two had a troublesome business relationship, Dickinson and Dresser became great friends and Dresser was godfather to Dickinson's son.[35] Dresser received letters from his family and friends on 8 January 1864, the first contact he had had with them for five months.

Dresser left San Antonio on 23 January 1864, leaving a box of bird skins in Adolphus Heermann's care but taking seventy of the most valuable skins with him. He travelled to Eagle Pass with just one companion, an English merchant called Jones. Within a couple of days they were again having adventures:

> I only by a chance escaped being 'chawed up' by the Comanches the day before I came in here and thus not losing only my horses but perhaps my hair [i.e. being scalped]. They passed up near us and attacked a Mexican train near us the next day and got clear off with more than 60 head of mules and horses.[36]

They reached Eagle Pass on 28 January 1864; there was no accommodation available so Dresser and Jones slept in the yard with their horses. Violence and unrest were high on the Mexican side of the border, so they moved into the fort (Fort Duncan) on 31 January for their safety, described in a letter to George Boardman:

> You cannot think what a case it is to be left afoot in these confounded sand plains. This place is a pretty queer hole I can assure you, not a decent house in the place nor in the town on the Mexican side, Pietras [sic] Negras. If one gets any sort of a place to spread one's blanket in and enough food (mostly chili) to keep body and soul together one is mighty lucky.... I and another English man are unusually lucky as we have hired the Powder Magazine in Fort Duncan and live there. We have certainly no windows but a latticed door which lets the light and air in at the same time and we actually possess furniture to the extent of two chairs and a small table. As for beds these are considered superfluities here and I for my part have not slept in one for a long time and do not feel the want of it now. We have also a Mexican woman who boards us and feeds us full of pepper [i.e. chilli] to her heart's content. The place is however very unhealthy and the small pox in particular is bad, scarcely a single house without it. Still I am thankful to say that as yet I have not had a reason to take medicine or consult a doctor since I have been out in this country, and indeed have never enjoyed better health.[37]

The powder magazine in which Dresser and his companion lived is still in existence and matches Dresser's description precisely; it is preserved as part of Fort Duncan Museum (see figure 2.3).

Within a few days of arriving at Eagle Pass, there was news of an uprising among the Mexicans, so Dresser volunteered to defend the fort along with

2.3 Powder magazine at Fort Duncan, Eagle Pass, Texas, where Henry Dresser once lived. Now part of the Fort Duncan Museum.

Jones and three others as sentinels, wrapped in blankets through the night, accompanied by a bottle of whisky. Dresser's goods arrived in Piedras Negras from Matamoros on 15 February and he tried to arrange their sale in Eagle Pass; the river was blockaded a number of times, adding to Dresser's difficulties.

He had one more adventure when two of his horses were stolen during the night. He and a companion rode hard down the valley of the Rio Grande on 22 February, travelling fifty-six kilometres on the first day and riding hard for ten hours the following, to try to retrieve the horses. Although unsuccessful, the two companions returned to Piedras Negras on 24 February to find that the blockade had been lifted. Dresser transported his cargo of fifty-two crates across the Rio Grande on 27 February to a store he had rented in Eagle Pass. He slept in his store, with his revolver under his pillow and a gun alongside him, in case anyone tried to interfere with his goods. He worked solidly for a week to unpack and repack goods. The lifting of the blockade meant that cotton was moving south into Mexico rapidly, just as other goods were travelling north; rumours of 'Yankees' approaching in early March served to increase the rate of passage still further.

On 17 March one of the teamsters ran off with Henry's goods when in 'a drunken fit'. Dresser got an order for the teamster's arrest and caught him after a forty-kilometre horse chase. The teamster was scared out of his wits and asked not to be harmed, so Henry had him take the temperance pledge. Dresser wrote how he 'played the devil generally and then let him slope [i.e. leave]'. The next day, Dresser was caught in a blockade when he was on the Piedras Negras side of the river. He jumped into the Rio Grande and swam across in sight of the Mexican guards, shouting to men on the Eagle Pass side to cover him should he be shot at by the Mexicans. Dresser crossed the river unhurt; he was also fortunate in that he did not lose any of his goods as a result of the blockade.

Last days in Texas

Dresser left Eagle Pass on 21 March for San Antonio, riding a wagon-train mule as he could not obtain a horse. Over the course of a week he and his companions hunted after peccaries and collected wild honey from bees' nests; he even saw Jaguar tracks. The first few days in San Antonio were spent trying to enter into trading agreements with merchants. Within a week of arriving there was news of more trouble as the Governor of Nuevo León and Coahuila, Santiago Vidaurri, had fled from the Mexican officials, destabilising the situation (political and business) still further (Tyler, 1973; Thompson and Jones, 2004). In the face of seemingly endless troubles, Dresser worked hard to finish up his accounts and to sell the last of his goods in order to leave the district. He travelled to Houston to settle matters with the Cotton Bureau, travelling with Santiago Vidaurri's group and Dickinson (by then a major), as well as Hiram Chamberlain, founder of the first protestant church on the lower Rio Grande, and another English merchant. The group travelled via Austin, crossing the Colorado River on horseback; they got into difficulty on the crossing and Dresser lost many of his business papers. In Houston, he could not settle his business with the authorities instantly, so he made a trip to Galveston Island to collect birds and eggs, having been granted special leave to do so by the authorities, as the islands were blockaded. On the first few islands he visited he saw many tern and Brown Pelican nests containing chicks. He and his companion had a dinner of fish cooked on a driftwood fire on the beach. The following day they visited an island that was home to enormous colonies of gulls and terns, which nested on the beach; on other islands they found colonies of Snowy Egrets and Tricolored Herons, and visited many of their nests. They took two bushels (seventy-two litres) of eggs, which formed the basis of their diet for several days afterwards.

There is little information on how Dresser settled matters with the Cotton Bureau. One contemporary report from Gideon Lincecum, a prominent naturalist, reveals that he had heard that Dresser was in Houston: 'Their [Dresser and his associates] aim now seems to be to purchase all the cotton they can get. They seem to be in funds too and as the stock of cotton on hand will soon be exhausted and the coming crop is comparatively nothing, this English firm is putting into action all their resources' (quoted in Burkhalter, 1965: 160n). After settling matters, Dresser returned to San Antonio in mid-June. He left Texas at the end of July, travelling to Matamoros via Laredo, as Brownsville was not passable. Matamoros had changed beyond recognition in the eleven months since Dresser had left the town. The two lagoons where he used to enjoy shooting had been shot out by pot-hunters (people who supplied birds for food). The town itself had swollen in size and up to 300 ships lay in anchor at the mouth of the Rio Grande (Bernath, 1970: 35). Dresser checked over his bird collections, which he had stored in the town: 510 skins were in good condition, but fifty skins of wading birds had been destroyed by damp, which had rotted the birds' wings. Dresser shipped most of these to England, keeping 200 of the best with him as well as eighty eggs and some insects and shells, in case the bulk of the collection was lost in transit: given the volatile situation, their was a reasonable chance of the collection being lost, seized or destroyed. He left Matamoros in mid-August; there were rumours of an outbreak of fever in Havana so he rode across the Union lines at Brazos Santiago, travelling north on the Mississippi. He was on a Federal steamer that was shot at by Confederate troops at Cypress Bend, close to Vicksburg on the Arkansas side of the river.[38] He travelled as 'a naturalist' to evade questions from the authorities. On reaching New York Dresser was amused to find that many of the merchants were Southerners and Confederate supporters. He had time to pay a tourist visit to Niagara Falls and to visit the bird collections at the Smithsonian Institution in Washington, DC, and at Philadelphia, where he met the naturalists John Cassin and John Krider.[39] Dresser sailed from New York, reaching England in the first week of October 1864, seventeen months after leaving England; he was still only twenty-six years of age.

Conclusion

Dresser remained a staunch Confederate supporter after he returned to England, and he continued to correspond with many of his friends in the South: 'I often think of Texas and one's friends there and think the better of it now I am so far away and [have] not forgotten how kind you Southerners were to me. I always stick up for the Confed. side of the question, which is the one I have always taken the most interest in.'[40] He tried to send some Derringer pistols to one of the officers in San Antonio and obtained more weapons for a cotton man; he must have known that these were contraband.[41] He wrote to Samuel Simpson to deplore Lincoln's assassination in April 1865,[42] but his political leanings are clearest from a letter written some time later, in May 1869, again to Boardman: 'What do you think of the rumours of war between us and your country? I myself don't think there is any fear of a rupture, as there are on both sides plenty of people who have good sense enough to work against it. There is no doubt we sympathised with the South (I rather more I think)'.[43]

Dresser frequently enquired about Adolphus Heermann when writing to his correspondents in Texas. Towards the end of 1865, Theodore Heermann wrote to tell Dresser that his brother had been killed, as his gun went off accidentally while he was out bird collecting (Stone, 1907: 5). Santiago Vidaurri, whom Dresser had travelled with as he left Texas, also died, in July 1867; he had been executed as a traitor in Mexico City, made to stand in a pile of horse dung and shot in the back by a firing squad to a cheering crowd (Tyler, 1973; Thompson and Jones, 2004: 130).[44]

The impact of Dresser's time in Texas would be far-reaching. His bird collecting exploits helped establish his reputation as an ornithologist. Closer to home, the impact of the Civil War would have a profound effect on his family in London, explored in the following chapter.

Notes

1. Manchester Museum (MM hereafter), ZDH/7/4.
2. MM, ZDH/1/2.
3. For discussion on the blockade and Confederate business relations with Europe, especially Great Britain and Liverpool, see Watson (1892), Thompson (1935), Diamond (1940), Owsley (1969) and Wise (1988).
4. Smithsonian Institution Archives, Record Unit 7071, Box 1, Folder 12: Dresser, Henry Eeles, 1862–71 (SIA, RU7071 hereafter), letter from Henry Dresser (HED hereafter) to George A. Boardman (GAB hereafter), 18 March 1863.
5. See Waller (1981, 2004) and Killick (2004). Archives relating to the firm of Leech, Harrison and Forwood are preserved in Hampshire Records Office, Winchester (HRO hereafter), 19M62, Arthur Bower Forwood Papers.
6. Arthur Forwood complained to the British government that seizure of ships close to British Crown Colonies was an invasion of British territory. HRO, 19M62/15/21, duplicate letter from Arthur Forwood to the Principal Under-Secretary of State to the Foreign Department, 18 August 1862.

7   Report of Commander J. R. M. Mullany, USS *Bienville*, on the capture of the *Stettin*, 24 May 1862, *Official Records of the Union and Confederate Navies in the War of the Rebellion*, 30 vols, Washington, DC, 1894–1922, series 1 (*OR* hereafter), vol. 13, pp. 29–30 (1901). Report of Commander Marchand, USS *James Adger*, on the capture of the *Stettin*, off Charleston, 25 May 1862, *OR*, vol. 13, pp. 30–5 (1901).

8   Reported in 'Austria', *The Times*, 23 May 1853, p. 6, col. b; 'Austria', *The Times*, 6 December 1853, p. 7, col. b; and 'Foreign Office, Dec. 28', *The Times*, 31 December 1853, p. 5, col. d. See also Hayes (1905: 186–7) and Keegan (2005).

9   MM, ZDH/7/4, 23 April 1863.

10 MM, ZDH/7/4, 30 April 1863.

11 MM, ZDH/7/4, 22 June 1863.

12 MM, ZDH/7/4, 26 June 1863.

13 HRO, 19M62/27/1, letter from Samuel Simpson (onboard USS *Bienville* off Charlestown) to Leech, Harrison and Forwood, 27 May 1862; Report of Commander J. R. M. Mullany, USS *Bienville*, on the capture of the *Stettin*, off Charleston, 24 May 1862, *OR*, vol. 13, pp. 29–30 (1901). Report of Commander Marchand, USS *James Adger*, on the capture of the *Stettin*, off Charleston, 25 May 1862, *OR*, vol. 13, pp. 30–5 (1901).

14 MM, ZDH/7/4, 28 June and 1 July 1863.

15 MM, ZDH/7/4, 3 July 1863.

16 MM, ZDH/7/4, 4 July 1863.

17 SIA, RU7071, letter from HED to GAB, 1 August 1863.

18 MM, ZDH/7/4, 18–19 August 1863.

19 SIA, RU7071, letter from HED to GAB, 26 August 1863.

20 Cambridge University Library (CUL hereafter), MS.Add.9839.1D.245, C732, letter from HED to Alfred Newton (AN hereafter), 6 November 1865. Casto (1995a) considered that McCormick must have been a Federal officer but this is incorrect as HED clearly states that McCormick was stationed at Fort Brown and gave him the collection of birds from Duffy.

21 MM, ZDH/7/4, 30 June 1863.

22 SIA, RU7071, letter from HED to GAB, 6 February 1864.

23 MM, ZDH/7/4, 6 and 9 September 1863.

24 MM, ZDH/7/4, 12 September 1863.

25 MM, ZDH/7/4, 19 September 1863.

26 MM, ZDH/7/4, 26 September 1863.

27 MM, ZDH/7/4, 23 September and 10 October 1863.

28 Information on Heermann's ill-health probably came to HED through GAB. See CUL, MS.Add.9839.1D.217, B867, letter from HED to AN, 13 March 1863.

29 MM, ZDH/7/4, 7 October 1863.

30 MM, ZDH/7/4, 28 October 1863.

31 SMI, RU7071, letter from HED to GAB, 6 February 1864.

32 MM, ZDH/7/4, 14 October 1863.

33 MM, ZDH/7/4, 8–14 November 1863.

34 MM, ZDH/7/4, 18 December 1863.

35 MM, ZDH/1/2, duplicate letter from HED to Major Dickinson, 5 January 1865, pp. 103–4.

36 SIA, RU7071, letter from HED to GAB, 6 February 1864.

37 SIA, RU7071, letter from HED to GAB, 6 February 1864.

38 SIA, RU7071, letter from HED (in Buffalo) to GAB, 1 September 1864.

39 MM, ZDH/1/2, duplicate letter from HED to Adolphus Heermann (San

Antonio), pp. 27–8; MM, ZDH/1/2, duplicate letter from HED to John Krider (Philadelphia), 17 March 1865, pp. 127–8.

40 MM, ZDH/1/2, duplicate letter from HED to Duncan Ogden (San Antonio), 31 December 1864, pp. 81–2.

41 MM, ZDH/1/2, duplicate letter from HED to Enrique D'Hamel (Matamoros), 31 December 1864, pp. 85–6.

42 MM, ZDH/1/2, duplicate letter from HED to Samuel Simpson, 12 May 1865, pp. 225–30.

43 SIA, RU7071, letter from HED to GAB, 12 May 1869.

44 SIA, RU7071, letter from HED to GAB, 29 August 1867.

3 Settling down to business

When Henry Dresser returned home from Texas in October 1864 he found his family's situation had changed drastically as his father had gone bust in the 'money panic' caused by the Civil War.[1] His father had liabilities and debts amounting to £157,521 against assets and property worth £99,425; he was owed £68,000 by one firm alone.[2] Henry Dresser senior was eventually discharged as a bankrupt in June 1866.[3] He had to give up the town house in Westbourne Terrace, the farm and Farnborough Lodge,[4] and two London villas (possibly purchased as homes for Dresser's unmarried sisters).[5] The Lancaster Mills in New Brunswick were advertised for sale by auction in May 1868 but remained unsold until 1872 (Thompson, 1980: 6).[6] The mills were completely destroyed by a great fire in 1873 along with forty of the workers' houses, as was widely reported in the US press.[7]

Henry senior managed to continue trading and he kept his business premises at Great St Helens. His bankruptcy must have been scandalous enough for the family, but was eclipsed by the activities of Henry Dresser's cousin Joseph Dresser Tetley in New Zealand, who embezzled £23,000 of his investors' money in 1863. Dresser Tetley's investors belonged to leading families around Thirsk and Topcliffe in North Yorkshire, who must have been known to Dresser's family. Dresser Tetley 'did the Pacific Slope' – an American and New Zealand term for those who fled the country to escape arrest – fleeing to Uruguay, where he set up as a farmer (Loftus, 1997; Hunt, 2001). The affair must have been a great source of embarrassment to Dresser's family, although it was kept out of the press.

Despite Henry Dresser senior's bankruptcy, the family (both parents and six of their children, including Henry Dresser himself) could move into a grand villa called The Firs on South Norwood Hill, Upper Norwood, in September 1865 (see figures 3.1 and 3.2).[8] Upper Norwood was 'upper' in both geographical and social terms, with a number of grandiose villas nestled between even grander aristocratic properties (Warwick, 1972; Wilson and Wilson, 1990). Directly opposite The Firs lived Admiral Plantagenet Cary (later eleventh Viscount Falkland) in a mansion in extensive parkland; the stupendous Beaulieu Heights lay next door, amid wooded grounds. Other neighbours were wealthy merchants and civil servants who commuted into the City from the nearby railway stations. Upper Norwood seems a most

3.1 The Firs, South Norwood Hill, photographed by Arthur Dresser.

unusual move for a bankrupt and his family: it was an expensive place to live and most of Henry senior's assets would have been required to pay his debts. Henry Dresser and his brothers may have clubbed together to provide for the family or, perhaps more likely, Henry Dresser senior had somehow managed to retain some wealth. He remained at his business premises at Great St Helens for many years.[9]

When he arrived back in England in 1864, Dresser had his own money worries, as his employers in the *Orizava* venture – both Leech, Harrison and Forwood of Liverpool, and Sichel, Alexander and Co. of Manchester – were in financial difficulties. He was supposed to earn one-ninth of the profits as commission, but both companies attempted to break their agreements. Sichel, Alexander and Co. were also involved in Dresser's father's finances, which strained their relations still further.[10] This led to a flurry of correspondence

3.2 Whitehall (formerly The Firs), South Norwood Hill, 1960, by F. Merton Atkins.

with the Liverpool and Manchester firms, and Dresser visited Liverpool many times to try to resolve the matter.[11] He took any work that came his way, travelling to Serbia in December 1864 and to Sweden the following month on business for banks and merchants. Arthur Forwood invited Dresser to take a cargo to the Confederacy in March 1865, but he turned down the offer because of his previous experiences.[12]

While Dresser was struggling to keep his business afloat, the *Orizava* affair dragged on, as Samuel Simpson (Dresser's business partner in Texas) could not be located. It transpired that Simpson had suffered a nervous breakdown and his brother had made a mess of the bookkeeping.[13] It took a year of disagreement and dispute to resolve the matter. Dresser finally got his commission (but not his travelling expenses) of £300 for his Texan adventure, having been almost drowned (twice) and nearly shot, but having successfully avoided epidemics of fever.[14] The Forwood brothers went on to become leading figures in Liverpool, both brothers serving as mayor of Liverpool (and both knighted for their contributions to business and politics); Arthur Forwood's

3.3 Statue of Arthur Forwood, in front of St George's Hall, Liverpool.

statue stands next to William Gladstone's in front of Liverpool's St George's Hall, reflecting his high standing in business and in society (Cavanagh, 1997: 175–6) (see figure 3.3).

Business was very slow in 1865 so Henry Dresser planned to depart England, partly to leave more work for his brothers Joseph and Arthur (as he worried about their business prospects).[15] He considered relocating to one of America, Paris, Berlin or Japan; he was offered a position in Archangel (north Russia) but turned it down, as he could not speak Russian.[16] The state of business soon improved, however, and Henry and Joseph remained at 7 New Broad Street. Henry also did some work (presumably translations of Baltic languages) for Morgan, Gellibrand and Co., a nearby firm of merchants who traded with Russia. Joseph travelled to the Lancaster Mills as an assignee during the resolution of his father's bankruptcy, and his younger brother Arthur worked at the Lancaster Mills from 1864 to 1871/72.[17] Frederick trained as an engineer at Deptford and Greenock; he worked in the West Indies then entered the Indian Public Works Department, working in Bombay (Mumbai) and Karwar (south-west India) during 1868–71. He set up a large rice mill in central Liverpool, which he ran for many years; he was a prominent figure in Liverpool mercantile society and married the daughter of Sir Thomas Brocklebank, a leading shipowner (Anon., 1904a).

Business travels through Europe, 1864–67

A great deal is known about Dresser's work during the late 1860s from his diaries and letters in Manchester Museum. He undertook private work for mercantile firms and banks, travelling to Europe to secure contracts and evaluate assets. He also secured a contract to supply the government with Baltic timber in 1866, and his business flourished.[18] Dresser's work involved two trips abroad each year, in the spring and autumn, to inspect timber yards and assess the value of trees in forests. When abroad, he would negotiate business terms on behalf of his English clients, draw up contracts (usually in German or French) and send translations of these to his employers; he would also act as an interpreter during disputes. This line of work suited him perfectly, as he could speak most of the major European languages. He clearly enjoyed the travel associated with his work, as he was still a young man with a taste for travel and field sports (see figure 3.4). He wrote to George Boardman in April 1865:

> I don't think I shall add the wife to the other things needful yet a bit, for here in England wives are generally luxuries that only rich men can afford. I think if I had one she would keep me quietly at home and not let me travel about which by the way would not be such a bad thing, for one must settle down some time or another.[19]

Dresser's diaries of his business trips record his views of places and people he encountered on his travels. They are filled with complaints about the

3.4 Portrait of Henry Dresser aged twenty-six.

lack of soap abroad and the standard of personal hygiene of other people he encountered, rather contradicting his oft-stated enjoyment of 'roughing it' himself. He held very prejudicial views against the Jews, Turks and Muslims he encountered, and only really mixed with naturalists, aristocrats, army officers and middle-class men (often merchants and engineers) when travelling. He often had large amounts of spare time while travelling and between business meetings. As an upper-middle-class man he enjoyed the high life, regularly attending operas and operettas, and paying visits to archaeological ruins. The bulk of his spare time was spent scouring markets for bird specimens, visiting local naturalists and hunting for birds whenever and wherever possible. How he managed to travel with the ever-increasing number of bird specimens is unclear, but he may have prepared study skins and sent them directly to England, or he may have skinned birds and kept the unmounted skins with him, to be fully prepared when he was back at home.

His first business trip of this period was to Serbia in November 1864.[20] There is an element of mystery to this, as he travelled with two Confederate officers through Austria and it seems probable that he was arranging for supplies (most likely of arms) for the Confederacy.[21] He travelled down the Danube by steamer, stopping in Pesth (now part of Budapest) to visit Emerich Frivaldsky (1799–1870), the bird curator at the Hungarian National Museum. As Dresser continued his journey downriver, the bird eggs and skins he acquired on his travels piqued the interest of customs officials:

> I had my things as carefully examined as if I were going to smuggle diamonds, by an old fellow who looked decidedly Turkish but was a Servian [sic, i.e. Serbian]. He opened a small box containing some eggs given me by the curator of the Vienna museum and after fingering an owl's egg very gingerly wanted to know the use of it. He spoke German so I informed him that it was a charm of considerable value and all the rest the same.[22]

Dresser travelled onward to Serbia, where he was to investigate the goings-on at a timber operation on behalf of a London accountancy firm. In Belgrade, he met with a man involved in the operation named Van Cleef; Dresser claimed (falsely) that he was there on behalf of Van Cleef's own associate, while his client was in fact a firm of accountants and liquidators, Coleman, Turquand, Youngs and Co. of London.[23] Dresser travelled on to Turnu Severin (Dobreta-Turnu Severin, in Romania) and to Majdanpek (in Serbia) to inspect the timber operations and forest for his London client.[24] The great forest was home to many rare birds and Dresser was quick to hunt out the forest keeper and give him a list of the birds and eggs that he wanted, including birds of prey and the rare Black Stork.[25] The true purpose of Dresser's visit was later revealed in a court case: two London businessmen had defrauded the Leeds Bank of £100,000 – equivalent to £9 million today – forming a fictitious timber company and writing bills and cheques in the company's name.[26] Dresser was a witness for the prosecution in the court case in London in May 1867, which was reported in *The Times* under the headline 'Extraordinary bank fraud'.[27] The episode also reveals that Dresser was more than just a businessman and that he held a very privileged and trusted position with his English clients.

In 1866, Dresser made two trips to the Continent: he spent a couple of weeks in January on a trip to Barcelona (also visiting Marseilles); and then March–May were spent on a trip that took in Paris, Marseilles, Barcelona, Milan, Pesth, Galatz, Turnu Severin, Trieste and Venice, and then back to Barcelona, Madrid and England. Dresser's diaries of these trips are full of accounts of the people he encountered, for whom he generally had a fairly low regard. For example, when travelling from Barcelona to Marseilles by train in January he saw the Moorish ambassador and his cortege, and noted in his diary 'such a queer lot of white linen covered top-heavy looking fellows'.[28] In Marseilles he associated with two Irish brothers, enjoying a tourist life; visiting an operetta, they 'were woefully disappointed for it was very badly given and the singing was of the very worst description, the prima donna as ugly as sin and without any voice worth reckoning'.[29] On one occasion a ferryman tried to overcharge Dresser, leading to a fight:

> he got cheeky and got to trying force so I squared up and the end of it was that he got a licking [a beating] and I was untouched, keeping well off him and giving him it straight from the shoulder. Some others came to help him and I took up a belaying pin and luckily kept myself clear of them and my friend [the attacker] went off as he said to fetch the police.[30]

During his second business trip of 1866, Dresser spent part of March in Barcelona, working on behalf of Arthur Forwood of Liverpool (of the *Orizava* venture). Travelling onwards to Marseilles he met another naturalist, G. H. White of Chelsea, who had collected birds in Mexico City, and the two became great friends. In Marseilles, Dresser once again saw a performance that was not quite to English tastes, although his diary reveals his usual sense of humour:

> when the dancing commenced I could not help laughing for the ultra style of dancing was given to perfection, the amount of leg shown was terrible and lastly a Can-Can was performed such as I should never have supposed could have been done outside Paris. I was much amused in watching the different effect it had on the audience. The English ladies looked as if they wished they were well away and did not know whether they should laugh or look starched…. The old gentlemen looked uneasy and the young ones tried to look as if they did not know what the Jardin Mabille [the name of the dance hall] was about…. As for the Frenchmen, they were perfectly wild, stamping and applauding and indeed to judge from the applause and encores it suited the audience's taste excellently.[31]

Leaving Marseilles, Dresser travelled onwards to Galatz (Galaţi, Romania), at one point with a French army contractor who was alleged to be the illegitimate brother of Napoleon III (Dresser was greatly amused at the resemblance his companion bore to the emperor). His diary of this journey reveals that Dresser was obsessed with washing and personal hygiene; his travelling companions did not always meet with his refined tastes. On a train journey towards Buda he wrote how he 'was the only Englishman on the train and consequently the only person who washed',[32] while on a steamer for Galatz he wrote: 'I slept but very little, the smell being so bad, dirty feet and perspiration seeming to be the chief component parts. I did venture to put my nose into the sleeping saloon for one moment and was very nearly sick, the smell being so bad.'[33]

In Galatz he again had lots of spare time between business meetings: after seven days of travel, business was concluded in a single day. During his free time, Dresser collected birds at the lagoons and forests, which were very rich in birdlife. He shot birds with a special single-barrel gun that looked like a walking-stick, given to him by Edmund Harting. These weapons aroused less suspicion among customs officials and police: the American naturalist Elliott Coues recommended them to collectors who 'approve of shooting on Sunday and yet scruple to shock popular prejudice', and to those shooting birds where the law forbade it (Coues, 1874: 7; see also Dike, 1983; Gabriel, 2011).[34] Once business was completed in Galatz he travelled to Turnu Severin and the forest at Majdanpek, where he had been two years previously. He spent a week investigating timber and hunting after birds, sending local lads up high trees to check eagle nests. He returned to Pesth on 14 April and spent a couple of days collecting at Cilli (in Styria; now known as Celje, Slovenia) with another ornithologist, Edouard Seidensacher, who was assisted by his local handyman. Dresser suffered badly from a complaint he referred to as 'rheumatism' at this time, a complaint that dated back to his teenage years. He wrote how he was 'almost bent double with rheumatism' when with Seidensacher.[35] This reveals a different side to Dresser's character from the vigorous, seemingly invincible young man who could endure all manner of hardships.

In Venice, Dresser paid a guide to show him round the main tourist attractions: 'I did not read [*Murray's* guidebook] until afterwards … I find I went into raptures at just where I ought to have done so, which is satisfactory.'[36] The following day (26 April) he travelled to Milan; he visited the local museum

Count. Ercole. Turati of Milan.

3.5 Count Ercole Turati, from Henry Dresser's album of correspondents.

and obtained a letter of introduction from the curator to two wealthy private collectors who lived nearby, Counts Ercole and Ernesto Turati (see figure 3.5). The Counts' collection of stuffed birds and skins filled eleven rooms on one floor of their palace, arranged in display cabinets and drawers. They showed Dresser their treasures and gave him a dozen extremely rare birds; they also agreed to exchange specimens with him (see also Martorelli, 1898).[37]

Dresser travelled from Milan to Marseilles, where he found a letter from his employer, asking that he return to Barcelona to conclude his business. He set off the following day, spending the night sleeping under the luggage rack on the roof of a stagecoach. Between business meetings in Barcelona, he searched the market for birds. He saw lots of smaller common birds for sale, such as Nightingales and Robins, sold as delicacies. Dresser also noticed many migratory birds in the market: on 9 May (Dresser's twenty-eighth birthday), he saw many Black Terns among the smaller birds; the following day he found lots of Curlew Sandpipers in the market (see plate 6). Identifying and collecting migratory birds was crucial to unravelling the timing and routes of migration. Ever on the lookout for good specimens to add to his collection, he could not control his collecting urge:

> Knew I had a busy day before me but could not resist the temptation and bought 5 of the better Curlew Sandpipers and 2 Little Stints. Called up at Devesa's [a natural history dealer] after breakfast on my way to Plancholet's [one of the businessmen] and he shewed me an exquisite specimen of *Falcinellus ignens* [Glossy Ibis, see plate 7] so I bought that also for 5/- and don't in the least know when I shall get them skinned.
> … Just before lunch was skinning a bird in the space of quarter of an hour I had when I felt a tap on the shoulder and on turning round saw the head porter of the Vicida Suria who asked me what on earth I was after and seemed highly amused when I told him but thought it rather stupid work for a man.[38]

After concluding some business in Valencia, Dresser travelled to Madrid by coach on 13 May. He met with the curator at the National Museum of Natural History, who introduced him to Manuel de la Torre, the keeper of the King's forests at the Casa de Campo, a hunting estate in Madrid (now a public park and the Madrid Zoo). Dresser paid de la Torre to show him the birds in the grounds and help him collect specimens. De la Torre also showed him nests of the uncommon Azure-winged Magpie, throwing the whole nest from the tree, eggs and all, as his collecting method. The following day they visited the nest of a Booted Eagle; Dresser climbed to the nest and, finding it contained two eggs, rested by sitting in the nest itself while he took notes.

In January 1867, Dresser's business partnership with his brother Joseph and John Buckland in Dresser Bros. and Buckland was dissolved.[39] Dresser worked independently from this time on (although, in practical terms, he had been working independently for some years). He made a business trip to St Petersburg, Moscow and the Volga in November 1867, in order to price timber in the forests on Prince Orlov's estate near Nizhny Novgorod. The Prince looked after Dresser and his business companion well, providing them with a French cook from Moscow and a valet each. In Moscow, Dresser paid tourist trips to the Kremlin, gilded churches and to the markets in search of birds. He witnessed the funeral procession of the head of the Russian Orthodox Church (later canonised as St Philaret), complete with icons, 400 priests with long hair and beards, and monks carrying the deceased's belongings on silk and velvet cushions. Then followed the body on a litter of golden cloth, and dignitaries in coaches and sleighs.[40]

Henry Dresser's illness of 1869

Dresser's life took a turn in 1869 when, aged thirty-one, he developed a mysterious illness. He was unwell for almost a year but it had one singular lasting effect, described to George Boardman in October 1870:

> I have recovered my health and strength as well as I possibly could do and am indeed stronger and healthier than I have been for years, but the result of the effort of nature is that I am <u>totally</u> hairless and it is questionable if I shall ever recover it again. An awful nuisance is it not, but still better than being in permanent bad health. I have consulted almost every good physician here and the general opinion is that it resulted from too sedentary a life [!] after so active a one as I had led. However to use the words of one of our most eminent men it will have no bad effect on general health or longevity but rather the contrary – though it alters my appearance totally and what is odd the loss is <u>all over</u> my body.[41]

This condition, alopecia universalis, is considered to be an autoimmune disorder. Dresser wrote to a correspondent: 'It would puzzle a Comanche to raise my hair now [i.e. to scalp him] ... he might catch a Tartar if he tried for all my bodily strength has returned to its fullest extent.'[42] Dresser's hair never grew back and he wore a wig for the remainder of his life. Dresser – very wiry and with prominent pointed features – became downright extraordinary in appearance with the addition of what was later (and unkindly) described as an 'ill-fitting wig' (Manson-Bahr, 1959: 59).

Cannon Street and the iron trade

Dresser gave up the timber-pricing business in 1869, just before the time of his illness. He took up work as an agent for buying and selling metal, working in the City at 1 St Helen's Place and then at 114 Cannon Street. He entered into partnership with Sir Antonio Brady, a very prominent figure who had been the superintendent of the Admiralty's purchase and contracts department (see figure 3.6).[43] On retiring from the Admiralty in 1870, Brady established himself as the sole London agent for the Bowling Iron Company of Bowling,

3.6 Sir Antonio Brady, from Henry
Dresser's album of correspondents.

near Bradford, in Yorkshire (Boase, 2004). Dresser settled in well with Brady
and he worked at Cannon Street for many years, one of many commuters
who travelled daily to the spectacular Italianate Cannon Street train station
(Kynaston, 1994). Brady, Dresser and another man (Marshall Mason Harris,
also of 114 Cannon Street) lodged a patent for metal couplings in 1871, a rare
reference to the detailed work of the firm.[44] Initially, when he was the junior
partner, Dresser was often detained for fairly long periods in the office while
Antonio Brady was away – something that interfered with Dresser's scientific
and social life.[45]

The agency of the Bowling Iron Company passed to Dresser from Antonio
Brady in 1873, and they moved offices to 110 Cannon Street in 1875, Dresser
taking an office of his own on the first floor, leaving Brady upstairs on the
second. Neighbouring merchants dealt in the products and futures of various
industries, principally coal, railways and metal. Dresser's brother Arthur took
over Brady's office in 1876, when he too worked in the iron trade (although
for only four years). Incredibly, the contract between Dresser and the company
still survives, beautifully written and carrying the wax seals of Dresser and the
company, and the directors' signatures; it is dated 30 April 1873.[46] The contract
established Dresser as the London agent for the company. He was to keep
normal business hours and could not engage in the iron or steel trade for any
other firm. In return, he was remunerated to the tune of £400 per year and a
1 per cent commission on all business, and a further commission on all foreign
orders. The company covered his office costs, fees and telegrams, and gave

THE

BOWLING IRON C°
— LIMITED —

BRADFORD, YORKSHIRE.

Established 1776.

London Office, 110, Cannon Street, E.C.

Makers of Best Yorkshire Iron, of
"Bowling" quality.
Bars, Boiler Plates, Bowling Hoops,
Rivets, Chains, &c.
Cold Blast Pig Iron.
Steel Manufacturers by the Siemens and
Crucible Processes,
of Steel Castings, Tyres, Expansion rings,
Machine Moulded Wheels,
Steel & Iron Forgings.
Engineers, Iron Founders & Boiler Makers
Sole Licensees of the
"Capell" Patent Mine Fan.

3.7 Trade advertisement for the Bowling Iron Company, from *Post Office London Trades' Directory* (1891).

him £100 for other expenses. His wage of £400 equates to around £200,000 in modern terms. Although the contract was initially for two years, Dresser remained with the company for over twenty years, until 1893.[47]

The activities of the Bowling Iron Company have a great bearing on Dresser's story as they provided the hard cash that enabled him to achieve what he did both in ornithology and in society. The company, established in 1788, mined iron-rich rock and coal from its own mines in the Yorkshire coalfield, mostly within a few miles of the works. By the mid–late nineteenth century it was one of the two largest ironworks in Bradford, extracting the iron ore from 4,000 square metres of land each week. Rock and coal were transported to the ironworks by narrow-gauge railway. Coal was converted into coke in 164 coke ovens on site and used to fuel six blast furnaces, each fifteen metres in height, which extracted the iron from the rock. Around 500 tons of iron was produced a week, most of which was converted into steel to produce boiler plates. All manner of machine-works turned the iron and steel into finished products, including engines and machinery for the local mills and factories. The ironworks employed 3,500 people in 1876.[48] A description from 1891 provides a good impression of the ironworks:

> The whole area, enclosed by a high stone wall, is somewhat more than a mile and a half round. Looking from the counting-house at the entrance, on the right is a large waste space, with the steaming lake and cinder hills behind. At night, when live scoria and ashes glow from the sides of the latter, and the lake is lighted up by vivid and fitful gleams emitted from the blast furnaces, the scene is strange

and weird-like. Leaving out of consideration the knowledge that one is in the neighbourhood of an immense ironworks, one might almost fancy himself in immediate proximity to an active volcano. (Cudworth, 1891: 222)

According to a trade directory advertisement of 1891, the company produced a wide variety of 'bars, boiler plates, bowling hoops, rivets, chains … cold blast pig iron … steel manufacturers by the Siemens and crucible processes, of steel castings, tyres, expansion rings, machine moulded wheels, steel and iron forgings' (see figure 3.7).[49] These are the sorts of products that Dresser would have sought customers for among the merchants and government agencies in London and Europe. Virtually nothing has come to light of his activities during this time beyond his natural history and some stray comments regarding the iron trade from his ornithological correspondents. This suggests two things: firstly, the dull nature of the work; and secondly that Dresser's clerks – he employed several at least – bore the brunt of the tedium.[50] Many of Dresser's ornithological letters were addressed from 110 Cannon Street, but whether they were only posted there or written during office hours is not known.

Conclusion

This chapter has sought to explore how Dresser established himself in business. Settling into 110 Cannon Street was a major turning point in his life. He remained there until 1911 (when he was aged seventy-three), a period of almost forty years. This firmly rooted position presents a strong contrast to his earlier life, yet it brought Dresser a number of advantages that helped further the development of his collections and his place in scientific society.

Notes

1 Smithsonian Institution Archives, Record Unit 7071, Box 1, Folder 12: Dresser, Henry Eeles, 1862–71 (SIA, RU7071 hereafter), letter from Henry Dresser (HED hereafter) to George A. Boardman (GAB hereafter), 12 October 1864; Manchester Museum (MM hereafter), ZDH/1/2, duplicate letter from HED to Jonathan Nocher (Matamoros), 1 April 1865, pp. 149–50.

2 See 'Court of Bankruptcy, Basinghall-street, Aug. 23 – In Re Henry Dresser', *The Times*, 24 August 1864, p. 11, col. d; 'Court of Bankruptcy, Basinghall-street, April 19 – In Re Henry Dresser', *The Times*, 20 April 1866, p.12, col. f.

3 Details of Henry Dresser senior's bankruptcy come from the Public Records Office, Kew, London District General Docket Book (June 1864–July 1865, A–K), no. 395, B/6/108.

4 Advertised in 'In Bankruptcy – Farnborough, Kent', *The Times*, 11 August 1866, p. 15, col. f.

5 Advertised in 'Sales by auction: Sutton, Surrey, Two elegant detached villa residencies', *The Times*, 25 January 1865, p. 16, col. c.

6   Advertised in 'Sales by auction: In Chancery – New Brunswick – Important and valuable freehold property', *The Times*, 21 May 1868, p. 16, col. c.

7   'Great fire near St John', *Bangor Daily Whig and Courier* (Bangor, Maine), 23 May 1873, 40(122), p. 1, col. e; 'Destructive fires', *Daily Evening Bulletin* (San Francisco, California), 22 May 1873, p. 2, col. d; 'Miscellaneous', *Sacramento Daily Union*, 23 May 1873, 45(6907), p. 3, col. e. See also Cropley (1874: 34), who reported to the Diocesan Church Society that the Musquash library had been burnt along with the mill and houses.

8   The Firs had an interesting history. It became a hotel named Court Royal in the 1890s, then the Court Royal Convalescent Home in 1901. In 1907 it was known as White Hall and served as a private nursing home specialising in 'nerve trouble' (being advertised in the *Journal of Mental Science*, July 1907, p. iii, for example). During 1915–19 it was the Princess Christian Auxiliary Hospital. It was converted into flats in 1922 and named Whitehall. It was demolished around 1980.

9   London Metropolitan Archives, MS11316, *London Land Tax Records 1692–1932*.

10   MM, ZDH/1/2, duplicate letter from HED to Arthur B. Forwood (ABF hereafter), 25 March 1865, p. 133.

11   Copies of HED's correspondence to ABF and to Sichel, Alexander and Co. are preserved in MM, ZDH/1/2.

12   MM, ZDH/1/2, duplicate letter from HED to ABF, 30 March 1865, pp. 145–6; MM, ZDH/1/2, duplicate letter from HED to Samuel Simpson (SS hereafter), 12 May 1865, pp. 225–30.

13   MM, ZDH/1/2, duplicate letter from HED to SS, 1 April 1865, pp. 153–4; MM, ZDH/1/2, duplicate letter from HED to SS, 12 May 1865, pp. 225–30.

14   MM, ZDH/1/2, duplicate letter from HED to William Forwood, 9 September 1865, pp. 398–9.

15   MM, ZDH/1/2, duplicate letter from HED to Arthur Dresser (Lancaster Mills), 4 January 1865, pp. 95–6.

16   MM, ZDH/1/2, duplicate letter from HED to SS, 20 October 1865, pp. 445–8. The firm that offered HED a position in Archangel was presumably Morgan, Gellibrand and Co., as that firm had an Archangel office and he was involved with the Morgans in London for a number of years.

17   MM, ZDH/1/2, duplicate letter from HED to Joseph W. Dresser (JWD hereafter) (Lancaster Mill), 4 August 1856, pp. 364–7. Information on Arthur Dresser's work in New Brunswick comes from his manuscript *Travels in New Brunswick, Canada and Manitoba*, National Archives of Canada, Ottawa, Microfilm A-1536.

18   SIA, RU7071, letter from HED to GAB, 12 October 1866.

19   SIA, RU7071, letter from HED to GAB, 27 April 1865.

20   MM, ZDH/1/2, duplicate letter from HED to SS (Houston), 1 November 1864, pp. 23–4.

21   MM, ZDH/1/2, duplicate letter from HED to JWD, 1 December 1864, pp. 59–62; MM, ZDH/1/2, duplicate letter from HED to JWD, 23 November 1864, p. 44; MM, ZDH/1/2, duplicate letter from HED to Enrique D'Hamel (Matamoros), 31 December 1864, pp. 85–6.

22   MM, ZDH/1/2, duplicate letter from HED to JWD, 1 December 1864, pp. 59–62.

23   MM, ZDH/1/2, duplicate letter from HED to Alexander Young, 30 November 1864, pp. 52–3. Coleman, Turquand, Youngs and Co. were public accountants and liquidators to the Leeds Bank.

24   MM, ZDH/1/2, duplicate letter from HED to Alexander Young, 12 December 1864, pp. 64–9.

25 MM, ZDH/1/2, duplicate letter from HED to Anton Wenns (Forstmeister, Ilovitza, Majdanpek), 14 December 1864, pp. 70–2.

26 The relative value was calculated as the purchasing power, using www.measuring worth.com, accessed 11 March 2017.

27 'Extraordinary bank fraud', *The Times*, 7 May 1867, p. 10, col. f.

28 MM, ZDH/7/5, 24 January 1866.

29 MM, ZDH/7/5, 26 January 1866.

30 MM, ZDH/7/5, 27 January 1866.

31 MM, ZDH/7/6, 12 March 1866.

32 MM, ZDH/7/6, 14 March 1866.

33 MM, ZDH/7/6, 21 March 1866.

34 See Ingram (1966: 133–40) for detailed accounts of collecting with a walking-stick gun in places where shooting was prohibited.

35 MM, ZDH/7/6, 16 April 1866.

36 MM, ZDH/7/6, 25 April 1866.

37 The Turatis' collection is now in the Museo Civico di Storia Naturale di Milano.

38 MM, ZDH/7/6, 10 May 1866.

39 *London Gazette*, issue 23,213, 29 January 1867, p. 514.

40 Cambridge University Library (CUL hereafter), MS.Add.9839.1D. 271, D562, letter from HED to Alfred Newton (AN hereafter), 4 November 1867; SIA, RU7071, letter from HED to GAB, 5 December 1867.

41 SIA, RU7071, letter from HED to GAB, 22 October 1870.

42 SIA, RU7071, letter from HED to GAB, 15 June 1871.

43 MM, ZDH/1/3, letter from Willy Hackman to HED, 21 February 1872.

44 See 'Abstracts of specifications (Miscellaneous)', no. 2788, *The Engineer*, 3 May 1872, p. 320, and the *London Gazette*, issue 23,791, 3 November 1871, p. 4477.

45 See, for example, CUL, MS.Add.9839.1D.365, G252, letter from HED to AN, 1 March 1873.

46 West Yorkshire Archive Service, Bradford, BIC/24/60, The Bowling Iron Company (Records of John Sturges and Company), Agreement.

47 CUL, MS.Add.9839.1D.502, L852, letter from HED to AN, 3 April 1893.

48 For information on the Bowling Ironworks and the Bowling Iron Company, see Cudworth (1891), Long (1968), Dodsworth (1969) and Richardson (1976: 61–78). Statistics on iron and steel production, and the number of employees, come from the *Journal of the Iron and Steel Institute* (1876), Appendix (report of visits at Leeds meeting), pp. xxiii–xxiv.

49 See www.gracesguide.co.uk/Bowling_Iron_Co, accessed 8 April 2017.

50 HED refers to his clerks in National Museums Scotland Library, GB 587, JHB 15/240, letter from HED to John A. Harvie-Brown, 2 February 1891.

4 Early exploits in ornithological society

This chapter further explores Henry Dresser's establishment in scientific society during the 1860s, when he was in his twenties. The separation between his life in scientific society, his business career and his personal life is in some senses artificial, as the different aspects were interwoven. He held great ambitions in natural history society, just as he did in business. His 'prospects' of advancing himself were determined by various factors: his position in society; what specimens he could bring to the table, both literally and metaphorically; his physical location; and his wealth, which ultimately financed his collecting and travelling.

As the previous chapters have demonstrated, Dresser had exceptional opportunities for collecting specimens and meeting naturalists in his early years, and he was ambitious enough to take full advantage of them. He could obtain birds and eggs himself through hunting, buying specimens and employing others to collect for him. His well-heeled background helped open doors for him, both in business and in ornithology; even as a young man he had the self-confidence to approach famous (usually much older) naturalists. For example, nineteen-year-old Dresser (writing as 'Harry') wrote to John Gould – the 'Bird Man' – in 1858, offering to obtain any bird skins Gould desired from Moulmein (Mawlamyein) in Burma through his brother Arthur, in a clear attempt to ingratiate himself with the great naturalist (see figure 4.1).[1] When in Musquash, he occupied the leading position in society, being effectively the squire of a country estate, so his social subordinates were practically obliged to provide him with birds and their eggs if they could.

Scientific socialising

Those with a taste for natural history came together in a variety of social settings. By the time he was seventeen, Dresser was attending natural history auctions at Stevens' Rooms in King Street, Covent Garden, the leading natural history auctioneers in Britain. Auctions were an important place for young collectors to get their faces known and to mix with their elders. Lord Walter Rothschild (whose life is explored in chapter 11), an avid collector and

4.1 John Gould, from Henry Dresser's album of correspondents.

member of the Rothschild banking dynasty, and who probably spent as much on natural history specimens at Stevens' as anyone, wrote:

> One of the less remembered functions of Stevens' rooms is the helpful opportunity they afford to collectors and men of science of meeting, of knowing one another and conversing, for, as is not the case in Scientific Congresses, the ground here is purely neutral as regards the rival theories of different factions of students, and men who meet in friendly competition for the possession of new or rare additions to their collections, have an almost unique opportunity of meeting their fellow-students with whom, but for the lure of the sale room, they might never become personally acquainted. (Rothschild, in Allingham, 1924: 6–7)

Whether we accept these magnanimous statements at face value or not, it is certain that younger collectors could at least come close to more eminent naturalists at auctions. At a particularly famous sale, at Stevens' on 13 July 1865, Dresser saw four eggs of the extinct Great Auk sell for a total of £122; three of the eggs were bought by his wealthy friends.[2]

Natural history societies were another important gathering place for those with tastes for natural history. In the 1860s the leading British societies were based in London: the Zoological Society of London held fortnightly meetings

4.2 Joseph Wolf and photographs of Wolf's tombstone – Henry Dresser's own composite in his album of correspondents. The inscription reads 'This tribute is offered by friends and admirers of his genius'. Dresser was one of the subscribers.

at its offices in Hanover Square; the Linnean Society, mainly concerned with botany, met at Burlington House, near Piccadilly; the British Ornithologists' Union (discussed further below) met officially once a year, initially at the offices of the Zoological Society of London. The societies had a hierarchical structure reflecting mainstream society. Royalty, aristocrats, military men and wealthy businessmen held the senior positions on committees; women were entirely excluded from many societies. Prince Albert had been the President of the Zoological Society; Alfred Newton and John Gould were Vice-Presidents (there were several at one time). Two posts of the Zoological Society were particularly distinctive: the Prosector, responsible for dissecting unusual animals that died at the Society's Zoological Gardens in Regent's Park (now London Zoo); and an official artist, at that time Joseph Wolf (1820–99), who produced illustrations of the animals when they were alive (Schulze-Hagen, 2001). Dresser first met Wolf at the Zoological Gardens in the late 1850s, and the two became longstanding friends (Palmer, 1895: 57) (see figure 4.2).

By 1861, Dresser was a regular attendee at the fortnightly meetings of the Zoological Society of London (he usually referred to it as the 'Zoo Soc' in correspondence).[3] He later wrote how 'I go out this evening to dinner but shall also attend the Zoological Society's meeting which is held this evening for that I would never miss if I could possibly attend.'[4] The fortnightly meetings were *the* place to hear about the latest zoological discoveries. They were held at 8 p.m. on alternate Tuesdays through October to June (Scherren, 1905; Mitchell, 1929; Zuckerman, 1976). The Society's Secretary (manager), Philip Sclater (see plate 8), announced the latest arrivals at the Zoological Gardens. Travelling naturalists described 'their' new species, and many rare and interesting specimens were exhibited to the audience.

Dresser used the Society's offices as a dropping-off and collecting point for specimens. He also used the occasion of the meetings to meet with Newton, who would travel down from Cambridge.[5] Philip Sclater gave him a card and letter with a general invitation to the Society's meetings in 1865 that he used to gain entrance.[6] He became a Life Fellow of the Society the following year. It was almost certainly at the Zoological Society that Dresser first met Richard Sharpe (1847–1909), then only nineteen years of age (see plate 9). Sharpe had worked for W. H. Smith and the book-dealer Quaritch before being appointed as the Zoological Society's librarian in 1867. He was working on a magnificent illustrated book on kingfishers, begun when he was only seventeen. Dresser and Sharpe would go on to become partners in an enormous project covering the birds of Europe, explored in chapters 6–8.

The importance of correspondence

In the nineteenth century, pretty much any naturalist could write and expect a reply from even the most prominent people in the field, and some naturalists were great letter writers.[7] Regular postal services ran through the British

Empire and elsewhere through trade links. These communication routes were, for the most part, reasonably reliable, encouraging people to send information and valuable (scientifically and financially) specimens long distances. Many of the leading ornithologists and naturalists maintained enormous correspondence networks. Spencer Baird of the Smithsonian Institution wrote over 3,000 letters in 1860 alone (Deiss, 1980: 639). Among the British ornithologists, Alfred Newton ran a huge correspondence network from his rooms at the University of Cambridge. Both Lord Lilford (1833–96) and John Harvie-Brown (1844–1916) – keen ornithologists and wealthy landowners in England and Scotland respectively – were often housebound due to illness but continued their ornithology through these correspondence networks (see figure 4.3).

In the United States, the importance of these networks led to the production of *The Naturalists' Directory*. This publication included the names and addresses of subscribing naturalists in search of correspondents, and listed what specimens they had to offer and what they were in search of in exchange. The *Directory* was first issued in 1865 and included 682 US naturalists, with some from Canada and the West Indies; a revised edition the following year included 1,603 naturalists. The project was taken up a decade later by another publisher, Samuel Cassino (1856–1937).[8] Many natural history societies published the names and addresses of their members in their journals (including the *Ibis* produced by the British Ornithologists' Union, discussed below), in order

4.3 Lord Lilford, from Henry Dresser's album of correspondents.

to promote communication among them. Dresser's own correspondence network increased with every one of his travels, adding new people to his network for exchanging information and specimens, which by the end of 1864 included ornithologists in America, Canada, Prussia, Finland and Sweden, as well as the leading ornithologists in Britain.

The British Ornithologists' Union

The British Ornithologists' Union (BOU) was established by a small group of wealthy ornithologists (for which also read 'hunters' and 'collectors') in 1858 to produce their own journal devoted to ornithology. Articles on birds had previously appeared in either the *Zoologist* (a popular natural history periodical) or, for more technical papers, in the *Proceedings of the Zoological Society of London* (Allen, 1976: 190–1; Bourne, 1988; Johnson, 2004, 2005). The Union adopted the Sacred Ibis as its emblem and named its journal the *Ibis*, which was issued quarterly from 1859.[9] Initially, authorship of papers in the *Ibis* was confined to the band of twenty BOU members; articles were mainly accounts of their hunting trips in search of birds and eggs.

The BOU consisted of the hard-core British bird enthusiasts and collectors, a sub-set of naturalists who attended the Zoological Society of London. For the most part, they had little (or, at least, much less) interest in other animals: Dresser once wrote 'the soft place in my head runs only on birds and eggs and I have often hard work to prevent my spending too much time over them'.[10] Dresser knew many of the BOU members by the end of 1864, and was corresponding and exchanging specimens with them. Alfred Newton was his closest and most influential friend among the BOU membership. He came from a wealthy Suffolk family and had travelled widely before becoming the University of Cambridge's first Professor of Zoology and Comparative Anatomy, in 1866. Newton was especially interested in extinct birds and the birds of islands, notably the birds of Mauritius, where his brother Edward was a diplomat. He encouraged students with an interest in birds by hosting meetings for them in his rooms at Magdalene College on Sunday evenings (Wollaston, 1921: 261–9). Newton was instrumental in establishing the BOU, which consisted of his closest colleagues and former students (Wollaston, 1921; Birkhead and Gallivan, 2012). Newton's closest friend had been John Wolley (1823–59), a wealthy English collector who died when he was very young.[11] Dresser's acquaintance with Newton was a significant advantage, as Newton's name could open doors to museum curators and collectors at home and abroad. Indeed, he once described himself as Alfred Newton's 'man for exchanging' bird skins and eggs, a bold claim and not strictly true, as Newton dealt with many other collectors.[12]

Philip Sclater (1829–1913) was another leading BOU member from its earliest days. He trained as a lawyer and became the Secretary of the Zoological Society in 1860, a position he held for over forty years. He was on the

4.4 Arthur Hay as Viscount Walden, from Henry Dresser's album of correspondents.

4.5 Frederick Godman, from Henry Dresser's album of correspondents.

committees of practically all of the leading zoological societies and edited the first series of the *Ibis*.[13] Thomas Littleton Powys, Lord Lilford, was the President of the BOU from 1867 until his death. He lived in Northamptonshire and spent relatively little time in London due to illness. Arthur Hay (1824–78) was another aristocrat – hailing from East Lothian.[14] Following a career in the army in India and the Crimea, he retired to Chislehurst in 1866 and played a leading role in several natural history societies; he became a member of the BOU in 1864 (Keene, 2004) (see figure 4.4).

One of the most prominent BOU members was Henry Tristram (1822–1906); as he was a churchman, he was referred to as the 'Sacred Ibis' by the other members (they often called one another 'Ibises', after the emblem of the BOU). He came from the north of England and worked in Bermuda as a naval and military chaplain during the late 1840s. His health deteriorated due to tuberculosis so he returned to County Durham in 1849 and served as rector of Castle Eden and of Greatham Hospital, and as vicar of Greatham. Tristram often spent the winter in the Mediterranean region and North Africa to escape the British winter (Mearns and Mearns, 1988: 384–93; 1992: 463–6; 1998: 238–41). He became Canon of Durham Cathedral in 1873, leading to a new nickname – the 'Great gun of Durham' (as a canon, not a cannon).

Other prominent Ibises were Osbert Salvin (1835–98) and Frederick DuCane Godman (1834–1919) (see figure 4.5). They both came from wealthy English families and had attended Newton's Sunday meetings when they were students at Cambridge. Salvin was the son of a famous architect, Anthony Salvin, while Frederick Godman's father was a partner in a brewing firm. Godman and Salvin collaborated closely with one another and were particularly associated with studies (and collections) of Central American birds; they visited Guatemala several times together during the 1850s and '60s (Sharpe, 1906: 461–2; Evans, 1909b,d; Fisher, 2004a).

A group of younger ornithologists, including Dresser, were especially friendly with one another. One of these was Howard Saunders (1835–1907, three years older than Dresser), whom Dresser knew from the early 1860s; Saunders had looked after Dresser's bird collection when he was in Texas.[15] Saunders, who came from a London merchant family, had worked at Callo (Callao) in Peru during Peru's 'Guano Era', an economic boom based on seabird droppings that were used for fertiliser (see figure 6.5, p. 108). He worked for the London-based firm Antony Gibbs and Sons: 'Antony Gibbs made his dibs selling the turds of foreign birds', as the saying went (see Miller, 2006: 37). Saunders gave up work and returned to England in 1862, after a long journey across the Andes and Amazon. He became a member of the BOU in 1870 (Evans, 1909f; Lloyd, 1985: 93–106). Henry Elwes (1842–1922) was an Eton-educated aristocrat who served in the Scots Guards from 1865; he became a member of the BOU in 1866 (see plate 11). Elwes became one of the most travelled naturalists of the day and formed large collections of birds, plants and insects (Balfour, 2004). Another member of this 'set' was Ernest Shelley (1840–1910), nephew of the poet; he too was a military man and travelled (and fought) in Africa.[16] Lastly, there was Edmund Harting, 'a great friend of mine … quite a young fellow, studying for the law', as Dresser wrote in August 1865.[17] Harting (1841–1928) trained as a lawyer, and became a natural history editor and librarian (Mullens and Swann, 1917: 272–6) (see plate 10).

Early in 1865, Newton told Dresser that he intended to propose him as a member of the BOU. Dresser wrote in reply:

> The first duplicate [egg] I receive of *Mergus cucullatus* [Hooded Merganser, an uncommon American duck] is yours, indeed you may rely on having any eggs I receive first at your disposal…. With respect to your offer to propose me for a member of the B.O.U., I am much obliged to you for it and need scarcely tell you it would give me great pleasure to become a member.[18]

Newton proposed Dresser at the Union's annual general meeting on 17 May 1865, held at the Zoological Society's offices in Hanover Square. Dresser's nomination was accepted and his position in ornithological society was confirmed; he entered into the leading fraternity of ornithologists, and of bird and egg collectors. Dresser's life contrasted strongly with that of most of the other Ibises: they were rich men who were able to take the summer months off to collect birds in warmer (and sometimes colder) climes. Dresser, by contrast, worked up to twelve hours a day in the City through the 1860s before commuting home to Norwood. He was the son of a bankrupt and, as the eldest son, had the responsibility of looking after his family. His collecting and corresponding had to be confined to evenings and weekends. He once wrote to his friend George Boardman, in 1867: 'I have not heard much now lately in the way of Natural History as almost everyone is out of town at this season of the year except one poor beggar in the City who must work hard for a living.'[19] Dresser was also one of the youngest members of the early BOU, although Godman and Salvin were near his age; indeed, it is remarkable how similar in age many of the early BOU members were. He did have some distinct advantages that he made the most of: he knew his birds well, travelled widely, he could translate well and he could obtain good specimens (such as the Hooded Merganser eggs mentioned above). All of these traits made him a very interesting contact for the collecting-minded ornithologists of the BOU.

As we have seen, the members of the BOU were a combination of wealthy aristocrats and landowners, members of the uppermost middle classes, church-men and military men. They were united in their love of field sports and the outdoor life, as much as by their interest in birds. Among the early BOU members, Colonel Henry Drummond-Hay, Lord Lilford (the Union's first and second Presidents, respectively), Frederick Godman, Philip Sclater, Arthur Knox and Edward Newcome were particularly noted for their keenness on field sports. Lilford and Newcome were famous for their interest in falconry, and Lilford was particularly fond of otter hunting (Drewitt, 1900: 219, 221). Most had travelled when they were young men, before settling down in business or in their country piles; practically all of them owned bird skin and egg collections, including some of the best of their kind in Britain (better in many cases than the collections of the British Museum itself).

The BOU members met once a year for a formal dinner. Dresser described the 1867 meeting to George Boardman: 'All knowing each other well we had a capital evening and kept it up late or rather early in the morning.'[20] They saw each other more frequently at the Zoological Society meetings, held parties for one another in their homes and went on holiday with one another to hunt after birds and eggs. Their alliances were not based solely on their interest in ornithological matters: they were united by their social class and many would have known one another before they mixed in scientific–social settings. Henry Elwes married Frederick Godman's sister; Tristram's and Salvin's families were also connected by marriage. Many were united by their political leanings, being confirmed Conservatives: Lilford sometimes signed himself 'Yours Torily', while Newton was described as 'a Tory of the Tories' (Drewitt, 1900: 247; Wollaston, 1921: 271); Dresser described himself as a 'red hot Tory'.[21] Lilford and Tristram were both prominent Freemasons; Tristram was appointed Grand Chaplain of England in 1884, one of the highest offices in British Freemasonry (Drewitt, 1900: 52, 249; Wollaston, 1921: 244–5; Buckland, 2004).

1 Gyr Falcon skins from Henry Dresser's collection, including two dark-plumaged 'Labrador Falcons'.

2 Female (left) and male Steller's Eider, from *A History of the Birds of Europe*.

3 Male Smew, Norfolk.

4 Buff-breasted Sandpiper from *A History of the Birds of Europe*, based on a bird collected by Henry Dresser at Matamoros.

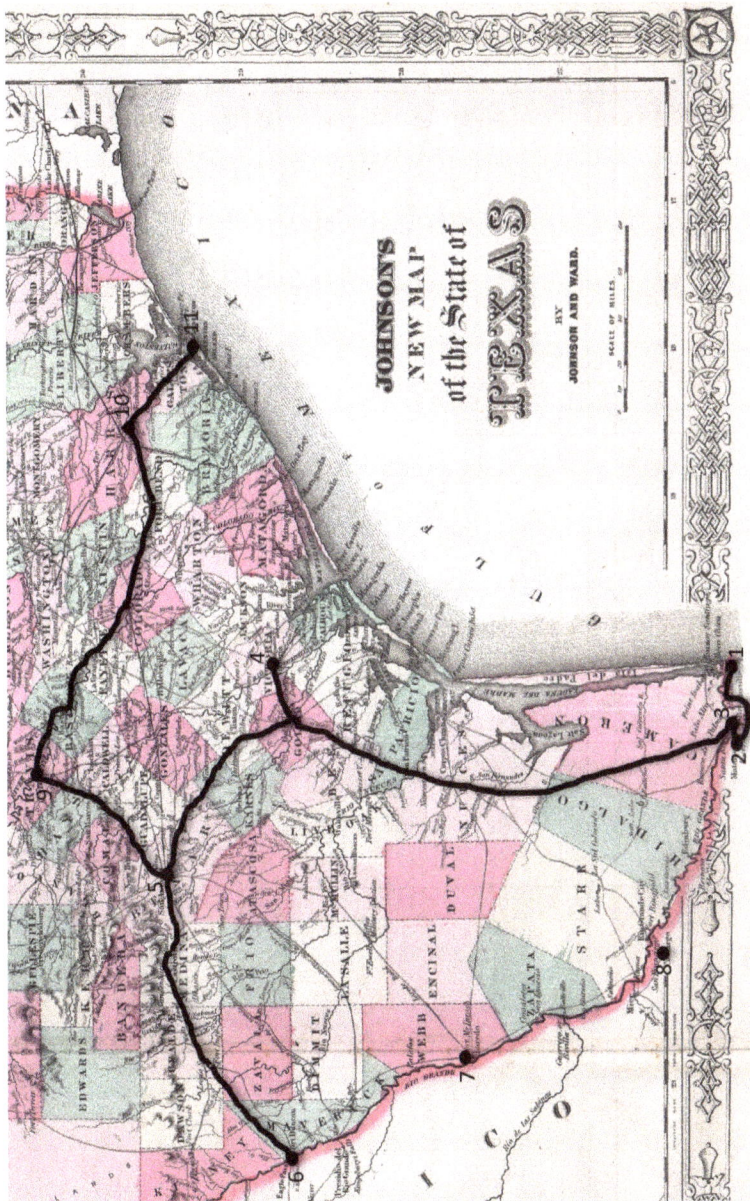

5 Map showing Henry Dresser's routes through Texas and places mentioned in the text: (1) Bagdad, (2) Matamoros, (3) Brownsville, (4) Victoria, (5) San Antonio, (6) Eagle Pass, (7) Laredo, (8) Camarga, (9) Austin, (10) Houston, (11) Galveston. Background map based on Johnson's New Map of the State of Texas (1866).

6 Curlew Sandpiper from *A History of the Birds of Europe*; the illustration of the bird in the foreground was based on a bird Henry Dresser bought in a Barcelona market.

7 Glossy Ibis from *A History of the Birds of Europe*, the illustration on the right is based on a bird Henry Dresser bought in Barcelona.

8 Philip Sclater, from Henry Dresser's album of correspondents.

9 Richard Bowdler Sharpe, from Henry Dresser's album of correspondents.

10 Edmund Harting, from Henry Dresser's album of correspondents.

11 Henry Elwes, from Henry Dresser's album of correspondents.

12 Waxwing adults, nest and chicks based on specimens collected by Henry Dresser, from John Gould's *The Birds of Great Britain*.

13 Golden-cheeked Warbler from an article by O. Salvin. The bird that formed the basis of the central figure was given to Henry Dresser while he was in Texas.

HOODED MERGANSER.
MERGUS CUCULLATUS.

14 Hooded Merganser male (foreground), female (swimming) and chicks, from *Supplement to A History of the Birds of Europe*, based on specimens sent to Henry Dresser by George Boardman.

15 Male (left) and female Surf Scoter from *A History of the Birds of Europe*, based on specimens shot by Henry Dresser in New Brunswick in 1862.

16 Flock of Knot flying in to roost, Norfolk.

Early publications

Getting one's name in print was a major achievement for any ambitious naturalist, and a serious step up the ladder of personal advancement: Charles Darwin once wrote how 'no poet ever felt more delight at seeing his first poem published' as when he saw his name in print for the first time (Darwin, 1958: 63). Dresser's name had first appeared in print in the *Ibis* in 1861, when Alfred Newton publicised his encounter with breeding Waxwings in Finland (described in chapter 1). Although the finding of an infertile egg of a small bird may seem like a small affair today, it was a major accomplishment to the avaricious, field-sport-oriented collectors who dominated English ornithology, as no other Englishman had achieved the feat. Newton wrote, with some gloss, how 'there was no bird whose egg was so longed for by the oologists of the world' (Newton, 1861: 93). The fame of this discovery remained with Dresser throughout his life among English ornithologists.[22] John Gould cited some of Dresser's observations on birds in his famous *The Birds of Great Britain* (Gould, 1862–73), and published an illustration of the nest of Waxwings Dresser had collected in Finland (see plate 12).[23] Dresser's name also featured in an article by Philip Sclater in the *Ibis* in 1865, on a bird Dresser had collected in Texas (it had been given to him by Duncan Ogden, a young lad). The bird represented the first record of the Golden-cheeked Warbler in North America (see plate 13; see also Salvin, 1876).[24]

The next step in Dresser's 'career' was to publish his discoveries under his own name. He had begun to prepare an article on the birds of New Brunswick for the *Ibis* in 1862, although he was not a member of the BOU at the time, but this plan was abandoned when he was called away to Texas.[25] His first publication was a more meagre offering: a short note on the 'Occurrence of a white Redwing in Norfolk', based on a bird (a kind of thrush) he had seen in a taxidermist's shop (Dresser, 1863). This appeared in the *Zoologist*, where many naturalists cut their teeth in publishing: Alfred Newton's first article appeared there when he was only fifteen years old, for example.[26]

Soon after he had joined the BOU, Dresser published his first major article (in two parts) in the *Ibis*, in 1865–66, on the birds of southern Texas. The paper detailed the occurrence and distribution of 272 species of birds encountered in the Rio Grande region, including supplementary information from Adolphus Heermann (Casto, 1995b, 1997). The paper was (and is) an important early record of the birds of the region, including the Ivory-billed Woodpecker, which was 'by no means rare', and Eskimo Curlew. Both of these species are now extinct (Fischer, 2001: 107–8; Lockwood and Freeman, 2014).[27] Following the success of his first major publication, Dresser published with increasing rapidity. An article on the breeding of the Booted Eagle in Spain (it should have been called 'the shooting of the bird and taking of its eggs') was rejected by Newton for the *Ibis*. Dresser presented it at the Zoological Society instead (Dresser, 1866b) and had the article reprinted in the *Zoologist*

(Dresser, 1867b); reports on the presentation also appeared in the *Gentleman's Magazine* and the *Scientific Review*.[28] Dresser put his skills with languages to good use in his early publications: he translated an article on the birds of Finnmark (in northern Norway) that had been published in Norwegian and published it in the *Zoologist* (Dresser, 1867a). A few years later, in 1869, Dresser began the first of a long series of exhibitions of specimens of eggs and birds at the Zoological Society (see Dresser, 1870). Also in that year, he had a letter published in *Nature* – the leading scientific journal – about the parasitic egg-laying behaviour of the Cuckoo, a particularly long-running debate among ornithologists (Dresser, 1869; see also Wollaston, 1921: 233–4).

The British Association and bird conservation

The British Association for the Advancement of Science – the BA for short – met annually as a congress of all forms of science. Dresser spent eight days at the annual meeting in 1868:

> I have just returned from Norwich where I have been to attend the meeting of the British Association for the Advancement of Science and have had a regular treat as fully 3000 people, English and foreigners were there. I went down with Professor [Thomas] Huxley on the 18th and the next day the proceedings opened by a grand speech by the president Dr. [Joseph] Hooker which touched chiefly on Darwin's theory and the relations now existing between science and religion.... The daily lectures and discussions were carried on in some 7 different section rooms, each devoted to one branch of science and ruled by a sectional committee under the general one. I was of course in section D, Zoology, and was on the committee. Our president was Berkeley a botanist but we had a good many bird papers.[29]

At the meeting, Alfred Newton complained that large numbers of seabirds were being shot at their breeding colonies by shooting parties on excursions from cities; many were also killed for feathers to decorate ladies' hats. Newton called for a scientific approach to the situation, and to explore the desirability of a close season, when birds could not be hunted. The Council of the BA established a committee 'for the purpose of collecting evidence as to the practicability of establishing a "close time" for the protection of indigenous animals'. Dresser was appointed to the committee, along with Henry Tristram and several other leading naturalists (British Association for the Advancement of Science, 1869; Wollaston, 1921: 136–40).

Following the Close-Time Committee's establishment, Dresser investigated the game laws of various countries, using his knowledge of European languages. The Committee met at the Zoological Society's offices in January 1869 and offered its support to the Yorkshire Association for the Protection of Sea Birds. Two months later, the Sea Birds Preservation Bill was brought before Parliament and implemented as law, establishing a close-time from 1 April until 1 August, with an amendment that allowed the people of St Kilda to collect birds and eggs for food. The Sea Birds Preservation Act was successful from the first and, at the 1869 BA meeting in Exeter, Dresser told the audience that the Committee was considering extending its remit to cover other birds that may be of benefit to agriculture (Close-Time Committee, 1870). The Sea Birds Preservation Act was a landmark in the development of nature conservation in Britain (see also Barclay-Smith, 1959: 115–16; Evans, 1992; Bircham, 2007: 314–16).

Birds and business

There are several indications that the two components of Dresser's 'career' – in business and in ornithology – did not coexist entirely smoothly. He complained to George Boardman in 1866 that he could not make use of the Zoological Society Library as it was open only between 10 a.m. and 4 p.m., when he was hard at work in the City.[30] Again, writing to Alfred Newton, regarding a meeting in January 1868: 'I expect to dine there [at Salvin's] to meet you tomorrow. This will suit better than meeting here in the city for one is always so busy that birds and eggs are quite driven out of one's head.'[31]

Dresser had to confine his ornithology to his spare time, after commuting home from the City. He gradually gave up hunting his own birds and acquired bird skins from others, and bird corpses in the London markets. In the evenings he would arrange his collection, prepare study skins from corpses, and relax and restuff skins he received from other collectors. In one evening alone he prepared twelve study skins of ducks after working a long day in the City, a considerable achievement by anyone's standards.[32] Dresser also found some escape from work through corresponding with collectors:

> I have been so hard at work myself that I have not had a single day for collecting
> – always at the desk in the City which after a time becomes dull work, but then
> one must look after the dollars and cents. I get lots of letters from men collecting
> in different parts of Europe and that keeps up my interest, but not so well as if I
> could be in the field myself.[33]

Pushing on

This chapter has explored Dresser's entrance into, and rapid advancement in, ornithological society. Through the 1860s, he established himself within the leading group of ornithologists in Britain, as a friend, a correspondent, as a competent ornithologist and as a collector. He valued his acquaintances with other naturalists greatly and collected signed letters and photographs from them from the mid-1860s onwards, which he kept in an album. He wrote to John Harvie-Brown in 1869: 'Please send me a photo so that the light of your countenance may shine in my "bird room" [his museum] when I open my book of collectors.'[34] His album of portraits and autograph letters of his scientific circle is now in John Rylands Library, Manchester; it represents fifty-odd years of social advancement in ornithological society. Dresser's name was also becoming more widely known as a result of his publications and references to his achievements in the publications of others. Dresser was on his way!

Notes

1   Natural History Museum Library and Archives, London, Z MSS GOU A/3, Zoology Manuscript Collection: John Gould Correspondence, letter from Henry Dresser (HED hereafter) to John Gould, 8 November 1858.

2   Smithsonian Institution Archives, Record Unit 7071, Box 1, Folder 12: Dresser, Henry Eeles, 1862–71 (SIA, RU7071 hereafter), HED to George A. Boardman (GAB hereafter), 22 July 1865. See also Fuller (1999). The three buyers whom HED knew were George Dawson Rowley, Arthur Crichton (Lord Lilford's brother-in-law) and Ernest Bidwell. HED placed a collection of duplicate eggs for sale at Stevens' in April 1867.

3   See Cambridge University Library (CUL hereafter), MS.Add.9839.1D.208, B287, letter from HED to Alfred Newton (AN hereafter), 25 August 1860, for the first mention of HED attending scientific meetings of the Zoological Society of London.

4   SIA, RU7071, letter from HED to GAB, 24 June 1862.

5   CUL, MS.Add.9839.1D.225, C641, letter from HED to AN, 13 May 1865, and CUL, MS.Add.9839.1D.244, C721, letter from HED to AN, 2 November 1865.

6   CUL, MS.Add.9839.1D.233, C700, letter from HED to AN, 28 June 1865.

7   See Barber (1980: 38–9) and Farber (1982) for discussion of correspondence networks. See www.darwinproject.ac.uk, accessed 8 April 2017, and Endersby (2008) for detailed discussion of the correspondence networks of Charles Darwin and Joseph Hooker respectively.

8   The *Naturalists' Directory* was expanded in 1880 to include some foreign (including European) naturalists. International editions (mostly called *The Scientists' International Directory*) were published between 1882 and 1914. Editions covering the United States and Canada were issued between these dates and well into the twentieth century. See Kisling (1994).

9   The account of the history of the BOU and biographical information of the early BOU members largely comes from the *Ibis* 1909 Jubilee Supplement (Anon., 1909a; Evans, 1909a–f; Sclater, 1909) and articles in the centenary issue of the *Ibis*, including Moreau (1959) and Mountfort (1959), as well as Johnson (2004, 2005) and Bircham (2007).

10   SIA, RU7071, letter from HED to GAB, 29 August 1867.

11   Following Wolley's death, Newton published a catalogue of his friend's collection (which he had continued to add to), entitled *Ootheca Wolleyana* (literally 'Wolley's eggshells'). This was published in four parts and is an important reference for understanding nineteenth-century egg collecting (Newton, 1864, 1902, 1905–7).

12   SIA, RU7071, letter from HED to GAB, 31 October 1862.

13   He held the post again from 1878 to 1912.

14   Arthur Hay was known as Viscount Walden from 1862 and ninth Marquess of Tweeddale from 1876.

15   CUL, MS.Add.9839.1D.219, B881, letter from HED to AN, 27 March 1863.

16 His name was George Ernest Shelley but he signed himself Ernest (Fredeman, 2006: 634). See Anon. (1911b) for biographical information on Shelley.

17 SIA, RU7071, letter from HED to GAB, 4 August 1865.

18 CUL, MS.Add.9839.1D.225, C641, letter from HED to AN, 13 May 1865.

19 SIA, RU7071, letter from HED to GAB, 14 September 1867.

20 SIA, RU7071, letter from HED to GAB, 29 March 1867.

21 SIA, RU7071, letter from HED to GAB, 7 November 1868.

22 HED's discovery of the breeding Waxwings was referred to in several of his obituaries (e.g. Rothschild, 1916: 194).

23 AN had seen the Waxwing specimens at the premises of Leadbeater's, the London natural history dealers, and wrote: 'I believe I am hardly divulging any confidence when I say they formed the subject of a beautiful picture, executed under Mr. Gould's superintendence, which I trust will before long be rendered more accessible to the public' (Newton, 1861: 104).

24 Manchester Museum, ZDH/1/2, duplicate letter from HED to GAB, 10 November 1864, pp. 40–1.

25 CUL, MS.Add.9839.1D.214, B657, letter from HED to AN, 4 June 1862.

26 Philip Sclater and John Harvie-Brown were also teenagers when their first publications appeared in the *Zoologist*. See Mullens and Kirke Swann (1917).

27 Griscom (1920), Griscom and Crosby (1925–26), and Graber and Graber (1954) also drew upon HED's records of Texan and Mexican birds.

28 'Zoological Society', *Gentleman's Magazine*, 221: 620 (November 1866); *Scientific Review*, 1 August 1866, p. 78.

29 SIA, RU7071, letter from HED to GAB, 23 August 1868.

30 SIA, RU7071, letter from HED to GAB, 31 August 1866.

31 CUL, MS.Add.9839.1D.274, D871, letter from HED to AN, 13 January 1868.

32 SIA, RU7071, letter from HED to GAB, 8 June 1867.

33 SIA, RU7071, letter from HED to GAB, 9 June 1868.

34 National Museums Scotland Library, GB 587, JHB 15/239, letter from HED to John A. Harvie-Brown, 21 July 1869.

5   Collecting

In previous chapters, we have seen how Dresser began his collecting career by shooting his own birds and searching for nests containing eggs,. This chapter explores the various sources, and tactics, that he and other ornithologist-collectors used to take their collecting to new heights. In the early days of his ornithological career, when Dresser was an avid field collector, he was usually keen that birds had 'a sporting chance', but he was not beyond killing birds on a grand scale or 'breaking the rules' of fair play in order to obtain a rare specimen, at least when he was a young man. When he was in Serbia in 1866, for example, he spent an idle evening shooting House Sparrows round the house where he was staying, shooting forty altogether.[1] His brother ornithologists in the BOU had similar tastes: in an early volume of the *Ibis* there is the extraordinary 'Quesal [Quetzal]-shooting in Vera Paz' from Osbert Salvin (Salvin, 1861), and many other *Ibis* articles were basically accounts of bird-hunting and egg-collecting forays (e.g. Salvin, 1859; Simpson, 1859). The popular *Zoologist* was also largely filled with records of rare birds, most of which had been shot, leading Alfred Newton to complain about the 'blood-stained pages of the Natural History Magazines' (Wollaston, 1921: 55).[2] Ornithologist-collectors preserved and retained specimens of interest, as mementos, bargaining chips, to fill gaps in their collections, and as evidence of the presence of particular species of birds, remembering the maxim 'what's hit is history and what is missed is mystery'.

The leading ornithologists took their collecting to a higher level by forming specialist collections, by buying and exchanging specimens from others, so-called 'cabinet collecting'. Many gave up fieldwork altogether, settling down and gathering specimens collected by others (Dresser's collecting 'career' followed this model). They each consolidated their place among their peers by having 'their' group of birds to collect and to study, an approach that avoided (or at least attempted to avoid) unnecessary competition between collectors, both in terms of subject areas and in terms of acquiring specimens themselves. While many people confined themselves to owning one or a few specimens of each species, more ambitious collectors built up series of specimens of their subjects. The American Elliott Coues wrote, in his characteristic emphatic way:

your own 'series' of skins of any species is incomplete until it contains at least one example of each sex, of every normal state of plumage, and every normal transition stage of plumage, and further illustrates at least the principal abnormal variations in size, form and color [sic] to which the species may be subject. (Coues, 1874: 28)

Some groups of birds were more collectable than others: birds of prey, waders, gamebirds and waterfowl were popular as skins and mounts (largely because of sporting interests). 'British birds' as a whole was a popular topic (among British ornithologist-collectors) for those who had more room to spare and many collectors sought to build a complete collection of eggs of British birds. This quest for 'British birds' meant skins of rare vagrants and visitors, and their eggs, had a particularly high value, and rare birds were harried as a result. Eggs of birds of prey and wader eggs were especially popular with collectors as they are particularly attractive and variably marked. Eggs of the parasitic Cuckoo and its hosts were another popular topic; more specialised subjects included the downy chicks of certain birds, nests, and albinos and other abnormally coloured birds.

The nature of collectors' specialisms varied from collector to collector but was usually linked with some period of travel in their younger years, entirely in keeping with the Victorian principles of self-help described in previous chapters. Philip Sclater, Osbert Salvin and Frederick Godman all collected Central and South American birds. Henry Gurney of Norwich formed a collection of 5,000 birds of prey; Howard Saunders concentrated on collecting gulls and terns, and Edmund Harting collected waders. Henry Tristram had varied tastes; he collected specimens from around the world from contacts he had in the Church Missionary Society, of which he was a governor. Lord Lilford kept live birds at his family pile in Northamptonshire, including pelicans, birds of prey and cranes in aviaries, and free-flying vultures (Drewitt, 1900; Trevor-Battye, 1903; Mearns and Mearns, 1998). Through the 1860s, Dresser specialised in birds and eggs from Europe and North America, reflecting his business travels in these regions. He was particularly keen on Arctic birds, eggs of sandpipers, and eggs of the Cuckoo collected along with the eggs of its hosts.[3]

The availability of specimens

Collectors had abundant opportunities to acquire birds and eggs in the mid-nineteenth century, both in Britain and when travelling abroad, and even while office-bound in London. Immense numbers of wild birds were sold for food, in the days before factory farming, and poulterers' markets could be an important source of specimens. In Britain, the most famous of these was Leadenhall Market in London (Collinson, 2012; McGhie, 2012). Huge numbers of gamebirds, waders and waterfowl were sold there: there were over 8,000 Black Grouse in the market in early 1883, for example (Gurney, 1883).

However, all manner of other birds could be found there – alive and dead – as well as eggs:

> Beyond all other places, Leadenhall Market is the emporium to which the purchaser of rare birds and animals, living or dead, should betake himself ... every kind of bird or beast that was ever yet made a pet of is here to be bought, sold, and exchanged, and frequently the collector may obtain very rare and valuable specimens. (St John, 1849: vol. 2, 264–5)

Being close to the Bank of England and Cannon Street, Leadenhall Market became a Mecca for City-working ornithologists, especially Dresser and John Henry Gurney junior, who searched for rare birds among the more common kinds during their lunch breaks (it was also an easy walk from New Broad Street). Some collectors also obtained large quantities of rare bird eggs from the market (Newton, 1905–7: 274). Gurney encouraged ornithologists to visit Leadenhall Market in an article in the *Zoologist* in 1870. The market produced a number of notable records of vagrant birds found among wildfowl (e.g. the first American Wigeon to be found in Europe) (Collinson, 2012). Howard Saunders discovered two Lesser Snow Geese (from America) in the market in 1871 that had been shot in Ireland, the first records for Europe; he gave one of the birds to Dresser (Saunders, 1872). Dresser purchased a wide variety of waterfowl, waders, rare seabirds and hybrid grouse at Leadenhall Market in the 1860s and '70s (see figure 5.1). The best specimen was an Pacific Golden

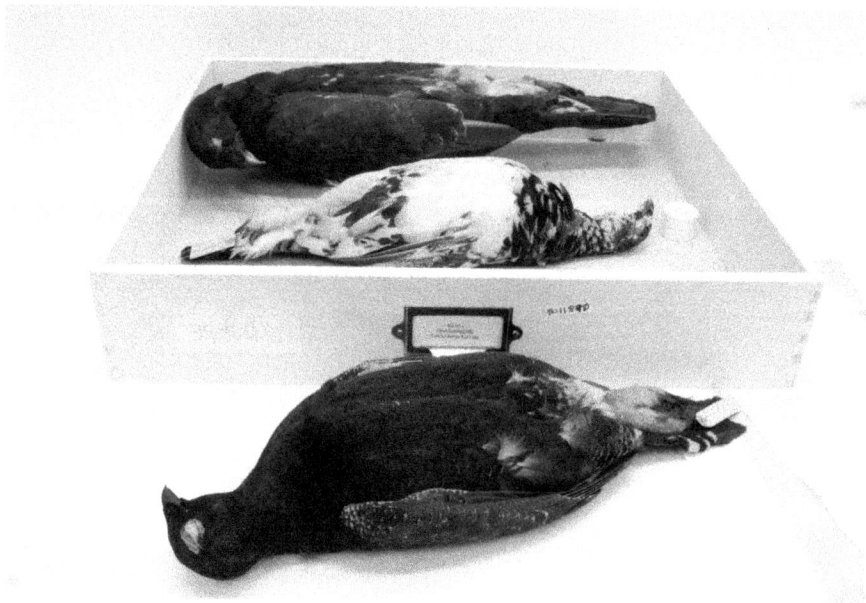

5.1 Rare hybrid grouse from Henry Dresser's collection. The bird in the foreground is a Black Grouse x Red Grouse hybrid purchased among gamebirds in Leadenhall Market.

Plover, a rare vagrant from Siberia and a 'first' for Britain, which turned up in the market in December 1874:

> Mr. [Ernest] Bidwell, a gentleman who visits Leadenhall market regularly, to pick up specimens of rare birds and eggs at the game-dealers' shops, told me that he had seen an odd variety of the Golden Plover. I immediately went to the market, and found amongst a lot of Golden Plovers from Norfolk the specimen in question. It was badly damaged, and having been kept in the shop during mild weather for ten days, it had already become tainted, so that it was only with the greatest difficulty that it could be preserved; but I have succeeded in getting it made into a passable skin. It closely resembles *Charadrius fulvus* [Pacific Golden Plover] from Asia. (Dresser, 1875c: 514)[4]

In Britain, gamekeeping reached a new high during the nineteenth century. Predatory birds and mammals of all kinds – including birds of prey and owls, herons, cormorants, gulls, crows, weasels of all kinds, badgers, foxes and hedgehogs – were deliberately killed in the belief that doing so would increase the amount of gamebirds, mammals and fish available for hunting and fishing. Woodpeckers and squirrels were killed for causing damage to trees. Traps and poisons laid to kill these 'vermin' also inadvertently killed many other animals (Yalden, 1999; Lovegrove, 2008; Yalden and Albarella, 2009). Moreover, gamekeepers and other country people (such as lighthouse keepers, farmers and labourers) supplemented their wages by selling the corpses of birds to taxidermists and collectors; Kingfishers, Barn Owls, Bitterns, hawks of all kinds, colourful birds and rare birds were in particular demand (Lascelles, 1915: 56; Morris, 2010: 147–53). Collectors often wrote hypocritically, berating gamekeepers and other country people for persecuting birds of prey, while shooting them themselves or taking their eggs for their own collections (e.g. 'Oologicus', 1863: 478–9).

Preserved birds were in great demand as household decorations, while their feathers (and to a lesser extent their preserved wings or whole bodies) were used extensively for the millinery trade. Almost unimaginable numbers of birds, alive and dead, came to Britain from overseas for the plume trade and for collectors: 'as for prepared specimens [of birds] we have the pick of the world', Dresser once wrote,[5] echoing the assertion that London was like 'a kind of emporium for the whole earth' (see p. 26). A single feather auction in London in 1876 accounted for the feathers from 10,000 Indian egrets and herons, and over 15,000 hummingbirds from South America; such sales were held almost weekly.[6]

The demand for preserved birds and other animals was serviced by an industry of taxidermists throughout Britain, Europe and North America, and there was a taxidermy firm in practically every town in Britain (MacKenzie, 1988: 41; Morris, 1993, 2010; Barrow, 1998). Specialist natural history dealers existed in much smaller numbers.[7] In London, the largest firms included Edward Gerrard and Sons (Camden Town), and Leadbeater and Son (Piccadilly) (Mearns and Mearns, 1998: 95–6; Morris, 2004). Natural history dealers could also be found in the larger northern European cities. In Paris there was the Maison Verreaux, 'one of the greatest, if not the greatest, emporium of natural history that the world has ever seen' (Sharpe, 1906: 503), which had 40,000 bird specimens in its storehouses at one point. The firm, run by the Verreaux family, received collections from many famous travellers (Mearns and Mearns, 1992: 403–7).

Commercial dealers obtained their wares by a variety of methods: some retained private collectors while others acted as collectors' agents; others purchased from auctions. Dealers were in competition with each other so their prices tended to be similar, with a 'going rate' for each species: during the 1870s an egg of the Golden Eagle was worth 24s (equivalent to roughly £100 today), for example.[8] The skins and eggs of birds of prey, Arctic birds and waders fetched high prices when compared with other birds.

Natural history auctions were another, if rather niche, source of specimens for collectors. Stevens' Rooms in King Street in London was the pre-eminent venue, selling natural history specimens (dead and alive) as well as all manner of curios, from tattooed Maori heads to archaeological artefacts (Allingham, 1924). Its regular natural history sales encouraged travellers to collect, as they could be sure of a place to sell any specimens they gathered on their return to England. There was a remarkable interplay between scientific societies, Stevens' and individual collectors. For example, Alfred Newton's friend John Wolley (see figure 5.2) received Waxwing eggs from his collectors in Lapland in 1856. Edward Newton (Alfred's brother) announced this at the Zoological Society of London in March 1857, later reported in the society's *Proceedings* (see Wolley, 1857). The eggs were subsequently sold at Stevens', the first going for more than £5 (Newton, 1860c, 1902: 212–39; Wollaston, 1921: 17).

5.2 John Wolley, from Henry Dresser's album of correspondents.

5.3 Henry Dresser's auction catalogues, marked with prices paid at auction for specimens.

Stevens' was particularly associated with sales of collections of eggs; sales of these were of popular enough interest to be reported in the *Ibis*, *The Times* and in other magazines and newspapers, serving to heighten the competition between potential buyers. Some of the sales became almost legendary among collectors, notably John Wolley's sales of eggs from Lapland (Wollaston, 1921: 17). Among the audiences, many collectors recorded the selling prices and buyers' names in their sale catalogues: auctions could be a spectator sport.[9] Dresser attended sales at Stevens' from the age of seventeen and was still visiting the rooms fifty years later, in search of rare eggs (see figure 5.3). He only offered a collection for sale at Stevens' once, in 1867, mostly consisting of eggs from Lapland (discussed further below; Chalmers-Hunt, 1976: 105).

Fortunately for Dresser, he was collecting at a time when increasing numbers of Europeans travelled abroad, whether for business or pleasure. Many private travellers acquired specimens that were made available to cabinet collectors when they returned to Britain.

Attitudes to commercial specimen dealers

Collectors – including Dresser – regarded commercial dealers with great suspicion.[10] While bird skins could be identified easily enough, the eggs of many species were indistinguishable, so the egg of an extremely rare bird could look identical to the egg of a common species. Commercial dealers were often suspected of selling the eggs of common species as rare ones; even the reputation of the Verreaux dealership in Paris was tainted by the poor quality of information with specimens, with vague (or, worse still, incorrect) collection localities and unreliable identifications of eggs. Lord Lilford commented on the business's practices:

> I was with Edward Verreaux (egg dealer) in Paris when there arrived a large consignment of skins and eggs from South Russia.... A big note-book was produced, and the two brothers proceeded to separate and name the eggs in the book, as it seemed to me, purely as fancy dictated.... The naïve way in which the brothers confessed their entire ignorance, and shot at probabilities, was most amusing, and gave me a lesson about buying eggs that I have never forgotten. (Trevor-Battye, 1903: 263)

In spite of Lilford's comment, the Verreaux brothers were otherwise excellent naturalists. They were constrained by the poor quality of information that collectors sometimes provided them with, and some misidentifications certainly occurred. Dresser had the same experience with the firm (Trevor-Battye, 1903: 263).

Alfred Newton considered egg dealers to be 'an evil' (Wollaston, 1921: 149), failing to appreciate that without a ready market from collectors there would be no dealers.[11] He published notes on the formation of 'scientific' collections of eggs (Anon., 1860; Newton, 1860a,b; Lindsay, 1993: 33–4), emphasising the importance of two points: identification and authentication. Identification meant that the parent birds often needed to be killed, either by shooting or (for small birds) catching them in nooses made of horse-hair placed over nests, to verify the identification of eggs. Many eggs in Dresser's collection were 'sent with bird' (some were sent with the parent birds' heads or feet) to verify their identification (see Tristram, 1860a, as a further example). Authentication was a way of assessing the character of collectors and dealers. It had little to do with bird biology but everything to do with whose word one was prepared to accept. It meant establishing a chain of certainty right back to the person who first discovered an egg. Newton encouraged collectors to write on eggshells, giving their name and a unique number to authenticate them. Dresser followed this practice: when he received eggs, he recorded them in a hardback catalogue, with details on whom they came from, and other notes on authenticity; the eggs were marked with the page number of the catalogue and a unique letter or other mark. When an egg proved to have been misidentified, Dresser removed it from his collection. He once sent a series of eggs back to the dealer as he did not believe them to be genuine; he recorded that the source was 'tainted' in his catalogue.[12]

In spite of collectors' best efforts, many eggs were deliberately passed off as being more interesting (and valuable) than they were. The commonest ruse, in Britain, was to pass off birds or eggs that had come from the Continent as having been collected in Britain, making them very desirable as 'British-taken'. Alfred

Newton was once duped with some eggs of Pallas's Sandgrouse (a Central Asian bird that occasionally spreads far to the west), which were alleged to have been collected in Norfolk; a number of ornithologists, including Dresser, investigated the matter and proved that the eggs had been imported from Central Asia (Newton, 1905–7: 12). The eggs of common species were often passed off as being those of rare species and collectors' marks were sometimes removed from eggs and falsified. Supposed 'clutches' of eggs were sometimes made up of eggs from separate nests (Cole, 2006: 96–100). Dealers' practices were criticised in the *Ibis* (e.g. Anon., 1860; 'Oologicus', 1863), including one particularly vitriolic attack:

> It has ever been the fate of true science to be attended by the false maiden who travesties her every step and parodies all her discoveries…. These [egg collectors] are the victims of a system of imposture as gross, and far less ingenious, than the fictitious antiquities of Italy and Egypt…. In oological [i.e. egg collections] beyond all other collections, dealers' specimens are most unsatisfactory; and from long acquaintance with the frauds of the trade, I would urge upon every young collector never to admit into his cabinet an egg purchased from a dealer. ('Oophilus', 1863: 372–4)

Because of the difficulty in authenticating specimens, many collectors – particularly egg collectors – avoided commercial dealers and only dealt directly with one another, and even then only once they had established mutual trust.

Bird skins were also passed off fraudulently on a number of occasions: there were the 'Tadcaster Rarities', an unlikely cluster of extremely rare birds (three 'firsts' for Britain) supposedly found near Tadcaster in Yorkshire during the 1840s–50s and all associated with a particular taxidermist (Melling, 2005). The most notorious incident was the 'Hastings Rarities' affair late in the century (and into the first two decades of the next), where a high concentration of vagrant birds were supposedly shot around Hastings in Sussex, many for the first time in Britain. Most of these records are now considered to be fraudulent, and to have involved a local taxidermist (see chapter 11; see also Collinson, 2012; Harrop *et al.*, 2012).

Setting up a 'system of collecting'

Elliott Coues, in his 'directions for field-work' (1874), wrote:

> How many birds of the same kind do you want? – All you can get – with some reasonable limitations; say fifty or a hundred of any but the most abundant and widely diffused species. You may often be provoked with your friend for speaking of some bird he shot, but did not bring you, because, he says, 'Why, you've got one like that!' This is just as reasonable as to suppose that because you have got one dollar you would not like to have another dollar. Birdskins are capital; capital unemployed may be useless but can never be worthless. Birdskins are a medium of exchange among ornithologists the world over; they represent value – money

value and scientific value. If you have more of one kind than you can use exchange with some one for species you lack; both parties to the transaction are equally benefited. (Coues, 1874: 27)

Even as a young man, Dresser had great ambitions for his collection, and his actions were a close match with those that Coues would advocate. In order to expand his collection, Dresser set up a 'system of collecting', first described in a letter to George Boardman in 1862:

> I think, on the whole, you and I will do a great deal of exchanging and shall in turn increase our own collections vastly, so please get together all you care and I will do the same…. If I were you I would never throw away a bird if I had time to skin it for I can always get exchange, for skins of all sorts, particularly those of hawks, owls, petrels.[13]

The system was to buy up all duplicate specimens of particular birds and eggs, establish a network of people who may want such things (notably collectors in other countries) and then exchange specimens directly with one another. This system reduced the risk of misidentifications (accidental and deliberate) from commercial dealers, and kept the price of specimens low by avoiding paying commercial dealers' profit margins; it also reduced the cost of specimens at source by buying in bulk. This system would have been a familiar approach to a commission merchant and Dresser once wrote that he thought he 'could do some commission business in the egg line'.[14]

Dresser also encouraged his youngest brother, Arthur, to collect for him while he worked at the Lancaster Mills in New Brunswick from 1863 to 1871:

> You speak about sending me mounted birds but I would much rather have skins such as Albert [Dunn] sent to me for the mounted birds are so difficult to keep and to transport. As for eggs, collect all you can of the [Purple] Martin, [Belted] Kingfisher, [American] Bittern, any snipes or sandpipers, Yellow headed Woodpecker [i.e. American Three-toed Woodpecker], all the Hawks and of the common birds ten or a dozen of each and if you don't know the birds shoot them or snare them and skin them. When you send the small birds eggs send the nest also. I find there are lots of the very small colored birds build [i.e. breed] there as Boardman has collected a great variety of considerable value. In fact you may collect all you like and I am sure I will get you good pay for them here and quite enough to keep you in pocket money.[15]

When Henry Dresser returned to England from America in 1864 (see chapter 2), there were boxes of bird skins and eggs waiting for him from his collectors in Finland, Berlin and Stettin, five boxes of bird skins and eggs from Russia, boxes from George Boardman in New Brunswick, and from his brother Arthur and the employees at the Lancaster Mills.[16] He also had 400-odd bird skins and 100 eggs that he had collected himself in Texas, as well as collections he had built himself on his travels. These specimens were the capital he used to further his collecting network. Dresser quickly exchanged many of his duplicate American bird skins and eggs, with most going to Osbert Salvin (smaller numbers went to Frederick Godman, Henry Tristram, Philip Sclater and others). Dresser recorded his exchanges in an account ledger so that he could readily calculate a 'balance' and 'exchange value' of birds and eggs sent to, or received from, each person, again using the working practices of business (this ledger is preserved in Manchester Museum).[17]

Through the 1860s, Dresser placed himself at the centre of a rapidly growing network of collectors (field and cabinet collectors) in America, Europe and Russia, established through face-to-face contact and correspondence. He used his business trips abroad to establish new relationships with collectors, curators and dealers, further growing his collecting network. For example, Dresser was held up in Copenhagen early in 1865 due to sea ice. Not one to miss an opportunity, he sought out the bird collectors and dealers, as he later described in a letter to George Boardman in Maine:

> During my stay there [I] was delighted to find some collectors of American eggs who have as yet but few, so I can do a good trade with common eggs. I traded off all my American duplicates and such as I think I can again replace through you and have got a whole lot of the rarest things for America.... Thus you see I can send you some good things and have a chance to get rid of any duplicates of all sorts of American eggs you can send me up to 20 of each sort or even more so please do not refuse any for I can send you [i.e. Boardman] full exchange in good things.[18]

This was a most important development for Dresser as it brought him a reliable source of large numbers of birds and eggs from Iceland and Greenland. Many of the unusual birds found there, such as Gyr Falcons and Snow Buntings, were found in both America and Europe and were in high demand with collectors there as being 'their' birds. The Copenhagen collectors were 'desperate' for eggs and took all the specimens that Dresser could supply them with from John Krider in Philadelphia and George Boardman in Maine. Dresser continued to exchange specimens with the Copenhagen collectors for many years.[19] Dresser began paying Manuel de la Torre (in Madrid) to send him specimens in 1866; he and some of the other BOU members paid the wages of a collector in Labrador from 1867.[20] Through the 1860s Dresser set up exchanging relationships with dealers he considered to be reputable, often setting up the relationships while travelling on business. These included three firms in Hamburg: H. F. Möschler (see figure 5.4), J. G. W. Brandt (who received duplicate specimens from his brother at the Zoological Museum in St Petersburg) and Gustaf Martens. Dresser also arranged to exchange specimens with Wilhelm Schlüter of Halle and Tancré of Anclam, as well as several Paris-based dealers: Edmond Fairmaire, the taxidermist Parzudaki, and the Maison Verreaux (which he visited in 1866).[21]

As well as meeting ornithologists and collectors face to face when abroad, Dresser wrote to others who might join (and be useful to) his collecting network. A typical example is a letter to the collector James Hepburn (see figure 5.5), then in San Francisco, written in 1865:

5.4 Heinrich Möschler, from Henry Dresser's album of correspondents.

5.5 James Hepburn, from Henry Dresser's album of correspondents.

By the introduction of our mutual friend Mr. Ed Heatley and as a co-jockey of the same hobby, Natural History, I take the liberty of addressing you…. I have been a most zealous collector for many years and having led a roving life have made a pretty fair collection of European birds and an egg collection rather better than usual … 1863 and 64 I spent in Texas and Mexico and having leisure time collected quite a number of birds and eggs in company with Dr. A.L. Heermann….

I see from your letter to Mr. Heatley that you are more anxious to get specimens of the arctic birds occurring in the western coast. Of these I can offer you those that are found in Greenland and Iceland having regular sources from which I can procure them, getting them yearly….

I should be very glad if we could do some exchanging and would quite as soon have eggs as birds indeed sooner as I should be sure of the authenticity of the eggs coming from you and one should be very careful from whom one receives eggs whereas birds one cannot mistake.[22]

This letter illustrates many of the methods of collecting ornithologists: Dresser used the name of a mutual acquaintance by way of introduction, and he name-checked the well-known Adolphus Heermann (establishing Dresser's social and scientific credentials). He made a clear offer and a clear request to set up an honest bargain, just as a merchant would. Dresser's letters to other naturalists often included lists of his desiderata (lists of specimens that collectors wished to acquire) and lists of specimens on offer. Like many other collectors, he often used a shorthand, giving a reference number for each species instead of writing the name out in full, based on published checklists of names of birds.

Specialities and monopolies

Members of Dresser's collecting network each brought their own 'specialities' to the network: birds and eggs they had special access to and that other collectors desired. George Boardman, who lived in Maine, was a reliable source of eggs of the Hooded Merganser (see plate 14) and Leach's Petrel, both extremely rare in British collections. The latter was (and is) certainly a 'British' bird, while the Hooded Merganser had supposedly occurred in Britain on several occasions (with rather little evidence) and was subsequently included in various books on British birds.[23] This 'Britishness' made both species particularly desirable to those who collected British birds and wanted to complete their collections. Boardman was the only source of eggs of the Hooded Merganser, so Dresser could exchange these for fine specimens for both himself and Boardman: 'As for the Hooded Merganser I can do what I like in the way of getting exchange for them as I have the monopoly. Please try and secure those from your friend at any hazard for I would like to keep the monopoly if possible and I will repay him in made-up plumes.'[24] Dresser sold an egg of the Hooded Merganser at Stevens' in April 1867 along with some eggs from Lapland; the egg was sold for £2 12s 6d.[25]

One of Dresser's own specialities was the Swallow-tailed Kite, a beautiful black-and-white hawk with a long forked tail (see figure 5.6). The bird came from southern North America and South America but was another of those species supposed to have occurred in Britain (a singularly unlikely occurrence, but a record that made this a 'British' bird). Dresser saw these birds when he was in Texas; he encouraged his friends there to hunt after them after he returned to England: 'Do please do all you can to get a lot of the Swallow-tailed Hawk by next year and send all you can to me and I will get rich exchange.'[26] A cotton planter (plantation owner) named Iglehart obtained some of the kite's eggs for Dresser – the first ever to arrive in Britain – in 1865. Dresser tried hard to obtain more eggs of this bird, eventually meeting with success in 1867, as he told Boardman:

> I have had a letter from my Texan friend who says he has found out the best locality for the Swallow-tailed Hawk by writing to the planters and would be pretty certain to get me the eggs. I send a letter to him enclosed and should be much obliged if you would post it and mark on it the best route. Please don't name his name to any of your northern collectors as I don't want anyone else to get hold of him.[27]

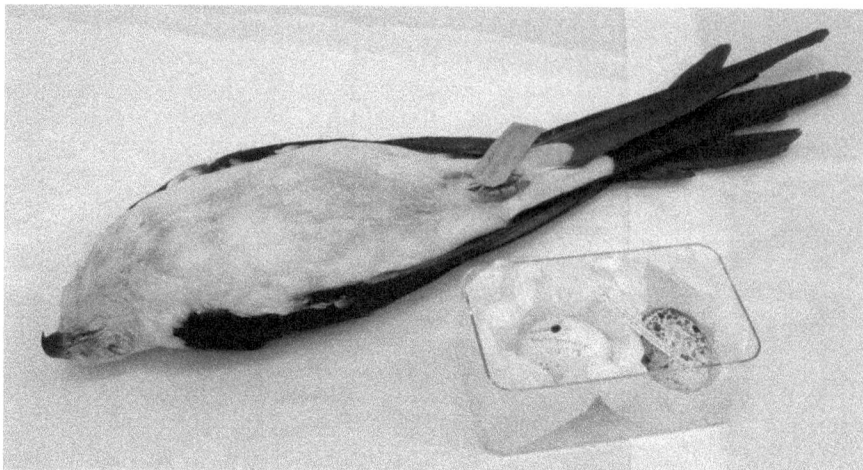

5.6 Swallow-tailed Kite study skin and eggs from Henry Dresser's collection.

Dresser wanted to keep Iglehart to himself, so that he could retain the monopoly on the Swallow-tailed Kite's eggs, and thus ensure that he could exchange them for good specimens; at the same time, he needed Boardman's assistance in posting his letter to Iglehart.

Dresser and museums

Outside of Britain, the best bird collections were generally to be found in museums, and curators were the most highly esteemed scientific ornithologists of the day. Some curators had begun their careers as field collectors before becoming the ultimate cabinet collectors. Museums could be very reluctant to part with their treasures and Dresser worked hard at establishing exchanging relationships with curators. When he was travelling home from Texas in 1864 he visited the curators of the Academy of Natural Sciences in Philadelphia and travelled to the Smithsonian Institution in Washington, DC, in the hope of meeting Professor Spencer Baird, the leading figure in the natural sciences in the United States (see figure 5.7); his biography is explored in the following chapter. The Smithsonian had built up enormous collections from state-sponsored geographical explorations of the United States. Dresser did not want the Smithsonian's bird skins (he thought they were of very poor quality, 'not worth a dollar') but was very envious of its eggs of Arctic birds. He tried to gain an introduction to Baird through George Boardman: 'He [Osbert Salvin] quite made me break the tenth commandment when he showed me the eggs he has just got from Baird ... cannot one get some of those rare eggs out of Baird before he disposes of all he has in duplicate? I would do anything I possibly

5.7 Spencer Baird, from Henry Dresser's album of correspondents.

5.8 Robert Collett, from Henry Dresser's album of correspondents.

can to get some!'[28] It took Dresser five years of persuasion and letter-writing to arrange some regular exchanging with Baird, with a good word on Dresser's behalf from Boardman.[29]

Dresser had entered into exchanging relationships with the curators of most of the major European museums by the end of the 1860s. Each country's museums specialised in different geographical areas through links developed as a result of trade, and imperial and colonial expansion. For example, Professor Johann Brandt (1802–79) was Director of the Zoological Museum of the Imperial Academy of Sciences in St Petersburg during a period of Russian imperial expansion in Siberia and Central Asia. Dresser visited him in December 1867 and January 1868, when on a business trip, and set up an exchanging relationship with him, trading American duplicates from Boardman and Krider for rare specimens from the Russian Far East.[30] Other museum ornithologists, including Theobald Krüper (at Athens

Museum), Count Tomasso Salvadori (at Turin Museum) and Robert Collett (at Christiania, now Oslo, University) (see figure 5.8), provided Dresser with many hundreds of birds from their respective parts of Europe.

Growth of Dresser's collection

As a result of his efforts and negotiations, Dresser's collection expanded rapidly through the 1860s. At the beginning of the decade his collection had been fairly small, with about 100 eggs.[31] Five years later he possessed a mounted specimen of almost every European bird (approximately 430 species) and an egg collection covering about 400 European and America birds.[32] He gave up collecting mounted specimens soon afterwards as they took up too much room, and focused his attention on acquiring study skins and eggs of European and North American birds (but keeping a selection of well-mounted birds of prey).[33] When Dresser's family moved into The Firs in 1865 (see chapter 3), he fitted out a room as his museum.[34] By 1866 he had filled 200 large glass-topped display boxes with the eggs of almost every species of European bird and many from North America; this must have run to several thousand eggs.[35] He wrote to Boardman, 'thus by degrees my collection mounts up and I hope someday to have a pretty good one'.[36] By 1868 his bird skin collection had grown to about 1,100–1,200 study skins stored in drawers in four large, low mahogany cabinets that formed a work surface in his museum room.[37] His family were less excited at these developments and referred to his collection as his 'rubbish'.[38]

Dresser's growing collection required considerable input and maintenance: preparing study skins from bird corpses, remodelling and restuffing old skins, labelling specimens, arranging them in collections cabinets, and checking preserved skins for damage from insects, soot and mould. He used a variety of chemicals to help preserve his collection, including arsenical soap, pyrogens, benzole to kill clothes moths, 'sublimate of lime' to preserve bird nests, alcohol (he sometimes received small birds pickled in alcohol) and mothballs: his collection must have produced quite an odour.[39] Dresser's exchanging activities also involved a lot of work, writing out letters and lists of desiderata, packing specimens and arranging to send them off. Dresser was pleased when Jules Verreaux visited him in 1870 when he was an exile fleeing from the Siege of Paris (Stresemann, 1975: 162–3). Dresser and his fellow ornithologists gave Verreaux boarding for a week each, as he described in a letter to Boardman: 'The old fellow is well worth his keep for he is the quickest worker in arranging a collection I ever saw and will do in a week what would take all my spare time for months.'[40] Apart from Verreaux's visit, Dresser curated his own collection after spending the day working in the City.

The birds and eggs in Dresser's collection were only a portion of the specimens that passed through his hands. 'His collection' referred to the birds and eggs he wished to keep; specimens for exchange were not considered to be part of this (and were not catalogued, for instance). It is difficult to establish exactly how many specimens passed through his hands but rarer birds and eggs would have required heavier exchanges, meaning that Dresser would have to supply larger numbers of commoner species, to exchange equally rare specimens, or to pay higher prices. Dresser once had Elliott Coues place an advert in an American journal in search of male and female specimens of the Labrador Duck, offering £40. The advert was placed in 1876, the year after the last sighting of the bird, which became extinct for unknown reasons (possibly habitat change or over-hunting) (Coues, 1876; Chilton, 1997). During 1865–66, Dresser received 609 eggs of a variety of species and sent out 790 eggs, most of which had come from other collectors. From the late 1860s, he seems to have acquired considerably more eggs than he disposed of and the same was probably true of skins. There are hints of the extent of his exchanging from his early correspondence:

> I have quite a large quantity of skins of the rarer water birds from North America: Harlequins, Surf Scoters, Fork-tailed Petrels [Leach's Petrel], Buffle-headed Ducks etc. etc. in all nearly two hundred, and am exchanging away the duplicates by degrees.... I am getting almost too many skins to stow them away with any degree of comfort.[41]

Most of the North American water birds mentioned above came from George Thomas, the lighthouse keeper at Point Lepreaux in New Brunswick. Among them were thirty-four Surf Scoter skins, large ducks that would form a sizeable consignment on their own (see plate 15).[42] On other occasions Dresser requested up to thirty Cedar Waxwing eggs and thirty-odd Spruce Grouse eggs from Maine and New Brunswick, and Swallow-tailed Kites (from Texas) by the dozen.[43] He traded away over thirty Azure-winged Magpie eggs received from Manuel de la Torre during 1865–69. Leach's Petrel eggs collected by George Boardman on Grand Manan Island (New Brunswick) during 1865–68 were sent on to Alfred Benzon (seven), Kammerråd H. C. Erichsen (two) and Pastor Theobald (two) in Copenhagen, and to Alfred Newton (ten), Henry Tristram (four), Osbert Salvin (three), Samuel Stafford Allen (two), Rev. Herbert Hawkins (two at 7s each), William Bridger (two), Alfred Crowley (one), Henry Elwes (one) and Robert Kent (two) in England.[44] Captain Rowland Sperling sent Dresser a batch of thirty-two Sooty Tern eggs from the Bahamas in 1868 with the instruction: 'You and Godman take half a dozen each and the remainder send to Newton for distribution amongst members of the BOU.'[45]

Dresser as a source of specimens

Dresser provided many collectors with rare bird skins and eggs, usually in return for specimens for his own collection, or which could be used to further his collection. He was quick to point out to correspondents that he was not a 'dealer', writing to one correspondent, 'I don't want to make any bargain with you for what assistance I may give. I do it willingly as one naturalist should do to another and am no dealer but an amateur.'[46] The key difference was that Dresser's motivation was to improve the quality of his collection (and those of his collaborators) rather than financial gain:

> I am much bothered with letters from all sorts of people but here in Europe we have so many private collectors that I make a rule never to let anything I have get into a dealer's hands and find that by so doing and only swapping with a few good collectors that I can do much better than otherwise and therefore if a dealer writes to me I just put his letter into the fire without further ceremony.[47]

While he did not trade specimens for money, he would acquire birds and eggs that he knew he could exchange for specimens for his own collection. For example, he accepted Chinese eggs from Henry Tristram that he then exchanged with Adolphus Heermann; he also acquired birds-of-paradise for Tristram, who then exchanged them with other collectors.[48] By any definition, this amounted to trading of some kind, albeit not for money. Dresser and others like him would have had a real impact on the business of commercial dealers by exchanging with each other directly and cutting dealers out of the loop (and because they had privileged access to first-rate collections). Dresser was resolutely opposed to commercial dealers throughout his career, writing to a correspondent in 1906:

> I would much rather exchange for eggs and not sell as I have always set my face against anything like dealing and work at science for the love of it and not for anything I may gain by it, but if you cannot give exchange I must in your case take money and spend it on eggs as soon as I can.[49]

Dresser did sometimes sell eggs and bird skins (possibly when his contacts could not make reasonable exchanges) and he paid for many specimens when he could not negotiate exchanges. The English collector E. J. Rhodes received 240 eggs of eighty species from Dresser during 1868–71, as well as many other eggs from a variety of dealers and from Stevens'. Rhodes' catalogue (which records the prices he paid for specimens) reveals that Dresser sold eggs to Rhodes at more or less the 'going rate' (i.e. the rate charged by commercial dealers such as the Rev. Herbert Hawkins, Dunn of Stromness and the London-based firms);[50] however, Dresser also supplied Rhodes with many eggs for free, so he provided Rhodes with eggs at a substantially lower cost overall. These included eggs of rare birds such as the Black Stork from Denmark, Ospreys from near Stettin, and Gyr Falcons and Peregrines from Lapland.

As well as dead and preserved birds, Dresser tried to obtain live birds on several occasions on behalf of his contacts. These included Hazel Grouse from Scandinavia for Lord Lorne to release on his estate in Scotland (Palmer, 1895: 224), and chicks of Ruffed and Spruce Grouse from New Brunswick for someone (probably Philip Sclater) who wanted to introduce them to Britain, presumably for sporting purposes.[51]

Collecting books and paintings

For serious ornithologists, 'ornithology' was as much a literary pastime as one concerned with specimens, as they needed to keep abreast of the latest discoveries and descriptions of expeditions to far-flung places. Access to a good library was essential to make the most of a good collection: understanding what the birds were, how they varied, and the circumstances of their collection. Many ornithological books and journals were produced in short runs and were collectors' items in their own right, particularly those with illustrations. Some journals provided authors with unbound 'reprints' of their articles to give away (without coloured plates); these were highly collectable in their own right, particularly when they included personal dedications from the authors themselves. In his younger days, Dresser usually bought his books second-hand, in order to save money (as he proudly explained to his friend George Boardman).[52] He was wealthy enough to subscribe to some of the great bird books of the day, including Gould's *The Birds of Great Britain* and Sharpe's book on kingfishers (*A Monograph of the Alcedinidae*). Over time, Dresser built up a fine ornithological library; this included a large collection of reprints – many of which related to birds and eggs in his own collection – that he had bound in leather-covered volumes.[53] He also collected paintings of birds from leading artists, including John Gerrard Keulemans and Joseph Wolf (see, e.g., plates 13 and 47).

Conclusion

This chapter has sought to explore some of the diverse means by which collectors built their collections. Dresser's own collection developed rapidly and productively due to his development of his exchanging network, which he carefully built up over many years by cultivating relationships with other collectors and sources of specimens. His tactics of dealing directly with other collectors – making clear bargains and constantly adding to the quality of his collection (and relationships) – has clear parallels with his mercantile work. Even by the late 1860s, he had one of the finest bird skin and egg collections in Britain, as a direct result of his personal travel, business links, wealth and determination. He continued to build on this strong foundation through the remainder of the nineteenth century and the early twentieth century.

Notes

1   Manchester Museum (MM hereafter), ZDH/7/6, 8 April 1866.
2   See also 'Oophilus' (1863).
3   Smithsonian Institution Archives, Record Unit 7071, Box 1, Folder 12: Dresser,

Henry Eeles, 1862–71 (SIA, RU7071 hereafter), letters from Henry Dresser (HED hereafter) to George A. Boardman (GAB hereafter), 30 September 1865 and 12 October 1866.

4   This specimen was formerly in MM but was destroyed after it suffered from attack by pests in the mid-twentieth century, its significance being unappreciated. A slightly older record, of a bird shot at Epsom Downs Racecourse in 1870 (in Charterhouse Museum) represents the first accepted occurrence of the species in Britain (Harrop *et al.*, 2013; Self, 2014: 160).

5   SIA, RU7071, letter from HED to GAB, 19 August 1865.

6   A. Newton, 'Borrowed plumes' [letter], *The Times*, 28 January 1876, p. 10, col. b.

7   See Frost (1987) and Morris (2010) for a history of taxidermy in Britain, and Cole (2006) for a history of commercial egg-dealing in Britain.

8   The prices charged for the eggs of a selection of species by the main commercial dealers are compared by Cole (2006: 110–11). The relative value was calculated as the purchasing power, using www.measuringworth.com, accessed 11 March 2017.

9   Many of these annotated catalogues are preserved in museums. Locations of these can be found in Chalmers-Hunt (1976). Some of HED's annotated sales catalogues are preserved in MM, ZDH/7/10.

10 'I would like to have eggs from the best sources [including Newton] having as yet kept quite clear of dealers' eggs.' Cambridge University Library (CUL hereafter), MS.Add.9839.1D.223, C613, letter from HED to Alfred Newton (AN hereafter), 18 April 1865.

11 AN wrote out his views on egg-collecting in 1893, following an attempt to introduce a Bill to protect wild birds and their eggs. These were circulated among politicians before being reprinted in the *Annals of Scottish Natural History* the following year (Newton, 1894). They are quoted in Wollaston (1921: 146–50).

12 MM, ZDH/11/1, 'Catalogue of eggs', p. 372: 'All these eggs returned to Bamberg [a dealer based in Weimar] as being doubtful', and p. 374: 'These two eggs I have retained [from Bamberg], as though from a tainted source they appear to be correctly named.'

13 SIA, RU7071, letter from HED to GAB, 24 June 1862.

14 SIA, RU7071, letter from HED to GAB, 28 September 1866.

15 MM, ZDH/1/2, duplicate letter from HED to Arthur Dresser, 10 November 1864, pp. 42–3.

16 MM, ZDH/1/2, HED's 1864–65 copybook, and SIA, RU7071, letter from HED to GAB, 12 October 1864.

17 MM, ZDH/1/1. HED'S 'Record of Exchanges'.

18 MM, ZDH/1/2, duplicate letter from HED to GAB, 4 March 1865, pp. 110–13.

19 SIA, RU7071, letters from HED to GAB, 26 May and 24 November 1865.

20 SIA, RU7071, letter from HED to GAB, 3 August 1867.

21 MM, ZDH/7/5, 23 January 1866.

22 MM, ZDH/1/2, duplicate letter from HED to James Hepburn (San Francisco), 11 November 1865, pp. 476–80.

23 See Dresser (1896: 296–7) for a list of occurrences and published references to these.

24 SIA, RU7071, letter from HED to GAB, 19 August 1865.

25 Advertisement – Eggs, bird-skins, Euplectellas, *The Athenaeum*, 2057, 30 March 1867, p. 404. The price of the Hooded Merganser egg was recorded by HED, see MM, ZDH/11/1, Dresser's 'Catalogue of Eggs', p. 194.

26 MM, ZDH/1/2, duplicate letter from HED to Adolphus Heermann, 1 November

1864, pp. 27–8. HED also advised Heermann to spend a month collecting the birds, eggs and chicks: MM, ZDH/1/2, letter from HED to AN, 26 April 1865, pp. 198–9.

27   SIA, RU7071, letter from HED to GAB, 8 June 1867.

28   SIA, RU7071, letter from HED to GAB, 7 November 1865.

29   SIA, RU7071, letter from HED to GAB, 15 February 1868; see also letters dated 14 December 1865, 1 February and 29 March 1867.

30   CUL, MS.Add.9839.1D.272, D866, letter from HED to AN, 7 January 1868.

31   CUL, MS.Add.9839.1D.263, D443, letter from HED to AN, 26 January 1867.

32   SIA, RU7071, letter from HED to GAB, 4 August 1865.

33   SIA, RU7071, letters from HED to GAB, 1 March 1867 and 13 March 1868.

34   SIA, RU7071, letter from HED to GAB, 17 March 1865.

35   SIA, RU7071, letter from HED to GAB, 17 August 1866.

36   SIA, RU7071, letter from HED to GAB, 31 August 1866.

37   SIA, RU7071, letter from HED to GAB, 13 March 1868.

38   CUL, MS.Add.9839.1D.235, C476, letter from HED to AN, 27 July 1865.

39   SIA, RU7071, letters from HED to GAB, 14 October 1865, 29 August and 14 September 1867.

40   SIA, RU7071, letter from HED to GAB, 22 October 1870.

41   MM, ZDH/1/2, duplicate letter from HED to William Culverwell, 9 November 1865, pp. 464–5.

42   SMI, RU7071, letter from HED to GAB, 13 September 1865.

43   SIA, RU7071, letters from HED to GAB, 17 March 1865, 13 September 1865 and 17 August 1867 respectively.

44   CUL, MS.Add.9839.1D.269, D516, letter from HED to AN, 5 October 1867, and information in MM, ZDH/1/1.

45   CUL, MS.Add.9839.1D.277, D986, letter from HED to AN, 2 May 1868.

46   ZDH/1/2, duplicate letter from HED to an unnamed correspondent, 5 January 1864, pp. 108–9.

47   SIA, RU7071, letter from HED to GAB, 13 April 1865.

48   MM, ZDH/1/2, duplicate letter from HED to Henry Tristram, 13 July 1865, pp. 334–5; McGill University Library, Rare Books and Special Collections, Blacker-Wood Autograph Letter Collection, letter from HED to Robert Ridgway, 4 July 1892.

49   Museum of Comparative Zoology Archives, Harvard University, John Thayer Collection, letter from HED to John Thayer, 16 August 1906.

50   E. J. Rhodes' catalogue is preserved in Chelmsford Museum along with his collection. Prices charged by commercial dealers are taken from Cole (2006: 110–11) and HED's annotated sales catalogues in MM (ZDH/10).

51   SIA, RU7071, letter from Joseph Dresser to GAB, 28 April 1866.

52   SIA, RU7071, letter from HED to GAB, 28 September 1867.

53   These volumes are now in MM.

6   Discovering the birds of Europe, I

We have seen how Dresser secured his position in ornithological society in chapter 4, and how he began to expand his collection in chapter 5. The most ambitious collectors directed their collecting impulse beyond simply acquiring specimens, but collected with a view to publication. From 1870 onwards, Dresser's collecting was directed towards one purpose: to review the bird species of Europe, including their distribution and variation around the world, in order to produce a great book, *A History of the Birds of Europe*. This collecting–publishing project was a mammoth undertaking: it took thirteen years to complete the eight large volumes, which were published between 1871 and 1882. The project – Dresser's largest and best-known book – forms the subject of the following three chapters. This and the following chapter explore the activities of some of the people who provided the specimens and information – the raw materials – that Dresser needed to produce the book, while chapter 8 explores how the book itself was produced. So, for the following two chapters, we will extend out from England to visit the far reaches of Dresser's collecting network, to discover who was there, for what reasons, and what they provided Dresser with in terms of specimens and information. The means by which Dresser acquired birds from particular geographical areas reflected the political and social relationships between parts of the world and the West; birds and eggs, and information on these, were effectively commodities in a global market, obeying laws of supply and demand. The people involved in Dresser's endeavour represent an incredibly diverse set of people who, for a wide variety of reasons, put themselves or found themselves around the globe.

Background to *A History of the Birds of Europe*

The instigator of the *History of the Birds of Europe* was Richard Sharpe, who, early in 1870, proposed to Dresser that they go into partnership to produce a great encyclopaedia on the birds of Europe.[1] The project required access to the very best specimen collections and libraries, whether in museums or private hands, as they were the most basic source of information on what species occurred where. New collections would also need to be acquired from explorers and other collectors. Sharpe had neither the opportunity nor the financial ability to produce such a collection, so he needed a 'collecting partner' if he hoped to produce his encyclopaedia.[2] Dresser fitted the bill perfectly: he owned a good collection of European birds and their eggs; he could acquire more good specimens from his contacts and access the collections of other BOU members, and he could make sense of the literature on birds published in other European languages. Just as importantly, he was ambitious and able enough to take on the project.

Sharpe and Dresser intended the book to cover the latest stage of knowledge of the species that were found in Europe. The two partners had plenty to do: the last significant book on the subject in English had been John Gould's *The Birds of Europe*, published in five volumes during 1832–37 (his *The Birds of Great Britain*, 1862–73, covered many of the same species, but did not review the wider literature). Charles Bree's *A History of the Birds of Europe, Not Observed in the British Isles*, issued over 1859–67, was an inferior work. These books represented the 'competition'. Information on European birds was scattered among books and journals in different languages, in the unpublished notes of colonialists and travellers, and on the labels attached to the legs of bird specimens. Information was of variable quality and reliability, and many species were as good as unknown. Large quantities of information had been generated about birds, in Europe and elsewhere, which meant that an authoritative ornithology on European birds was much needed.

A History of the Birds of Europe was one of a genre of similar projects underway at the time, concerned with the production of reliable, standardised information and extending across all natural history subjects (other examples are discussed in chapter 8). Empirical description and cataloguing were the order of the day, coinciding with colonial and imperial discovery, when enormous numbers of novelties were first made known to Western science (see Pickstone, 2001). Lists of the birds of particular places formed a mainstay of the pages of the *Ibis* (Moreau, 1959). Standardisation in the natural sciences included understanding patterns of distribution of animals and plants, and ecological zones (Browne, 1983). Two English naturalists, Alfred Wallace and Philip Sclater, were particularly eminent in this area, defining biogeographical zones characterised by particular or peculiar fauna and flora (e.g. Sclater, 1858; Wallace, 1876).[3] *A History of the Birds of Europe* was based on the Western Palaearctic, one of the great biogeographical regions of the world.

The book was to cover all the birds of Europe, the Azores, Canary Islands and Madeira, North Africa, the Urals, Caucasus, Turkey, Syria, part of Persia and across to the Red Sea. It was also to include American vagrants that had been reliably observed in the area (even if there were only one or two records). Dresser excluded thirty-nine species that had occurred in the peripheral areas to the south of 'his' region but which were not felt to be 'European' enough. They were African birds that had a toehold in the margins of 'Europe', including colourful sunbirds, large vultures and even the Ostrich. In this we see that there was more to being a 'European bird' than just 'a bird from Europe'. As if this was not enough, *A History of the Birds of Europe* had an even greater scope. It aimed to provide the fullest account of these 'European' species by digesting the latest information on their distribution, movements and variation wherever else they occurred in the world. So little was known about many birds that the two partners had to collect and assess specimens from all parts of the world, to ascertain the true distribution of 'European' species elsewhere. This meant that *A History of the Birds of Europe* was to be more than a book about European birds, but was intended to provide a globalised knowledge of these species.

A History of the Birds of Europe was to be a great adventure for the two partners, connecting them with the exploits of travellers and explorers in the far reaches of the Empire and in remote wildernesses (something that Dresser had gone some way to achieving already through his collecting network). The project gave a kind of structure, direction and purpose to Dresser's collecting enterprise. He would have to identify which collectors, travellers and museums had the very best collections of birds from particular geographical areas, to cultivate relationships with them, and to arrange to acquire their collections – whether by gift, exchange or purchase – to fill in the blanks and ensure the completeness of the project and his collection. The project required collecting on a near-industrial scale: 'I have been getting lots of good things in lately and have been awfully extravagant in my purchases of birds and eggs, but we [Sharpe and Dresser] must have them if we intend our work [i.e. the book] to be up to the mark'.[4] Dresser gave up collecting American birds in the 1870s in order to focus his efforts on the work (literary and in collecting) required for *A History of the Birds of Europe*. Although it was Sharpe who suggested the project in 1870, the two ornithologists went their separate ways in 1872, leaving Dresser to complete the project alone through to the early 1880s (discussed in further detail in chapter 8).

Dresser described the project to Alfred Newton in February 1870: 'We don't want to make a pretty picture book, but if possible a good useful book embodying as much as possible the latest information on the ornithology of Europe.'[5] Newton lent his support, but at least some others, including Henry Tristram, were critical of the idea.[6] Charles Bree complained to Dresser that he had a poor choice of partner, who had subjected him to a 'vulgar attack' in *The Field*; Bree sent Dresser an 'intemperate and uncalled for tirade, prophesying downfall to the *Birds of Europe*'.[7] Lord Lilford, on the other hand, offered to help cover the costs of publication and the production of plates (Trevor-Battye, 1903: 254). A number of Dresser's European contacts promised to contribute information for the book.[8]

A word on scientific travellers

Field collectors were a hugely varied band: they could be acting on behalf of governments or scientific societies, or their activities could be more informal. Many were incidental travellers who went abroad for leisure, business or trade. Others were speculators, trying to earn a living by collecting (Mearns and Mearns, 1988, 1992, 1998). Travellers, especially British ones, were a particularly important source of specimens for Dresser. Official naturalists were attached to many scientific and military expeditions, but their specimens were often destined for museums, so were not readily available to private collectors. Far greater numbers of people found themselves, or put themselves, in situations far from home, whether as part of the machinery of trade and Empire – as colonialists, soldiers, doctors, missionaries or merchants – or as leisure travellers, escaping the straitjacket of British society and exploring the world with ever-increasing ease. Many of these became 'scientific travellers', gathering information that could be shared with others as part of a wider scientific endeavour, or simply for their amusement while far from home (Lloyd, 1985; Raby, 1986; Fan, 2004):

> Perhaps no other activity in natural history was more filled with intense action and experience than fieldwork.... Whether high in the Himalayas or deep in the Amazon jungles, Victorian naturalists' expeditions, often full of trials and dangers, demanded psychological fortitude and physical stamina – a situation at which the naturalists never tired of hinting in their travel accounts. The travelling naturalists valued the experiences of an expedition as much as the specimens shipped home.... There were encounters between bearers of the torch of civilization and savages, primitives, barbarians.... The mysterious, the unknown, the space devoid of Western presence might actually be some other people's backyard. What was a heroic adventure to the Westerners might be an everyday routine to others. (Fan, 2004: 122–3)

Travellers were encouraged to collect specimens (and other things) and information by a number of popular books and instruction manuals. The Royal Society and the Admiralty collaborated to produce the *Admiralty Manual of Scientific Enquiry: Prepared for the Use of Officers in Her Majesty's Navy; and Travellers in General* (first issued in 1849 and running to many editions). The *Manual* sought to encourage rigorous collecting of scientific data (and materials) that could be of use to experts; chapters were written by leading naturalists and scientists, including Charles Darwin (Herschel, 1849; Levere, 1993: 142–89).[9] The Royal Geographical Society's *Hints to Travellers*, first issued in 1854 and running to many editions, was produced with similar aims (Driver, 1998). Books such as Francis Galton's *The Art of Travel* (1855), 'a manual to all those who may have to "rough it"' (quoted in Driver, 1998: 27), encouraged resourceful travellers to find their own adventures and to make use of a notebook, gun, thermometer and barometer, all in the name of geographical and scientific endeavour, and as a fulfilling and morally uplifting activity.

Spencer Baird of the Smithsonian Institution issued a pamphlet with instructions for collecting and preserving natural history specimens in 1851. This was subsequently expanded and published as *Directions for Collecting, Preserving, and Transporting Specimens of Natural History* in 1852, with specialised instructions for collecting and preserving different types of animals (Baird,

1852; Lindsay, 1993: 27–35, 143). Baird's notes were 'such hints as may enable travellers and others to secure and preserve the different objects of Natural History with which they may meet.... Officers of the Army and Navy, Clerks of Trading Posts, Indian Agents, Land Surveyors, Missionaries, etc. very often have it in their power to procure specimens of the highest interest' (Baird, 1852: 3). Baird's aim was to establish a collecting network that could supply the Smithsonian with specimens that could be used to answer research questions on a continental scale (Deiss, 1980; Bruce, 1987: 188; Mearns and Mearns, 1992: 43–54; Lindsay, 1993; Binnema, 2014). Allan Hume, one of the leading ornithologist-collectors in India (discussed further in the following chapter), issued *The Indian Ornithological Collector's Vade Mecum*, with collecting and preserving instructions specially tailored for India, including local names for preservatives so they could be bought in markets. Hume's main advice was 'in preserving birds the main point is to use good arsenical soap, and plenty of it' (Hume, 1874: 1). This manual was intended to be carried by collectors when travelling (the Latin *vade mecum* translates as 'go with me'). Travellers to remote regions were thus encouraged to collect and make observations of all the birds they saw, not only rare ones or new species, as even this information helped naturalists understand the distribution of more widely distributed species, filling in the blanks of knowledge.

As well as promoting travel, the large numbers of published travel narratives encouraged people to consider becoming authors themselves; many tried to publish their accounts, only to have their hopes dashed (see Speake, 2003; Steinitz, 2003: 331–4). First-hand travel narratives typically emphasised the personal danger travellers had endured, and included descriptions of local customs (descriptions of marriages and funerals were *de rigeur*) and the local economy, government, agriculture and industry, as well as natural history; local people were often written about in a derogatory way.[10] Collectors often flagrantly violated local customs by killing protected birds and wrote boastfully about these encounters afterwards.[11] The important thing to recognise is not just that they disrespected local customs, but that they wrote about these events with such gusto. Once travellers had written up (or tried to write up) their travels, they often disposed of any collections, either privately or at auction.

Collecting the birds of Great Britain

Dresser had already built a good collection of British birds, with many specimens coming from Leadenhall Market (see chapter 5). Richard Sharpe provided him with hundreds of specimens of commoner birds from Berkshire and Sussex, including many that Sharpe shot at Pagham Harbour in Sussex, a popular shooting spot. A professional bird catcher, C. Davy of Kentish Town, provided hundreds of birds from Hampstead. Between them, these two collectors provided a representative series of 'English' birds for Dresser's collection and *A History of the Birds of Europe*.

6.1 John Harvie-Brown, from Henry
Dresser's album of correspondents.

Scotland was a popular holiday destination for those with a taste for field
sports and natural history, with rare and spectacular birds (many of which
were not found in England) in equally spectacular landscapes. Dresser took a
number of holidays to Scotland, visiting the breeding sites of rare birds in 1876
along with John Harvie-Brown, a leading expert on Scottish natural history
(see figure 6.1).[12] Dresser visited Sutherland alone in 1877 and obtained eggs
of some rare Scottish birds from gamekeepers, whom he 'paid' in fishing flies.[13]
Dresser visited the Scottish Highlands again in July 1886, well after the *Birds
of Europe* project was completed. He visited a small loch where the Slavonian
Grebe was supposed to breed, the only site in Britain. Dresser saw some
grebes on the water but could not be absolutely certain of their identity, so he
afterwards sent the gamekeeper illustrations of the birds from *A History of the
Birds of Europe*, so he could try to identify the birds.[14]

Collectors in northern Europe and the Arctic

Dresser had many opportunities to acquire northern birds to help him with *A
History of the Birds of Europe*. The region was a popular destination for British
scientific travellers, as many birds that were common in Britain in winter bred

farther north. The breeding grounds of some of these species, such as the Knot (see plate 16), Sanderling, Little Stint and Grey Plover (four species of wading bird), were either poorly known or unknown, providing the more ambitious collectors with a reason (or post-hoc justification) for venturing north (Seebohm, 1880b: 2–3; Harvie-Brown, 1905: v–vi; Nethersole-Thompson and Nethersole-Thompson, 1986; Vaughan, 1992; Ratcliffe, 2005). Dresser dreamt of tracking some of these birds down himself: 'I should like amazingly to hunt out the breeding places of the Knot and Sanderling and feel convinced they must take up their quarters for breeding somewhere in the far NW of Europe'.[15] Although his wish was never to be realised, he did get as far as the North Cape of Norway in 1881 on a pleasure trip.[16] Through the 1870s, he assisted a number of other British collectors who were travelling to northern Europe by putting them in touch with his business and scientific contacts; he was often repaid for his troubles with good bird and egg specimens on the travellers' return.

The birds of Greenland and Iceland were effectively dealt with by Dresser's contacts in Copenhagen, notably Alfred Benzon and his acquaintances. Benzon (1833–84) owned a large pharmaceutical firm and had a large egg and bird

6.2 Alfred Benzon, from Henry Dresser's album of correspondents.

6.3 Kammerråd H. C. Erichsen, from Henry Dresser's album of correspondents.

skin collection, with his own curator (Grouw and Bloch, 2015) (see figure 6.2). Another of Benzon's circle, Kammerråd (Counsellor) H. C. Erichsen, paid the wages of a collector in Greenland for many years (see figure 6.3). The birds of the Faeroes were dealt with by another member of the Copenhagen group, Sysellmann (Governor) Hans Müller of Streymoy (Strømø), the leading authority on Faeroese natural history (e.g. Müller, 1862). An English friend of Dresser, Henry Feilden (1838–1921), an English army officer and aristocrat, paid a six-week visit to the Faroe Islands in 1872 with an introduction from Dresser.[17] Feilden investigated the indigenous Pied Raven (a local aberration of the Raven, now extinct); witnessed the counting of the *naebbetold* or bill tax, which obliged each man to kill a number of predatory birds each year and submit their bills (beaks) as evidence; and studied the locals' harvesting of wild birds for food (including chicks of the Great Skua, a large, predatory gull-like bird) (Feilden, 1872; Bloch, 2012). Feilden gave Dresser a number of birds from his trip in return for his help.

Dresser acquired a representative collection of Norwegian birds from Robert Collett (1842–1913), a curator at the Zoological Museum in Christiania (Oslo) and the leading authority on Norwegian birds (see Collett, 1869). Collett gave Dresser some very rare Norwegian birds in 1881, when Dresser visited Christiania. Another museum curator, Wilhelm Meves (1814–91), provided Dresser with collections of birds from Sweden and the Urals, including both common and rare species. Meves, originally from Brunswick in Lower Saxony (now Germany), worked at Stockholm Museum during 1841–77 and travelled widely in northern Europe (see figure 6.4).[18] Dresser obtained another collection of birds from the Urals from Leonid Sabaneyev (1844–98; Dresser referred to him as Sabanaeff), a Russian aristocrat who wrote a number of books on natural history and hunting. Dresser helped to arrange for the publication of Sabaneyev's articles in Britain.[19]

Two English collectors supplied Dresser with eggs of the birds of Lapland to help with *A History of the Birds of Europe*. Alfred Newton gave Dresser eggs collected by his friend John Wolley, widely feted as one of the greatest English field ornithologists (Newton, 1860c, 1864).[20] Wolley built a vast collection of bird eggs collected around Muoniovara during 1853–57, assisted by upwards of 300 locals, mostly teenage lads and peasants (see e.g. Newton, 1864: 121–38; Newton, 1902: 212–39). One of Wolley's best-known discoveries was the nest of the Smew (plate 3), although this seems less remarkable when it is realised that the discovery was made in a nest box that the local people had erected for ducks in order to harvest their eggs (Wolley, 1859; Newton, 1905–7: 626). Dresser acquired more eggs from another English collector, Horatio Wheelwright (1815–65), who lived in Sweden for ten years, sending birds and eggs to auction in England during 1861–66. Wheelwright, who had led an adventurous life, wrote popular books and articles for *The Field* under the *nom de plume* of 'An Old Bushman' ('An Old Bushman', 1864, 1865; Chisholm, 1976, 1979). Dresser was a great admirer of Wheelwright's writings and purchased many of his specimens at Stevens'. He wrote to George Boardman:

6.4 Wilhelm Meves, from Henry Dresser's album of correspondents.

'He [Wheelwright] was originally a gentleman collector and by profession a lawyer but took to drinking.… I like his writings very much and I dare say should like him personally if I knew him personally. I have been often amused by odd tales the Swedes tell about him when on the [drinking] spree.'[21]

Dresser was particularly successful in obtaining birds and information on birds of northern Russia. His friend John Harvie-Brown visited Archangel (Arkangelsk) in 1872 along with another naturalist, Edward Alston, with letters of introduction from Dresser.[22] They were assisted there by a Polish exile named Ignati Piottuch.[23] Piottuch subsequently sent Dresser hundreds of birds towards *A History of the Birds of Europe*, although Piottuch could drive a hard bargain for rarer specimens: 'we must tell him [Piottuch] not to send too many common birds.… We must keep in with him', Dresser once wrote to Harvie-Brown.[24] Another English collector, Henry Seebohm (1832–95; a wealthy steel manufacturer from Sheffield), travelled to northern Russia armed with introductions from Dresser, help for which Dresser was repaid in valuable information and specimens for his book.[25] Seebohm and Harvie-Brown travelled to the Pechora River in European Russia in 1875, again accompanied by Piottuch, where they collected over 1,000 bird skins and 600 eggs, including some of the first Little Stint and Grey Plover eggs to be brought to Britain (Seebohm, 1880b). Seebohm made a more extensive trip to the Yenisei (the largest river flowing northwards through Siberia) in 1877 with Captain Joseph Wiggins, an English entrepreneur who hoped to establish a sea-trade link with Siberia via the Kara Sea (see Johnson, 1907; Stone, 1994).[26] On arriving on the Yenisei, Seebohm and Wiggins purchased a small schooner to sail on the river, naming the boat the *Ibis* after the journal of the BOU; this trip produced another 1,000 bird skins. Seebohm's accounts of his expeditions (Seebohm, 1880b, 1882a, 1901) are classic natural history travelogues, filled with adventure stories: shipwreck, mutiny, hunting and eating bears, and particularly derogatory accounts of the local people (see also Harvie-Brown, 1905; Lloyd, 1985: 107–19). Between them, Seebohm, Harvie-Brown and Piottuch 'covered' the birds of northern Russia for *A History of the Birds of Europe* by providing Dresser with many specimens and information from their travels.

Travelling naturalists in southern Europe and North Africa

Some of Dresser's closest associates regularly visited southern Europe and North Africa, usually for the benefit of their health. Among these travellers was Lord Lilford, who visited Spain and the Mediterranean between 1856 and 1882, hunting birds and mammals, and sailing in a yacht. Lilford wrote several lengthy papers on his experiences in the *Ibis* (Lilford, 1865, 1866, 1875, 1887) and sent many animals to the British Museum. Lilford gave Dresser a number of specimens, including skins and eggs of the rare Audouin's Gull, which he rediscovered, and specimens of two kinds of bird that Dresser named as new species.

Dresser's friend Howard Saunders provided him with a representative collection of the birds of Spain (see figure 6.5). After giving up work in 1862, Saunders visited Spain several times during 1863–70; having spent considerable time in Peru, Saunders was possibly attracted to Spain by the language.[27] He also travelled in search of relief from what he described as 'rheumatism'; while his condition was sometimes debilitating, he also managed to collect many birds and hunted large mammals. Among the more unusual trophies of Saunders' collecting exploits were two Bonelli's Eagle chicks that he took from their nest, alive, tied up in his braces; these birds were sent to London Zoo (Saunders, 1869: 184–5). Dresser received more Spanish birds from Howard Irby (1836–1905), an English soldier stationed at Gibraltar during 1865–72. Irby was a keen ornithologist and wrote up his observations on birds in his *Ornithology of the Straits of Gibraltar* (1875) (Mearns and Mearns, 1998: 190–2; Vibart, 2004). Irby gave some birds to Dresser and later sold his collection at Stevens' (Chalmers-Hunt, 1976: 125).

One of the other young 'Ibises', Frederick DuCane Godman, had spent time collecting birds on the Atlantic islands. He gave specimens of some of these to Dresser, including specimens of Bolle's Pigeon of the Canary Islands

6.5 Howard Saunders, from Henry Dresser's album of correspondents.

and the Azores Bullfinch, both species that Godman first made known to science. These were valuable additions to Dresser's collection, and to *A History of the Birds of Europe*.

Two Englishmen provided Dresser with first-rate collections of birds from Turkey, making an invaluable contribution to *A History of the Birds of Europe*. Charles Danford (1844–1927), a wealthy independent traveller, made two expeditions to Turkey during the 1870s, collecting all manner of natural history specimens and painting watercolours of the places he visited.[28] On his first expedition, in 1875–76, he encountered 185 species of bird and collected 156 species. While on his travels, Danford sent birds to Dresser, including some specimens of a snowcock (a gigantic mountain-living partridge) that were thought to belong to a new species, which Dresser named straight away (Dresser, 1876d). Following Danford's return to England, Dresser identified his birds to assist with publication (Danford, 1877, 1878), and bought the entire collection of several hundred bird skins. Danford's second expedition involved a 3,000-kilometre round trip in the Anti-Taurus Mountains and the Anatolian Peninsula in 1878–79. The party visited the town of Birecik on the Euphrates, where Danford encountered the rare Northern Bald Ibis breeding on the cliffs behind the town, an extraordinary bird with a long red beak, bald head, shaggy crest and black-bronze plumage. Although the bird had previously been seen by European travellers to the town, Danford was the first to announce to the scientific community the species' presence in Turkey. The bird was guarded by superstition but the town's governor shot one for Danford, 'a special act of courtesy' (Danford, 1880: 88); Danford shot another of the sacred birds the next day (with permission). He gave both of these birds to Dresser; one of them was illustrated in *A History of the Birds of Europe* (see plate 17). Danford's collections included several animals new to science, including a woodpecker, lamprey, lizard and small fish; he gave many of his specimens to the British Museum.[29]

The other Englishman to send bird collections from Turkey was Thomas Robson (1812–84), a Newcastle-born engineer who relocated to Constantinople during 1861–83 on the advice of his doctor (Turner, 2013). Robson formed a large collection of birds, mostly collected near the city. Some of these went to Robert College (now Bogazici University, Istanbul) but he also sent several collections to Britain. Robson provided Dresser with hundreds of birds; his letters to Dresser frequently refer to railways, as both of them were involved in metal and engineering.[30] Dresser had met another English railway engineer, Charles Farman, when he was travelling in Bulgaria in 1864. Farman subsequently sent Dresser many bird of prey eggs, including those of the Saker Falcon and Imperial Eagle. Dresser communicated a paper on these birds on Farman's behalf in the *Ibis* (Farman, 1868–69).

Dresser obtained a good series of Greek birds and their eggs from Theobald Krüper (1829–1921), the curator of the Athens Museum. Krüper, who came from Ueckermünde in Western Pomerania (now in Germany), collected in Greece and the Ionian Islands during 1858–77, where he made some important discoveries, including a new species of nuthatch that was named in his honour. He supplemented his meagre museum income by selling bird skins and eggs (Palmer, 1922; Mearns and Mearns, 1988: 219–22).[31] Krüper visited Smyrna in western Anatolia in 1862–63 and again during 1871–72. Henry Seebohm, travelling to recuperate from an attack of smallpox, visited Krüper there, with a letter of introduction from Dresser (Hobson, 1992: 63, 86). Another of Dresser's English friends, Colonel Hanbury Barclay (1836–1908), gave him a collection of birds that he had collected in Albania in 1871, filling in another 'blank' in *A History of the Birds of Europe*.

Russian expeditions to northern Asia

Dresser's business connections meant he was well placed to acquire birds and eggs from Siberia to help towards *A History of the Birds of Europe*.[32] The nineteenth century was a time of imperial expansion in Russia, which, after the Crimea, was focused eastwards. The Russian Geographical Society organised the Great Siberian Expedition of 1855–63 to explore the Amur region (Bassin, 1983). Gustav Radde (1831–1903), originally from Danzig in West Prussia, was the naturalist of one of the two sections of the Expedition. As well as producing reports in German (Radde, 1861, 1862–63), an account of the Expedition was published in English in the *Journal of the Royal Geographical Society* in 1858, which Dresser would have been familiar with. Dresser obtained some of Radde's specimens from the Expedition through the Zoological Museum of the Imperial Academy of Sciences in St Petersburg.[33] Radde relocated to Tiflis (now Tbilisi) in the Caucasus in 1864, where he established a natural history museum. He explored the Caucasus during 1864–65 and 1876–85, written up in many articles and a book entitled *Ornis Caucasica* (1884) (Kropotkin and Freshfield, 1903; Blasius, 1904; Bassin, 1983; Mearns and Mearns, 1988: 298–303). Radde provided Dresser with information on the Caucasus for *A History of the Birds of Europe*, as well as specimens of some of the endemic Caucasian bird species.

Another museum curator, Ladislas (Władysław) Taczanowski (1819–90) of the Warsaw University Zoological Museum, provided Dresser with larger numbers of birds and eggs from the Russian Far East. Taczanowski had close links with Polish collectors throughout the Russian Empire as well as with museum curators (see Mlíkovský, 2007a). One such collector was Ludwik Młokosiewicz (1831–1909), an aristocrat who had served in the Russian Army, explored the deserts of Persia, and been exiled for inciting political unrest. He settled in Lagodekhi (in the Georgian Caucasus), where he worked as an inspector of forests (Mearns and Mearns, 1988: 260–2). High in the mountains, close to Lagodekhi, Młokosiewicz discovered a new species of Black Grouse (now called the Caucasian Grouse) in 1875. He sent two specimens to Taczanowski in Warsaw, who named the bird in his honour. Młokosiewicz sent further specimens to Taczanowski and some of these were offered to Dresser, as he wrote to Alfred Newton:

> I have just heard from Taczanowski that he has received the eggs (8) of the new grouse <u>Tetrao (Lyrurus) mlokosiewiczi</u> and four skins. I am writing for an egg…. He asks me what he should charge for the skins and eggs – what do you say? I must have both at whatever price he fixes. He says that Mlokosiewicz is poor and he wants him to get as good a price as he can and that the bird is rare and only inhabits the most inaccessible places in the mountains. The result of his entire trip was eleven birds killed of which seven fell down the gorges and were too much damaged to preserve, so he only brought back 4 and one nest of eggs.[34]

Dresser acquired all four specimens of the Caucasian Grouse and some of the eggs. Others subsequently went to other collectors such as Henry Seebohm, presumably through Taczanowski (see plate 18).

Taczanowski was also a source of specimens collected by Polish exiles in Siberia. The most prolific of these was Benedykt Dybowski (1833–1930), one of the most extraordinary of all travelling naturalists.[35] Dybowski held two separate doctorates in lake biology. He was arrested in 1863 for his part in preparing for the Polish Revolt. He was sentenced to hanging but this sentence was commuted to twelve years' hard labour in Siberia. Dybowski travelled to Irkutsk in 1864 to begin his sentence and spent a number of years working as a doctor near Chita, treating scurvy victims. He was pardoned from hard labour in 1866 and spent the following ten years exploring the Russian

Far East, in the Baikal, Amur and Ussuri regions. Dybowski encouraged other Polish exiles to assist him, including Alfons Parvex (1833–90?), Wiktor Godlewski (1831–1900) and Michał Jankowski (1843–1903) (Dubrovin, 1890; Bassin, 1983). In 1876, Dybowski, Godlewski and Jankowski learnt that they were to be permitted to return to Poland. Instead, they explored Kamchatka in 1878–82 and spent time on the Commander Islands, where Dybowski assisted the local lepers and sufferers of syphilis, and helped them combat alcoholism. In 1883 he took up an offer from Lviv University to become head of the zoology department. Throughout his time in Siberia Dybowski sent bird collections to Taczanowski, possibly amounting to thousands of specimens, although it is difficult to imagine how they were safely transported across such vast distances. Taczanowski wrote up these collections in a large number of publications, culminating in the *Fauna Ornithologique de la Siberie Orientale* (1891–93). Dresser acquired several hundred bird skins and eggs from Dyboswki and his collaborators, received via Taczanowski and the Maison Verreaux in Paris.

'European' birds from North America

Although Dresser gave up collecting American birds and eggs in the 1870s, he continued to acquire American specimens of the species that also occurred in Europe, to contribute towards *A History of the Birds of Europe*. In addition to the birds he had gathered himself in New Brunswick and Texas, he was provided with many more from Adolphus Heermann, whom he had spent time with in Texas (see chapter 2). Heermann was one of the pioneer naturalists of the southern United States, a combination of medical doctor, army officer and naturalist. He took part in a number of expeditions, including a spell with one of the Pacific Railroad Survey Expeditions, before settling in San Antonio, where Dresser met him during the American Civil War.[36] As we have seen in chapters 4 and 5, George Boardman, of Calais, Maine, was an extremely important source of American birds as he was well-connected to other naturalists and to the Smithsonian Institution in Washington, DC. Boardman collected mounted birds and game trophies, which were housed in an eight-by-five metre museum built in the grounds of his house (Boardman, 1903; Barrow, 1998: 26–7). Dresser's youngest brother, Arthur, provided him with many birds and eggs from New Brunswick. Arthur clearly had his own aspirations as a writer-naturalist, producing two lengthy manuscripts on birds and his travels, but these were never published.

Dresser was given lots of eggs and some bird skins by another Englishman, James Hepburn (1811–69), who had emigrated to San Francisco in 1852 to work as a lawyer (see chapter 5).[37] Hepburn travelled extensively around San Francisco and along the north-west coast to British Columbia. Dresser encouraged him to write up his discoveries for English ornithologists but Hepburn was not keen on writing for the *Ibis* as it took too long to reach

Vancouver and San Francisco; instead, he planned to write a book on the birds of the north-west, but this was never completed.[38] When Hepburn died (aged only fifty-eight), his collection was passed to a friend in England who wished it to go intact to a museum. Dresser negotiated for the collection – which included over 1,500 birds, as well as many eggs, shells and reptiles – to go to the University of Cambridge, no doubt currying favour with Alfred Newton.[39]

Birds from the American Arctic were particularly poorly known, yet Dresser needed to ensure that he had access to the latest information – and ideally specimens – from the region, as many of the bird species found there were also found in Europe and would be included in *A History of the Birds of Europe*. Great strides had been made in uncovering their distribution under the aegis of Spencer Baird of the Smithsonian Institution.[40] Baird had sent a young naturalist, Robert Kennicott (1835–66), northwards in 1859 to set up collecting relationships with the Hudson's Bay Company. Kennicott's greatest success was with Roderick MacFarlane (1833–1920), the clerk in charge of Fort Anderson in what is now the Northwest Territories, close to the northern coast of Canada. MacFarlane, a Scot, had joined the Company in 1852 aged nineteen; he was soon placed in charge of Fort Good Hope on the Mackenzie River and established Fort Anderson in 1861 to trade with the Inuit and acquire furs. MacFarlane became quite obsessed with collecting specimens for Kennicott and the Smithsonian and had to defend himself against criticism that he was neglecting his Company duties in favour of ornithology. He encouraged people to bring birds, eggs and ethnographic objects to the Fort, and under-took a 150-kilometre egg-collecting trip in June and July each year from 1862 to 1865 through mosquito-infested marshes and an expanse of tundra known as the Barren Grounds towards the Arctic coast. His party included around twenty native hunters and canoe-bearers – local Athapaskans and Inuit – who did most of the work of collecting and preparation. MacFarlane spent the long, dark winter writing up his notes and packing specimens, sending them south to the Smithsonian along with the fur harvest in the late winter and early spring. Fort Anderson was beset by difficulties: distemper killed many of the sled dogs in 1864; the following year scarlet fever ravaged the Native American hunters at the Fort (killing many of MacFarlane's best local collectors) and measles killed many of the Inuit. Fort Anderson was abandoned in July 1866, having failed to make a profit for several years (Stager, 1967).

During his time at Fort Anderson, MacFarlane amassed around 5,000 specimens on his annual trips, and discovered the nests of a number of birds that were almost or entirely unknown. Most remarkably, he found thirty nests of the Eskimo Curlew, now almost certainly extinct. The numbers of specimens collected are mind-boggling: over 1,000 eggs of the Common Eider and 3,000 eggs of the Willow Ptarmigan, for example (see MacFarlane, 1891; Preble, 1922; Gollop *et al.*, 1986; Mearns and Mearns, 1992; Vaughan, 1992; Lindsay, 1993; Binnema, 2014). These huge collections were valuable to the Smithsonian Institution as they contributed directly to its collection, and any that were not required could be exchanged with other museums and collectors for specimens the Institution desired. To put these figures in context, the Hudson's Bay Company was handling (killing and receiving) approximately 60,000 geese annually during the same period for the food and feather industries (Barnston, 1860).

Kennicott's success was entirely dependent upon the co-operation and participation of a small number of Hudson's Bay Company workers. However, their successes were just as dependent upon the active collaboration, co-operation and coercion of the local people. Indigenous people did most of the actual collecting and preparation as, by their own admission, Kennicott and MacFarlane were poor at making specimens (Lindsay, 1993: 69–75). As Debra Lindsay notes, for many people Arctic ornithology was not so much a labour of love as just plain labour (Lindsay, 1993: 63).

The collections that the Smithsonian Institution received from the north were among the most desirable objects for collectors such as Dresser: 'Can one not get any of those rare waders eggs from him [Baird] by hook or crook for such eggs as those taken by a man like Kennicott are indeed a treasure', he once wrote to George Boardman.[41] Dresser managed to obtain some of MacFarlane's specimens, mostly through the Smithsonian. The Smithsonian also provided Dresser with two specimens of Ross's Gull – an almost mythical bird closely associated with Arctic exploration – after *A History of the Birds of Europe* was completed. Dresser had sought specimens of the rare Arctic gull from the Smithsonian since 1867, but they had so few specimens that there was no chance that they would give him any.[42] Larger numbers of the bird were collected at Point Barrow (Alaska) on one of the expeditions of

6.6 Ross's Gull specimens from Henry Dresser's collection, collected on the First International Polar Year Expedition to Point Barrow.

the First International Polar Year (1882–83, although work extended over 1881–84) (Murdoch, 1885; Barr, 1985; Levere, 1993; Todd, 2001). The gulls were discovered by one of the expedition naturalists, John Murdoch. He and his companions shot over 100 of the birds for the Smithsonian, so many that Spencer Baird forbade them from revealing the number collected, 'for fear he should be overwhelmed with requests for gifts or exchanges' (Murdoch, 1899: 152). Murdoch provided an account of the unusual challenges he faced in preparing the specimens:

> Arctic taxidermy has its drawbacks. The carpenter's shop, where I had to work, would not warm up in spite of the little Sibley stove in it, and by the time I had a skin turned inside out and the skull cleaned, the skin would be so stiff from freezing that it would not turn back, and I used to have to warm it at the stove before I could finish the skin. (Murdoch, 1899: 152–3)

Dresser persuaded the Smithsonian to give him two of the birds Murdoch collected in 1892, after twenty-five years of negotiations for specimens of the gull (see figure 6.6).[43]

Other collectors in the Arctic

Dresser received many specimens from Labrador through the dealer H. F. Möschler of Hamburg (see figure 5.4). These were collected by (or purchased from local people by) missionaries of the Moravian Church, a German Protestant episcopalian church that had been settled in Labrador since the eighteenth century.[44] The best specimens Dresser acquired from Möschler were three very rare dark Gyr Falcon study skins: even the Smithsonian Institution had only one such bird (see Dresser, 1875b, 1876b). Dresser considered (incorrectly) that these formed a separate species, the Labrador Falcon, and had two of the birds illustrated for an article (see plate 19). Dresser and some other English ornithologists paid the wages of a collector in Labrador for some time from 1867 onwards, probably one of the missionaries.[45]

Another of Dresser's friends, Henry Feilden, who had earlier given him specimens from the Faroe Islands, also provided him with the latest information on Arctic birds for *A History of the Birds of Europe*. Feilden served as one of the two naturalists on the British Arctic Expedition of 1875–76, the last official British assault on the North Pole of the nineteenth century (see Nares, 1878). The Expedition was a disaster, as it was beset by scurvy, of which four men died. However, Feilden found the nest of the Sanderling, then almost unknown, and the solution of the 'knotty tangle' (as Alfred Newton called it): the question of the breeding place of the Knot (Feilden, 1877; Sharpe, 1906: 350; Wollaston, 1921: 21–2). Feilden's collections from the Expedition went to museums but he did provide Dresser with one item that he treasured greatly. Dresser had lent Feilden a collection of plates of Arctic birds from *A History of the Birds of Europe*, to check that the coloration in the plates of Arctic birds' beaks, eyes and legs was correct. After the Expedition, Feilden returned these to Dresser along with his field notes, writing:

> Dear Dresser,
>
> I send you back as an Artic [sic] relic of the 1875–76 Expedition your book of plates, it wintered in lat 82°27' No., onboard H.M.S. Alert. I found the plates very useful as guides to my companions and in less than no time they were equally good ornithological observers as myself.[46]

Feilden remained closely associated with the Arctic (Caswell, 1977; Levere, 1988, 1993) and provided Dresser with some noteworthy Arctic specimens he received from other collectors. One of these was an Ivory Gull, collected on the German North Polar Expedition of 1869–70, which surveyed the hitherto unknown north-east coast of Greenland. This was a real adventure: one of the two ships, the *Hansa*, was crushed by ice and the crew drifted south on an ice floe for eight months, living in a hut made of coal briquettes. The other ship, the *Germania*, reached the coast of north-east Greenland in the summer of 1869 and gathered important natural history collections, including the Ivory Gull; some collections were subsequently sold in Bremen and acquired by collectors such as Feilden (Wollaston, 1921: 21; Venzke, 1990). Feilden travelled to Novaya Zemlya (north of Siberia) in 1895 and 1897 with another private English traveller, Henry Pearson (Feilden, 1898; Pearson, 1898; Pearson and Feilden, 1899). In 1897, they discovered Little Stints breeding in very large numbers on Waigats (Vaygach) Island, south of Novaya Zemlya, and collected almost 200 of their eggs, some of which subsequently passed into Dresser's hands. Sailing along the eastern coast of Novaya Zemlya, they reached a glacier they named the Ibis Glacier, 'in compliment to our brethren of the British Ornithologists' Union' (Feilden, 1898: 363), and they named branches of the glacier after one another.

Arguably the most interesting Arctic specimens in Dresser's collection are those that came from sailors and whalers. Unfortunately, few details are available on the lives or activities of these people. Among these is a specimen of a Snowy Owl nestling from Greenland, collected by sailors on a whaling ship, brought back to Dundee and sold on to John Harvie-Brown (see plate 20).

Notes

1 Cambridge University Library (CUL hereafter), MS.Add.9839.1D.301, E493, letter from Henry Dresser (HED hereafter) to Alfred Newton (AN hereafter), 10 February 1870.
2 Sharpe had a large personal collection, but this was mainly of African birds.
3 On a smaller scale, Britain was divided into provinces (vice-counties) for the purpose of recording plants by Hewett Watson and birds by Alexander More;

Scotland was similarly divided into faunal provinces by Frank Buchanan White (see Browne, 1983).

4   National Museums Scotland Library (NMSL hereafter), GB 587, JHB 15/239, letter from HED to John A. Harvie-Brown (JAHB hereafter), 18 January 1871.

5   CUL, MS.Add.9839.1D.301, E493, letter from HED to AN, 10 February 1870.

6   CUL, MS.Add.9839.1D.302, E499, letter from HED to AN, 14 February 1870.

7   CUL, MS.Add.9839.1D.328, I143, letter from HED to AN, 25 May 1871.

8   CUL, MS.Add.9839.1D.309, E640, letter from HED to AN, 26 October 1870.

9   Expeditions such as the British Arctic Expedition (1875–76) were sometimes provided with detailed instructions on what to observe and collect (Jones, 1875; Vaughan, 1992).

10   Henry Lansdell wrote with three audiences in mind: 'general readers', students and 'men of science and specialists' (Lansdell, 1885: vi).

11   This can be said of Howard Saunders, Charles Danford and Henry Tristram (Danford, 1880: 88; Tristram, 1882: 415–17; Lloyd, 1985: 95).

12   An egg JAHB sent HED from Perthshire (in 1871) was thought to be the first breeding record of the Goosander in Britain (although this record was 'beaten' by Edward Booth), NMSL, GB 587, JHB 15/240, letter from HED to JAHB, 16 September 1871.

13   NMSL, GB 587, JHB 15/240, letter from HED to JAHB, 19 June 1877.

14   CUL, MS.Add.9839.1D.497, L338, letter from HED to AN, 16 July 1886.

15   CUL, MS.Add.9839.1D.253, C927, letter from HED to AN, 6 August 1866.

16   CUL, MS.Add.9839.1D.465, J960, letter from HED to AN, 5 September 1882.

17   Henry Feilden is known to have visited Norwood in March 1871: see NMSL, JHB 15/239, letter from HED to JAHB, 4 March 1871.

18   He showed that the sound made by displaying male Snipe (a bleating noise produced in flight, usually referred to as 'drumming') was produced by the bird's tail rather than being its voice (Dresser once saw Meves demonstrate this). Meves wrote a book on the size of birds' eyes (649 species!), based on his long experience of taxidermy (Meves, 1886).

19   HED had a translation of Sabaneyev's paper on the birds of the Urals made when he was working on *A Birds of Europe* (so he could publish the details) and for Henry Seebohm (see Harvie-Brown, 1877).

20   Richard Sharpe described Wolley as 'This prince of field naturalists [who] laid the foundation of all that splendid method of collecting specimens of natural history, especially birds' eggs, which has been the distinguishing feature of the work done by British Ornithologists since his day' (Sharpe, 1906: 512).

21   Smithsonian Institution Archives, Record Unit 7071, Box 1, Folder 12: Dresser, Henry Eeles, 1862–71 (SIA, RU7071 hereafter), letter from HED to George A. Boardman (GAB hereafter), 14 October 1865.

22   See Alston and Harvie-Brown (1873: 55), and NMSL, GB 587, JHB 15/239, letter from HED to JAHB, 18 April 1872.

23   NMSL, GB 587, JHB 15/239, letter from HED to JAHB, 17 April 1872.

24   NMSL, GB 587, JHB 15/239, letter from HED to JAHB, 22 February 1873.

25   Seebohm belonged to a Quaker family. His father, a German, moved to Bradford; he was involved in the wool trade and travelled widely as a Quaker minister. Henry Seebohm's sister was married to the leading industrialist and philanthropist Joseph Rowntree.

26   Wiggins initially invited John Harvie-Brown to accompany him to the Yenisei, but, being detained by previous arrangements, Harvie-Brown recommended Seebohm instead (see Harvie-Brown, 1905: ix).

27 See Lloyd (1985: 93–106) for information on Howard Saunders. See Saunders (1869) for a particularly detailed description of a hunting expedition in Spain.

28 Danford had visited Transylvania in 1872 and again with John Harvie-Brown in 1874, with assistance from Dresser. CUL, MS.Add.9839.1D.366, G160, letter from HED to AN, 5 August 1873. His album of watercolours exists in private hands.

29 Danford wrote about his second expedition in the *Ibis* and, together with Edward Alston, wrote an important article on the mammals of South-West Asia (1877, 1880).

30 MM, ZDH/1/3, Henry Dresser's 1871–73 letterbook.

31 See Elwes and Buckley (1870), Kumerloeve (1975) and Kirwan *et al.* (2008) for fuller accounts of the history of ornithology in Greece and Turkey.

32 See Johansen (1952) and Vaughan (1992) for an overview of ornithology in Russia, and McGhie and Logunov (2005, 2006) for discussion on HED's relationships with Russian ornithologists.

33 CUL, MS.Add.9839.1D.272, D866, letter from HED to AN, 7 January 1868.

34 CUL, MS.Add.9839.1D.406, I175, letter from HED to AN, 4 August 1875.

35 Information on Dybowski comes from Wszolek *et al.* (1990); further information on his travelling companions comes from Mlíkovský (2007a).

36 Biographical information on Heermann's early life comes mainly from Stone (1907), Mearns and Mearns (1992: 225–33), Casto (1997) and Weintraub (2015). See Heermann (1853, 1854, 1859) for some details of his expeditions and ornithological discoveries.

37 Information on Hepburn comes from Swarth (1926) and Kinnear (1931). See also Mearns and Mearns (1998: 135–6). Hepburn's notebooks are preserved in the University of California and the University of Cambridge.

38 CUL, MS.Add.9839.1D.283, D770, letter from HED to AN, 28 December 1868. He did contribute many notes to Baird *et al.* (1874).

39 CUL, MS.Add.9839.1D.297, E071, letter from HED to AN, 13 July 1869. See also Mearns and Mearns (1998: 135–6).

40 See Levere (1993: 344–55) and Lindsay (1993). Additional details come from Mearns and Mearns (1992).

41 SIA, RU7071, letter from HED to GAB, 13 September 1865.

42 SIA, RU7071, letter from HED to GAB, 1 February 1867.

43 McGill University Library, Rare Books and Special Collections, Blacker-Wood Autograph Letter Collection, letters from HED to Robert Ridgway, 4 July and 7 September 1892. See Vaughan (1992), Densley (1999) and McGhie (2013) for detailed accounts of Ross's Gull and specimen collectors.

44 The Moravian missionaries provided ornithologists with specimens and information into the twentieth century (Anderson, 1928).

45 SIA, RU7071, letter from HED to GAB, 3 August 1867.

46 HED had this collection of plates and notes (together with the letter) bound up in a magnificent leather and gold binding. The book exists in private hands.

7 Discovering the birds of Europe, II

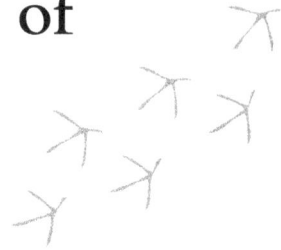

This chapter continues to explore the activities of those who supplied Dresser with birds and information for *A History of the Birds of Europe*, but pays special attention to those who were overseas as a result of imperialism and colonialism, in China, Japan, Central and South Asia, and Africa. They may have been stationed there as part of formal empire as paid 'professional' naturalists or geographers, or informally as private travellers or colonialists. Many of them had extraordinary first-hand experiences; some died at a young age, victims of disease and long-endured hardships. Only a small number of these people can be explored here at length. Far greater numbers of people – nearly 300 are known – provided Dresser with small numbers of specimens. Untold numbers of people did the bulk of the dirty work of specimen preparation but their names are not preserved for posterity: they were not of sufficient social standing to warrant recognition by collectors.

Whether they were part of formal or informal empire, scientific travellers' outlook was influenced by the political relations between the seats of Empire and the areas in which they found themselves, and this is often apparent from their writings. Travel narratives were frequently written from an imperialist point of view, deriding local people, even in supposedly scientific journals. To take one example, Allan Hume, a prominent government official and leading figure in Indian ornithology, wrote of one of his Indian servants in the *Ibis*: 'in season and out of season, with reason and without reason, he lies, lies, lies.... It would not do for one of Her Majesty's judges to be seen kicking one of Her Majesty's subjects about his premises, besides, I am a patient man, or else –' (Hume, 1869: 6).[1] The cabinet collectors in the West would have known of scientific travellers' exploits against the background of reports in scientific journals and the popular media, and the specimens that they sent would have been viewed as coming from the frontiers of the progress of civilisation.

Britain's 'Imperial Century' and natural history

The period between 1815 and the First World War has been described as Britain's 'Imperial Century', as Britain's Empire extended around the globe (Parsons, 1999). European powers entered a period of relative stability following the Franco-Prussian War (1870–71) and many of them turned their attention to political and economic expansion during the period of 'New Imperialism' (c.1870–1914), notably with the so-called 'Scramble for Africa' (1881–1914), in which Britain became heavily involved.

Many of the imperialists and colonialists who spread out from the West spent some of their time in natural history pursuits, whether hunting or collecting plants, insects or animals (their collections can still be found in museums in Britain and Europe). They sent collections back to auction houses in Europe or brought them home with them when on sick leave or following retirement, so their specimens entered collectors' exchanging networks in the West.[2] A particularly notable feature of this genre of collectors is that remote areas were particularly strongly identified with small numbers of individuals, or even single individuals. These were the 'lone voices' from remote stations that appealed so much to the Victorian imagination, obsessed as it was with the discovery and conquest of uncharted regions in the name of progress. They were, in effect, one-man institutions. Very few women were involved in these practices, at least on the face of it, although wives, daughters and servants surely provided unrecognised help. These scientific travellers relied heavily on assistance from naturalists in the West: they had access to museums and libraries that could help identify species that the travellers encountered. Alternatively, scientific travellers worked up their results during their own infrequent visits to Europe. Whether they wrote them up themselves, or had their discoveries written up for them, scientific travellers provided a vast mine of information and specimens that dominated many of the natural history journals through the nineteenth century (see, e.g. Moreau, 1959; Johnson, 2004, 2005).

Those who travelled as a result of the forces of imperialism and colonialism thus form a fascinating group of people: they navigated between different cultures in a variety of ways, ranging from respect and friendship to outright aggression and oppression, with many contradictions and ambiguities. They took advantage of their opportunities for ornithology, and many were rewarded with exciting and interesting discoveries as a result.

Robert Swinhoe and the birds of China

The majority of birds that Dresser received from China originated from Robert Swinhoe (1836–77), an English consular official (see figure 7.1).[3] Swinhoe's natural history achievements in China were enormous: he personally described over 200 new species of birds and two dozen mammals,[4] and was hailed as 'one of the most industrious and successful travelling naturalists that ever lived' (Sclater, 1875: 380; see also Wallace, 1880: 372–3). Swinhoe was educated in London before entering the consular service; he arrived in Hong Kong in 1854 and spent a year learning Mandarin (Bruner *et al.*, 1986: 9; Hall, 1987; Coates, 1988: 74–6, 501). At the time of his arrival, Europeans were not allowed to travel beyond Hong Kong and the five coastal Treaty Ports forced

7.1 Robert Swinhoe, from Henry Dresser's album of correspondents.

upon China following the First Opium War (1839–42). Consular officials ensured that trade agreements were honoured and oversaw the behaviour of European merchants and missionaries. They were expected to be the embodiment of upright Victorian social and moral values. Consuls had to keep office hours of nine until four, six days a week, but frequently had too little to do (Coates, 1988: 95–6; Fan, 2004: 77).

Swinhoe was based at Amoy (Xiamen) during 1855–59, a town with only a few dozen British subjects. Disease was rife and the three previous Consuls had died at the posting (Coates, 1988: 93–4, 115, 205). Swinhoe devoted himself to investigating the birds and mammals of Amoy, employing Chinese hunters (who roamed wherever they liked). He also corresponded with other Brits in China who shared his taste for natural history, as well as with naturalists back in Britain, including Charles Darwin himself.[5] Swinhoe paid two trips to Formosa (Taiwan) on consular business, in 1856 and in 1858, when his party searched for two Englishmen rumoured to have been abducted by the indigenous people. While in Formosa, Swinhoe hunted after birds and encountered some fantastic new species, including a large species of pheasant that now bears his name.

In 1860, Swinhoe acted as interpreter to the leaders of the British forces during the Second Opium War (1856–60), accompanying them on the assault on Tientsin (Tianjin) and Peking (Beijing). His published account of the campaign makes for rather curious reading, combining political and military observations with ornithological ones (Swinhoe, 1861). After the assault, Swinhoe was posted to Formosa as Vice-Consul, aged only twenty-four. He arrived there in 1861 and established the consulate in a Buddhist temple in the city of Taiwan-fu (Tainan). Trade was poor, so Swinhoe moved the consulate to the town of Tamsui, firstly in a moored opium-trading ship and then in an old Spanish fort. Trade prospects were just as poor in Tamsui: there were only three Europeans in the town (Takahashi, 1965; Fan 2004: 76)! Swinhoe's health deteriorated and he returned to England in 1862 on sick leave (unauthorised but subsequently granted for a year). While in England, he was extremely active in scientific society, exhibiting birds at scientific societies and at the London Exhibition, and writing a dozen or so scientific papers, including the first major checklist of Chinese birds. His achievements were recognised by fellowships of several of the leading societies, including the BOU (he was elected as an honorary member in 1862).

Swinhoe returned to Formosa in January 1864 to find his assistant, George Braune, completely exhausted. Braune died of a heart attack soon afterwards, aged twenty-six. Swinhoe moved the consulate once again, to the town of Takow, in August 1864, living in the hulk of another opium-trading ship with his wife, the daughter of an Amoy missionary. He was called back to Amoy in 1866 to cover the leave of the Consul there. Swinhoe subsequently made several expeditions to investigate trade prospects: to Hainan, beyond the Great Wall to Kalgan and up the Yangtze (Coates, 1988: 320–1). During all of these trips he collected birds and mammals, subsequently described in articles in the *Ibis*. These ornithological articles were also filled with tales of pirates, typhoons and near-drownings; they were often written in the present tense for dramatic effect.

While Swinhoe was a successful naturalist, his consular career was sometimes beset with difficulties: he criticised his superiors in print for not supporting his natural history researches (e.g. Swinhoe, 1863: 207) and tried to leave the diplomatic service, applying (unsuccessfully) for a job in the Chinese Maritime Customs in 1863 (Smith *et al.*, 1991: 58). In December 1867, Swinhoe and the acting Consul at Takow, Gibson, were present in a Chinese court that was trying people for anti-foreign or anti-Christian violence. They had called for the accused to be flogged; when one of the accused pleaded his innocence, torturers were brought in, at which point Swinhoe and Gibson withdrew from the court. Swinhoe's superiors were horrified and there was talk of dismissing both men. As this incident was concluded, Swinhoe took a second period of sick leave to England for eighteen months from July 1869, again producing large numbers of scientific papers. He returned to China in 1871 to serve as Acting Consul (and later Consul) at Ningpo (see Coates, 1988: 324–8). After less than two years there he suffered two 'paralytic strokes'.

Sulas P. Verreaux of Paris

7.2 Jules Verreaux, from Henry
Dresser's album of correspondents.

He was relieved of his post and spent three months in Shanghai while on his
way to Chefoo, collecting birds in the market while being transported in a
kind of wheelbarrow that the locals used as a taxi. Following a third 'paralytic
stroke' late in 1874 he returned to England and retired the following year (see
Swinhoe, 1873a: 362; Swinhoe, 1873b; Fan, 2004: 179). He was elected a
Fellow of the Royal Society in 1876, the petition being led by Charles Darwin.

Swinhoe has been described as 'the authentic voice of imperialism', with
good justification (Gittings, 1973: 57). He once ridiculed the Chinese for
venerating a fossil elephant tooth as being that of the Buddha (Swinhoe, 1862:
262). On another occasion, he tried to shoot a tiger in Amoy to satisfy the
locals, but had to beat a retreat to escape being attacked. The episode has clear
parallels with George Orwell's anti-imperialist essay 'Shooting an elephant'
(1936), where a British resident in Burma shoots an elephant 'solely to avoid
looking a fool' (Fan, 2004: 142).

Following his return to England, Swinhoe lived in Chelsea. His fellow
ornithologists flocked to visit him in order to study his collection of almost
4,000 bird skins of over 600 species (including many type specimens). Dresser
was one of these and he and Swinhoe became very friendly: they even began to
prepare a book on the birds of China together (discussed further in chapter 9).
Swinhoe gave Dresser several hundred bird skins from China and some rather

unreliably identified eggs. Dresser made very good use of these and Swinhoe's publications in *A History of the Birds of Europe*.[6] Swinhoe died in 1877, aged only forty-one; the cause of death was officially recorded as syphilis, presumably the reason for his long-endured paralysis (Coates, 1988: 328).

Another collector, Père Armand David (1826–1900), provided Dresser with a handful of birds from China for *A History of the Birds of Europe*. David was a French Lazarist missionary, so, unlike British subjects, he was free to travel in the interior of China. Père David was the first European to encounter the Giant Panda and discovered over sixty new species of birds and hundreds of species of plants. Most of David's specimens were destined for the Muséum National d'Histoire Naturelle in Paris but some of his specimens may have passed through the hands of the Verreaux firm in that city, as some of his specimens made their way into private collections (see figure 7.2).[7]

British colonialist collectors in Japan

Like China, Japan had been largely off limits to Europeans so its natural history was poorly known in the West; however, several British colonialists were actively studying Japanese birds (and other animals) at the time when Dresser was preparing *A History of the Birds of Europe*. Among these was Colin McVean (1838–1912), a Scottish engineer and surveyor who helped develop lighthouses in Japan for the Imperial Government from 1868 to 1876. He built up a good bird collection and published an article on the birds of Yedo (Tokyo) (McVean, 1877). McVean gave Dresser 250 skins in 1876 after he retired to Britain.

Thomas Blakiston (1832–91), an English aristocrat, provided Dresser with more birds from Japan.[8] After an adventurous career in the army and as an explorer, Blakiston settled at Hakodate on Yezo (Hokkaido) and initiated various engineering projects. He was clearly made of stern stuff and had a poor opinion of the bureaucrats and consuls he worked with. In 1873 he was tried in the consular court after a boy died in his custody in suspicious circumstances. Blakiston had detained a boy he suspected of theft; the boy was later found dead, having supposedly strangled himself. Blakiston got into trouble on another occasion when his firm was issuing trade coupons, which were illegal, leading to a diplomatic incident. Blakiston made a number of expeditions in Japan, beginning in 1869 with a journey of almost 1,500 kilometres round Yezo. He published a number of articles on birds in the *Ibis*, notably 'A catalogue of birds of Japan' (1878), with Harry Pryer (1850–88), an English merchant and naturalist based in Yokohama. Blakiston made an important discovery in recognising that the faunas of Hokkaido and Honshu were substantially different (Blakiston, 1883, 1884). 'Blakiston's Line' continues to be recognised as separating Siberian and tropical faunas in Japan. Blakiston gave the bulk of his Japanese collection (over 1,300 birds) to the Japanese authorities; he also sent many specimens to Swinhoe and some of these went to Dresser.[9]

Discovering the birds of India

India, the 'jewel in the crown' of the British Empire, was home to generations of British administrators and military men and their families. Hunting and collecting birds and eggs were just as popular in India as they were in Britain, with the added bonus of abundant colourful birds. Two Englishmen had dominated Indian ornithology: Thomas Jerdon (1811–72), an army surgeon and author of *The Birds of India* (1863–64), and Edward Blyth (1810–73) of the Indian Museum in Calcutta. Dresser knew both men after they retired to London (Jerdon lived close to Dresser's home in Upper Norwood).

By the time Dresser was gathering materials for *A History of the Birds of Europe* the central figure of Indian ornithology was Allan Hume (1829–1912).[10] Hume, a Scot, was posted to the Northwest Provinces in 1849, becoming Commissioner of Customs for the Northwest Provinces in 1867 and, in 1870, Director General of Agriculture, moving to Simla (now Shimla). During 1862–84 he amassed over 80,000 bird skins and 20,000 eggs, the largest collection in private hands at the time, kept in a purpose-built extension to his Simla home (Sharpe, 1885). His chief curator and collector, William Davison, had his own retinue of taxidermists, horses and even elephants. Hume also established the journal *Stray Feathers* to cover articles on Indian ornithology; he served as the journal's editor and was one of its chief contributors. Being a critical writer, Hume was nicknamed the 'Pope of Indian Ornithology' by his contemporaries (Charles Marshall, quoted in Wedderburn, 1913: 41). Hume worked in Allahabad between 1879 and 1882, having been demoted for disagreeing with his superiors. Following this he retired to Simla, where he planned to complete a great book on the birds of India that he had been working on for over twenty years. This was not to be realised as, in the winter of 1884, when he was away, his draft notes for the book were mysteriously removed from his museum. Hume's friend Charles Marshall wrote that the disappearance of these notes 'must have been the dastardly act of a discontented servant' (quoted in Wedderburn, 1913: 42). Hume practically gave up working on birds after this incident and concentrated on plants instead (see Wedderburn, 1913; Mearns and Mearns, 1988: 201–7; Moxham, 2001; Moulton, 2004).

Dresser corresponded with Hume and relied heavily on his information on birds for *A History of the Birds of Europe*, but he received only small numbers of specimens from him. Most of Dresser's Indian birds, and information about them, came from Edwin Brooks (1828–99), a Newcastle-born railway engineer who was based in Etawah (now in Uttar Pradesh) (Anon., 1899a).[11] Brooks collected vast numbers of birds himself and employed at least two people to hunt birds for him. He made one long-distance expedition, to Cashmere (Kashmir) (Brooks, 1871), and was particularly interested in groups of birds with similar species that were difficult to identify ('affined species' he called them), notably eagles and small warblers. Being a keen observer and a good draftsman, he was particularly skilled at noticing fine differences between birds (see plate 21). Brooks sent many boxes of birds to Dresser, including specimens of the rare Siberian Crane, eagles, falcons and all manner of other Indian birds, including supposedly new species. He corresponded frequently with Dresser during the 1870s (his letters to Dresser are preserved in Manchester Museum); their relationship is discussed further in chapter 9.

An Italian aristocrat, Louis Mandelli (1833–80), supplied Dresser with specimens and information on birds around Darjeeling, where he was the manager of three tea estates in the 1860s and 1870s. Mandelli began collecting birds around 1869 and paid local hunters to collect for him in Sikkim, Bhutan, Tibet and Nepal. He sent thousands of bird skins to other collectors, providing 5,000 birds for Hume alone and probably more than 20,000 skins in total. Mandelli was overworked and did not enjoy his situation, as described to a friend in 1876:

> I can assure you, the life of a Tea Planter is by far from being a pleasant one, especially this year: drought at first, incessant rain afterwards, and to crown all, cholera amongst coolies, beside the commission from home to inspect the gardens [i.e. a commission to examine the management of the tea plantations], all these combined are enough to drive any one mad.[12]

He hoped to retire in 1876 (when his wife nearly died of disease), but suffered financial problems and his health deteriorated. He was replaced by three managers – one on each tea estate – in 1879, indicating how overworked he was. Mandelli died in February 1880 aged forty-seven; he has been reported to have taken his own life (see Pinn, 1985), although there does not appear to be any firm evidence of this.

Dresser received specimens and information from several other 'members' of Hume's circle. Andrew Anderson (–1878), a judge at Futtehgurh (Fatehgarh, Uttar Pradesh), formed a large collection and wrote a number of articles on birds.[13] The Marshall brothers, Charles (1841–1927) and George (1843–1934), were prominent military men who mainly worked in the Punjab. Charles's collections of birds and eggs mostly passed to Hume (Mearns and Mearns, 1998: 195). The last person to provide Dresser with significant numbers of Indian birds was Eugene Oates (1845–1911), a civil servant in the Public Works Department in India and, more especially, Burma. He is perhaps best known for his *A Handbook to the Birds of British Burmah* (1883).

Russian explorers in Central Asia

Central Asia was one of the most poorly known areas of the Old World, a great blank space on Western maps of the period, yet Dresser needed to understand the distribution of birds there as many European species were also found further east (see plate 23 for a map of southern Central Asia). Most exploration was carried out by Russia as part of a programme of imperial expansion (see chapter 6; see also Bassin, 1983; Postnikov, 2003a–d). Russia expanded by military force into western Turkestan during 1865–76.[14] Russia's new territories were consolidated to form Russian Turkestan (now divided into Turkmenistan, Uzbekistan, Kazakhstan, Kyrgyzstan and Tajikistan); the land to the east (Chinese or East Turkestan, Xinjiang in Chinese) was under Chinese control. Following the establishment of Russian colonies in Turkestan, a number of expeditions explored the region's natural resources (Hooson, 1968; Postnikov, 2003a,c). These expeditions produced enormous amounts of data, but most of these results were published only in Russian, keeping them beyond the reach of Western naturalists and geographers.

The most famous Russian explorer of this period was (and is) undoubt-edly Nikolai Prjevalsky (1839–88) (see Rayfield, 1976; Mearns and Mearns, 1998: 266–74; Postnikov, 2003b).[15] He dreamed of reaching the fabled city of Lhasa, off limits to outsiders since 1792. To reach Lhasa would have been a great achievement both for him and for Russia (Hopkirk, 1982, 1992; Bishop, 2003). Prjevalsky made four unsuccessful attempts on Lhasa during 1870–83, being forced back by the impassable Yangtze River (twice), by the great Altyn Tagh mountain range, and by Tibetan officials who suspected that he intended to kidnap the Dalai Lama (when he was only 270 kilometres from Lhasa).[16] His first expedition involved a trip across the Gobi Desert; Dresser assisted with an English edition of the account of the expedition, published as *Mongolia, the Tangut Country and the Solitudes of Northern Tibet* (Prejevalski, 1876). During his second expedition, in 1876–77, Prjevalsky's party discovered the rare wild horse named in his honour (Prejevalski, 1879). His third attempt (1879–80) set out from Lake Zaysan (in present-day Kazakhstan) but was turned back near Lhasa; the fourth attempt (1883–85) explored the headwaters of the Yangtze and headed home across the Taklamakan Desert. Prjevalsky mounted a fifth expedition in 1888 but died of typhus at the outset, aged only forty-nine. He had travelled over 33,000 kilometres during his attempts to reach Lhasa, through some of the most arduous terrain on earth. His work was continued by a number of his assistants, who led expeditions of their own to Central Asia.

Prjevalsky's motives were as much political as they were scientific (see Kunakhovich, 2006). 'Scientific research will mask the political aims of the expedition and ward off the interference of our adversaries', he once wrote.[17] The reports that reached London and Calcutta of his travels must have aroused great suspicion among British officials, nervous of Russian expansion towards India. He often felt thwarted by local authorities and Central Asian peoples, whom he regarded as an interference: 'three things are necessary for the success of long and dangerous journeys in Central Asia – money, a gun and a whip', he wrote in his last book, *From Kyakhta to the Source of the Yellow River* (1888). This book called for open war with China so Russia could take control of Chinese Turkestan, on the basis of racial superiority. His empire-building motives were widely known, published even in *The Times*:

> This is scientific exploration with a vengeance, and goes beyond anything Mr. Stanley did with his 'six-shooter' among the negroes in Africa.... The *Viedomosti*

[a Russian newspaper], referring to this, says: – 'Among the natives visited by Colonel Prejevalsky [sic] there exists a deep conviction that sooner or later the Great White Czar will enter their country and take them under his domination.'[18]

During the course of his Central Asian travels, Prjevalsky collected over 5,000 birds, as well as thousands of other natural history specimens.[19] Around half of his collection went to the Zoological Museum of the Imperial Academy of Sciences in St Petersburg while the other half was exchanged with museums and private collectors. Dresser relied on Prjevalsky's information in *A History of the Birds of Europe*; he eventually acquired about thirty of his bird specimens (including species discovered by Prjevalsky) after the book was completed. Dresser acquired similar numbers of specimens from some of Prjevalsky's followers, including Mikhail Pevtsov (1842–1902), and Grigory Grum-Grzimailo (1860–1936),[20] mostly through the Zoological Museum in St Petersburg.

Dresser received larger Central Asian collections – several hundred birds – from Nikolai Severtzov (1827–85; the name is sometimes spelt Severtzoff, Severtzow, Severtsov or Severtsof), an influential Russian zoologist less well known in the West than Prjevalsky (see Todes, 1989: 144–56) (see figure 7.3). After failing to obtain a university post, Severtzov travelled through western Turkestan during 1857–79 with military support, gathering information on natural history, natural resources and military intelligence. He sometimes acted in a purely military capacity and once had the role of truce-envoy to one of the Khans, who had had the two previous envoys impaled on stakes.

7.3 Nikolai Severtzov, from Henry Dresser's album of correspondents.

During his first expedition (1857–58), to the Syr Darya (a great river), Severtzov's party were attacked by local forces (Kokands). His assistant was killed, while Severtzov received serious sabre wounds to the head. He was held hostage for a month before being released by a band of Kazakhs. When he returned to St Petersburg, late in 1859, 'the story of his captivity was in everybody's ears, and he had become the hero of the day'.[21] Severtzov sported a torn ear and scars to his face throughout his life as a result of his injuries. During 1864–67 he was back in Turkestan, accompanying a campaign against the locals (Kokands) and travelling through the Tian Shan mountains. Severtzov's group discovered important mineral resources (coal, iron and gold) and gathered huge numbers of natural history specimens, including several thousand bird skins (Severtsof and Michell, 1870). During 1869–73 he produced three major publications, in Russian. Of these, 'Vertical and horizontal distribution of Turkestan animals' (1873) was particularly influential. Dresser had the information on birds translated into English by Carl Craemers, one of his business contacts in Archangel (he also collected birds with Harvie-Brown in Archangel in 1872), and published in the *Ibis* for 1875–76 (Dresser, 1875a, 1876a). These articles provided detailed information on almost 400 birds species, including some that Severtzov described as new species. Severtzov visited Europe in 1875, where he received a gold medal from the International Geographical Congress in Paris. He visited London and met many of the English naturalists, including Dresser (discussed further in chapter 9). The information and specimens that Dresser received from Severtzov at this time were quite a coup, as they meant that *A History of the Birds of Europe* included the last word on birds of Russian Turkestan.

During 1877–78, Severtzov led an expedition to the Pamir Mountains, investigating mountain passes at over 4,000 metres altitude, visiting the Karakul (a saline lake that occupies a giant meteorite crater) and other little-explored regions (Lomonossof, 1880). A paper on the birds of the Pamir Mountains was prepared for the *Ibis* (Severtzow, 1883). Severtzov spent the rest of his life working up the results of his expeditions to throw light on the evolution of species, writing about migration, hybridisation and individual variation in eagles. He met an untimely end: while riding in a horse-carriage on a frozen tributary of the Don, the carriage fell through the ice. Although Severtzov, his companion and the driver managed to get themselves out of the freezing water, Severtzov and the driver died immediately (Anon., 1885b). His work was continued by his pupil Mikhail Menzbier, based at Moscow University.[22]

British naturalists of the 'Great Game' era

As the Russians expanded their territory in Central Asia, the British became increasingly uncomfortable at the narrowing gap between the two empires. A vast swathe of country, from the Caucasus eastwards across Persia, Afghanistan and Chinese Turkestan, became the location of a strategic conflict between the two powers, an early version of the Cold War. Diplomatic missions were sent to establish political boundaries, and set up trade agreements and peace treaties with remote Khanates. The more clandestine activities were those of the 'Great Game', immortalised in Kipling's *Kim* (Hopkirk, 1982, 1992).[23] Much of what was known in the West about the natural history of the region came as a direct result of the political anxiety about the spread of Russian influence there: the India Office Political and Secret Department employed an interpreter, Robert Michell, to review and translate Russian printed books, reports and articles for intelligence. Michell made many of the natural history discoveries of explorers such as Prjevalsky available in the West for the first time.

Dresser knew some of the most prominent British naturalists who explored and surveyed the lands between the British and Russian Empires. Of these, William Blanford (1832–1905) had spent fifteen years with the Geological Survey of India and served as zoologist and geologist to the punitive expedition to Magdala in Abyssinia (discussed below) (see figure 7.4).[24] He accompanied Major Oliver St John on one of the expedition teams of the Persian Boundary Commission (1870–72), which was to settle disputed sections of borders between Persia and Afghanistan in order to secure a barrier to the spread of Russian interest in the region (Mearns and Mearns, 1998: 168–9; Hopkins, 2007). The party spent two months travelling along dry riverbeds and ravines, mapping the region; Blanford then travelled alone to Tehran and the Elburz (Alburz) Mountains. Both he and St John formed large collections of birds and other vertebrates.

St John and Blanford returned to England to write up their results. Blanford spent two years working up the birds and mammals, and collaborated closely with Dresser. The Persian Boundary Commission expeditions were written up in two volumes (Blanford, 1876; Goldsmid *et al.*, 1876). Blanford's

7.4 William Blanford, from Henry Dresser's album of correspondents.

volume provided information on the distribution of 384 species of birds, as well as mammals, reptiles and amphibians. He gave Dresser approximately 100 birds from the expedition, which, together with the information he had published, were a valuable addition to Dresser's *A History of the Birds of Europe*. Blanford undertook further expeditions in India and Baluchistan, retiring in 1881 after a severe fever. He edited the seventeen volumes of the *Fauna of British India* series (writing three volumes himself) and was a leading figure in natural history societies.[25] Most of Blanford's bird skins went to the Calcutta Museum; his private collection of 1,344 birds went to the British Museum (Natural History) along with many mammal specimens (Anon., 1905a; 'TGB', 1907; Mearns and Mearns, 1998: 167–9; Moore, 2004).

John Biddulph (1840–1921) was another wealthy Englishman who had extraordinary first-hand experiences. He had served in the army during the Indian Rebellion and then entered the Political Department. During 1873–74 he participated in the Second Yarkand Mission, sent to negotiate diplomatic relations with the Muslim ruler at Kashgar in Chinese Turkestan.[26] After settling matters there the party investigated the Pamir Mountains, the 'roof of the world', where the Hindu Kush, Tian Shan, Kunlun and Karakoram mountain ranges meet in a great knot of mountains and remote valleys. Biddulph discovered that some of the passes could be easily approached from the north, raising serious concern that Russia could readily penetrate the region. The party returned to India via the Karakoram Pass, at an altitude of 5,580 metres.[27]

Biddulph was sent to the region again in 1876, ostensibly to investigate the murder six years earlier of a British explorer, George Hayward (1839–70). His real mission was to investigate the extension of Kashmiri control over the remote tribes, to defend the region against the spread of Russian influence. The mountainous region was divided between feudal kingdoms with wonderful names: Hunza, Nagar, Chitral, Wakhan, Yasin. He travelled through Hunza and Wakhan along nightmarish trails, using sheep as pack animals (and as food) as the 'road' was too steep for horses or mules. He found that the locals were unlikely to co-operate with the British, and he recommended that the Kashmiris should police the region under the guidance of a British official. After a brief home leave, Biddulph took up this position as 'one of the loneliest posts in the British Empire' (Keay, 1979: 97) in 1877. He was based in Gilgit (now in Gilgit-Baltistan, Pakistan), close to where Kashmir, Tibet, China, the Russian Empire and Afghanistan converged in the mountains. He gathered intelligence on the local people and listened for reports of any Russian spread into the region. At one point, Biddulph wrote that he had not seen another European for ten months. He focused his mind by building his bungalow and tennis court, writing on the languages of the local people (written up as *Tribes of the Hindoo Koosh*, published in 1880), shooting big game and collecting birds. The body of George Hayward, beheaded on the orders of a local ruler, lay buried in the orchard below his bungalow as a constant reminder of potential hostility with the locals. Infighting among the feudal rulers ultimately spelt the end of the Gilgit agency and Biddulph was recalled to India in 1881, where he continued to serve as a political agent or resident, before retiring to England in 1895 (Anon., 1922d; Gratzl, 1971).

Biddulph collected many birds to investigate the ornithology of Gilgit; he produced several articles, notably a list of 249 species from Gilgit (Biddulph, 1881–82). He sold 448 birds from Gilgit to the British Museum (Natural History) in 1881 and presented a further 3,194 birds in 1881 and 1897 (Sharpe, 1906: 310). He also collected coins, ancient weapons and armoury during his time in India and Gilgit (some of these coins are in the British Museum). Dresser knew Biddulph from his visits to the Zoological Society of London and received at least seventy birds from him. Dr John Scully, who served as Biddulph's locum at times, collected 1,543 birds during his time at the remote post. Scully consulted Dresser's and Seebohm's collections to help with the preparation of an article on his birds (Scully, 1881–82), and he gave Dresser some eggs in return for his help. Through their specimens and publications, Biddulph and Scully helped Dresser fill in the blanks in contemporary understanding of the distribution of birds in one of the most remote regions on earth.

Another English collector, Henry Lansdell (1841–1919), provided Dresser with information on and specimens of Central Asian birds through the 1870s and '80s, after *A History of the Birds of Europe* was completed. Lansdell was an evangelist based in Blackheath, London (Hopkirk, 1982: 75–8; Baigent, 2004). He distributed thousands of biblical tracts in prisons and hospitals in Europe and Siberia during the 1870s, written about in the hugely successful book *Through Siberia* (1882). On his next expedition, in 1882, he travelled 19,000 kilometres through Omsk, Kulja, Bokhara, Khiva and Merv, described in his book *Russian Central Asia* (1885; see also Lansdell, 1887). Dresser helped prepare the section on birds for this book and Lansdell gave him a collection of eggs from the expedition in return.[28]

For his most ambitious expedition, Lansdell attempted to reach Lhasa during 1888–91. Knowing the country to be closed to Westerners, he went with a letter from the Archbishop of Canterbury, mounted on yellow silk, which he hoped would gain him an audience with the Dalai Lama. He attempted to enter Tibet by way of Leh (Ladakh) in November 1888 and Kalimpong in the eastern Himalayas but was turned back each time. He took up an invitation to visit Kathmandu to distribute Christian texts. Lansdell travelled onwards to Peking, hoping to negotiate an entry to Tibet with the authorities there. Officials there told him that he was threatening diplomatic relations between Britain and Tibet, so he gave up his mission on patriotic grounds. A number of English naturalists, including Dresser and Alfred Newton, had encouraged Lansdell to collect specimens and information more rigorously than he had done on his previous expeditions. He and his servant Joseph, a Persian from the Strangers Home for Asiatics in East London, took taxidermy and collecting lessons at Edward Gerrard and Sons, the natural history dealers in Camden. Lansdell collected about 100 birds, but most specimen preparation was done by Joseph. Dresser examined Lansdell's collection on his return to Blackheath and classified the birds for Lansdell's published account; he was repaid with at least half of the specimens from the expedition.

Henry Tristram in South-West Asia

Dresser obtained information on and specimens of the birds of South-West Asia for *A History of the Birds of Europe* from Henry Tristram, one of his fellow members of the BOU. Tristram suffered from tuberculosis and spent the winters south of Britain for the benefit of his health (see Mearns and Mearns, 1988: 384–93) (see figure 7.5). He made six trips to South-West Asia during 1858–97, visiting biblical sites and studying natural history as well as local customs. Tristram was the first person to make a detailed study of the birds of the region, which had been less accessible to travellers during the Ottoman Empire. He published his discoveries in *The Fauna and Flora of Palestine* (1884) and in lengthy papers in the *Ibis*. His book *The Natural History of the Bible* (1867) ran to eleven editions and he also contributed to the immensely popular *Picturesque Palestine, Sinai and Egypt* (Wilson, 1881–83). Tristram formed large collections of birds and eggs on his travels, some of which were sold at Stevens' auction rooms. Dresser and Tristram exchanged many specimens, with Dresser providing Tristram with American birds in return for specimens from Tristram's own travels.

7.5 Henry Tristram, from Henry Dresser's album of correspondents.

Collecting birds in Africa

Africa was a subject of great interest to colonial and imperial powers during the mid–late nineteenth century, and the 'Scramble for Africa' (1876–1914) began in earnest as *A History of the Birds of Europe* was being produced. Dresser primarily obtained North African birds from Geronimo Olcese, an Italian naturalist and natural history dealer based in Tangier. A number of British travellers had also visited North Africa during the 1860s, including Henry Tristram, who spent the winters of 1855–56 and 1856–57 in Algeria and Tunisia as respite for his tuberculosis, sometimes travelling far into the desert, described in his book *The Great Sahara* (Tristram, 1860b) (see Mearns and Mearns, 1988: 384–93).

Two of Dresser's English contacts provided him with information on the birds of Egypt. Samuel Stafford Allen (1840–70) was senior partner in the family's business, which produced essential oils and pharmaceuticals. Allen gave Dresser a collection of about fifty birds from Egypt in 1866, including some rare birds he purchased in the food markets in Cairo and Alexandria: 'I have preserved four or five [White-tailed Plovers]; but as they mostly have their throats cut, according to the Mahommedan custom, it is troublesome

7.6 Ernest Shelley, from Henry Dresser's album of correspondents.

133

work' (quoted in Wright, 1865: 462). Ernest Shelley, one of Dresser's closest London-based colleagues, wrote a book entitled *A Handbook to the Birds of Egypt* (1872), as an account of his own hunting exploits. Dresser assisted him with the book and acquired some of Shelley's specimens (see figure 7.6).

Dresser received a much larger collection of bird skins from Charles Wright (1834–1907), an Englishman who was the son of the founder of the *Malta Times* (Sultana and Borg, 2015: 78–108). He maintained the definitive list of birds of the island; his collection included many rare birds (including vagrants from Africa) bought in the food market. Among these were two Slender-billed Curlews, now possibly extinct (see Wright, 1864a: 145). Wright gave his collection of bird skins to Dresser in 1873 for use in *A History of the Birds of Europe*.[29]

The Prussians Friedrich Hemprich (1796–1825) and Christian Ehrenberg (1795–1876) accompanied an archaeological expedition to Egypt from 1820, but set off on their own path to explore Egypt, Nubia and Abyssinia, sponsored by the Berlin Academy. In 1823 they travelled to the Sinai Peninsula, Red Sea and Lebanon. Their expeditions were beset with difficulties and Hemprich died of fever when he was twenty-nine. Their expeditions generated great collections that went to the University of Berlin, including 46,000 botanical specimens and 34,000 animals of 4,000 species, which were written up by Ehrenberg (Mearns and Mearns, 1988: 183–7). Dresser acquired a handful of Hemprich's and Ehrenberg's specimens from the Berlin Museum; these would have been regarded as great treasures, as Hemprich and Ehrenberg were well-known explorers.

In 1868, Queen Victoria ordered a punitive expedition to Abyssinia, after Emperor Tewodros imprisoned some British subjects. William Blanford was the expedition geologist (an important role needed to locate water supplies). Another naturalist, William Jesse, was sent by the Zoological Society of London, but he arrived too late to join the marching party. Instead, he remained near the coast, where he was beset by illness. After the expedition had achieved its aim (burning the fort at Magdala), Blanford and Jesse collected birds (750 specimens) and mammals in the Bogos country, along with two companions (Blanford, 1870). Jesse's collection was studied and written up by the German Otto Finsch, who was in the process of preparing a book on the birds of East Africa (Finsch, 1870; Finsch and Hartlaub, 1870). Dresser made extensive use of Jesse's notes in *A History of the Birds of Europe* and acquired some of the specimens from the expedition.

Herbert Ussher (1836–80) was Governor of the Gold Coast (a British colony, present-day Ghana) during 1867–72 and 1879–80 (also of Tobago 1872–75 and Labuan 1875–79). He formed a large collection of birds and sent many specimens to Richard Sharpe, described in a series of papers in the *Ibis*. Although some of Ussher's specimens went to collectors, including Dresser and Ernest Shelley, most of his collection was lost (Sharpe, 1906: 359). The specimens and information he provided for *A History of the Birds of Europe* were particularly helpful, in that many birds that spent the summer in Europe spent the rest of the year in Africa. Ussher's chief collector was a black African who styled himself 'Mr. St. David Thomas Aubinn, Esq., Royal Hunter to the King of Denkera' and as a gentleman. Sharpe gives the fullest account of this collector, as well as a reproduction of a sketch by Ussher showing Aubinn in a top hat, suit and boots, with a dead bird as a 'rarety from Denkera' (Sharpe, 1898: 172–3). Here is a rare (rather derogatory) glimpse of an individual who actually did the legwork of collecting, something that is missing from many travel narratives and ornithological accounts.

Conclusion

Through this and the previous chapters, we have met all manner of travellers, adventurers, private collectors, political exiles, aristocrats and colonial officials, to say nothing of countless unknown helpers, employees and servants. Their experiences and discoveries were as disparate as the individuals were as a group: most would have had no knowledge of the activities, or indeed the existence, of one another. Yet, they, collectively, provided the raw materials from which *A History of the Birds of Europe* was created. The standardised collecting and observational practices they followed, and that we have explored in chapters 5 and 6, meant that their separate experiences could contribute to a greater endeavour, namely the scientific description of the world's birds. Dresser knitted their results and experiences together in *A History of the Birds of Europe*, producing a globalised narrative of scientific and geographical knowledge, which forms the subject of the following chapter.

Notes

1   It is worth noting that this and similar quotes made their way into a small collection of 'buried treasure' selected from the *Ibis* during its centenary year in 1959. For further examples see Salvin (1860: 250) and St John (1876: 28).

2   See for example Fan (2004) on colonial natural history and MacKenzie (1988) on big-game hunting and imperialism. Lloyd (1985) and Raby (1986) discuss British scientific travellers.

3   See Takahashi (1965), Hall (1987), Mearns and Mearns (1988: 365–71), Yuteng (1993), Collar (2004) and Fisher (2004b).

4   See 'The Published Writings of Robert Swinhoe', http://home.gwi.net/~pineking/RS/MAINLIST.htm, accessed 8 January 2008.

5   See Fan (2004: 68, 74, 77–8) and Coates (1988: 98). See also the 'Darwin Correspondence Project', www.darwinproject.ac.uk, accessed 8 April 2017.

6   Henry Dresser (HED hereafter) wrote in his 'Egg Catalogue': '[Robert] Swinhoe was a very unreliable egg collector, and not like his brother Charles', Manchester Museum (MM hereafter), ZDH/11/2, p. 10.

7   Including, for example, Henry Tristram's collection in World Museum Liverpool.

8   The account on Blakiston is based on Cortazzi (1999). The name is pronounced 'Blackiston'.

9   Blakiston's collection is now in the Agricultural Department of Hokkaido University.
10 Biographical information comes from Wedderburn (1913), Mearns and Mearns (1988: 201–7) and Moulton (2004). See Moxham (2001) for discussion of Hume's role in protecting the 'Great Hedge of India', a 2,400-kilometre-long barrier hedge established as a part of the salt taxation system.
11 Brooks also provided large numbers of specimens to Henry Tristram and John Hancock (a Newcastle-based collector and naturalist).
12 This letter is reproduced as the frontispiece in Pinn (1985) and quoted on p. 8.
13 Henry Seebohm purchased his collection after his death (Sharpe, 1906: 297).
14 Tashkent, Khodzhent and Dzhizak were occupied in 1865–66 and Samarkand in 1868. The Khanates of Bokhara, Khiva and Kokand were annexed by Russia during 1868–76.
15 His name is pronounced 'prrjVALski', with a rolling 'r' and soft 'j', as in French 'Jean', or 'shiVALski'.
16 Incredibly (given the distances involved), this last incident was reported in Britain: 'Colonel Prejevalsky's [sic] travels in Tibet', *The Times*, 8 May 1880, p. 7, col. f.
17 Archival report from 25 August 1878, quoted in Rayfield (1976: 108) and Kunakhovich (2006: 5).
18 'Colonel Prejevalsky', *The Times* (3 February 1886), p. 5, col. d–e.
19 The bird collections were written up by Prjevalsky himself and by curators of the Zoological Museum of the Imperial Academy of Sciences in St Petersburg.
20 'Pevtsov' is spelt by Mearns and Mearns (1998) as 'Peltzov'.
21 As reported by E. F. Lunge, cited in Dement'ev (1948: 58) and Todes (1989: 145).
22 The bulk of Severtzov's collection is divided between St Petersburg and Moscow University Museum of Zoology. Menzbir sold nearly 200 of Severtzov's birds to the British Museum (Natural History) but the most important collection in the UK, including many types, went to HED.
23 Some writers (e.g. Morgan, 1973; Yapp, 2000) are critical of the concept of the 'Great Game', suggesting that it consisted of a small number of ad hoc instances, largely played out by the British. Others have emphasised the role of the indigenous population in driving political relations in the region (see Hopkins, 2007).
24 His name is sometimes incorrectly spelt 'Blandford'.
25 He was elected a Fellow of the Royal Society in 1874, President of the Geological Society (1888) and Vice-President of each of the Royal Society, Royal Geographical Society and the Zoological Society of London.
26 Information on Biddulph and the Second Yarkand Mission comes from Forsyth (1875) (the official report of the expedition), Gratzl (1971), Keay (1979: 81–112), Mearns and Mearns (1998: 257–66), and Waller (2004: 147–68). Information on the Gilgit agency can be found in Chohan (1984).
27 See Hruby (2005). The Second Yarkand Mission produced many valuable scientific results, but the work on birds was beset by difficulties. Richard Sharpe finally published a description of the birds in 1891.
28 Forty birds from the expedition were eventually purchased by the British Museum (Natural History) (Sharpe, 1906: 360).
29 MM, ZDH/1/3, letter from C. A. Wright to HED, 6 November 1871, p. 23.

17 Northern Bald Ibis from *A History of the Birds of Europe*, illustration based on a specimen collected by Charles Danford.

18 Male (foreground) and female Caucasian Grouse from *A History of the Birds of Europe*, illustration based on birds collected by Ludwik Młokosiewicz.

19 'Labrador Falcon' (dark-phase Gyr Falcon) by J. G. Keulemans, based on a bird in Henry Dresser's collection (shown in plate 1, see also plate 47).

20 Snowy Owl from *A History of the Birds of Europe*, illustration based on a well-grown nestling taken from a nest on the shores of Davis Strait, Greenland, by a sailor on the Dundee whaling ship *Polynia* in 1871.

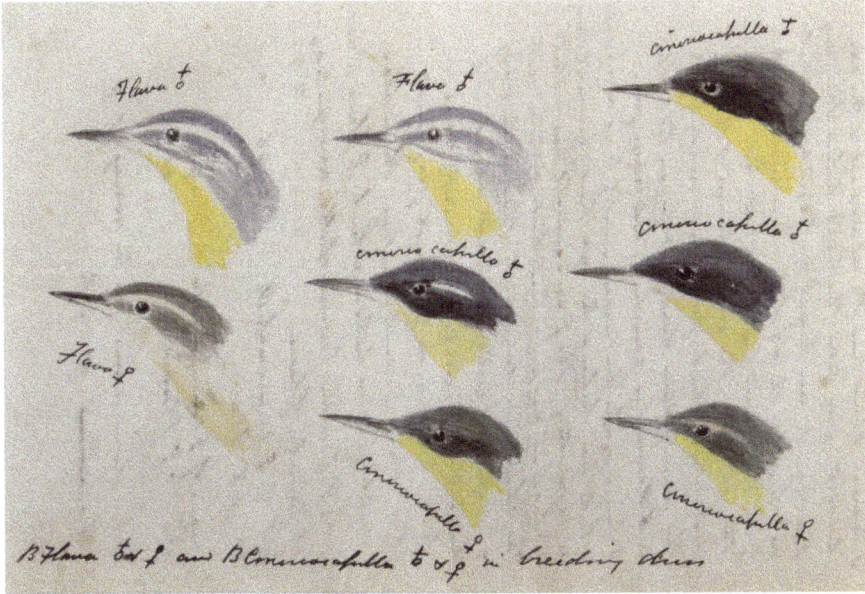

21 Illustration of the heads of Yellow Wagtails, drawn by Edwin Brooks and sent to Henry Dresser.

22 Crossbills, a particularly animated illustration from *A History of the Birds of Europe*, based on birds in Henry Dresser's collection.

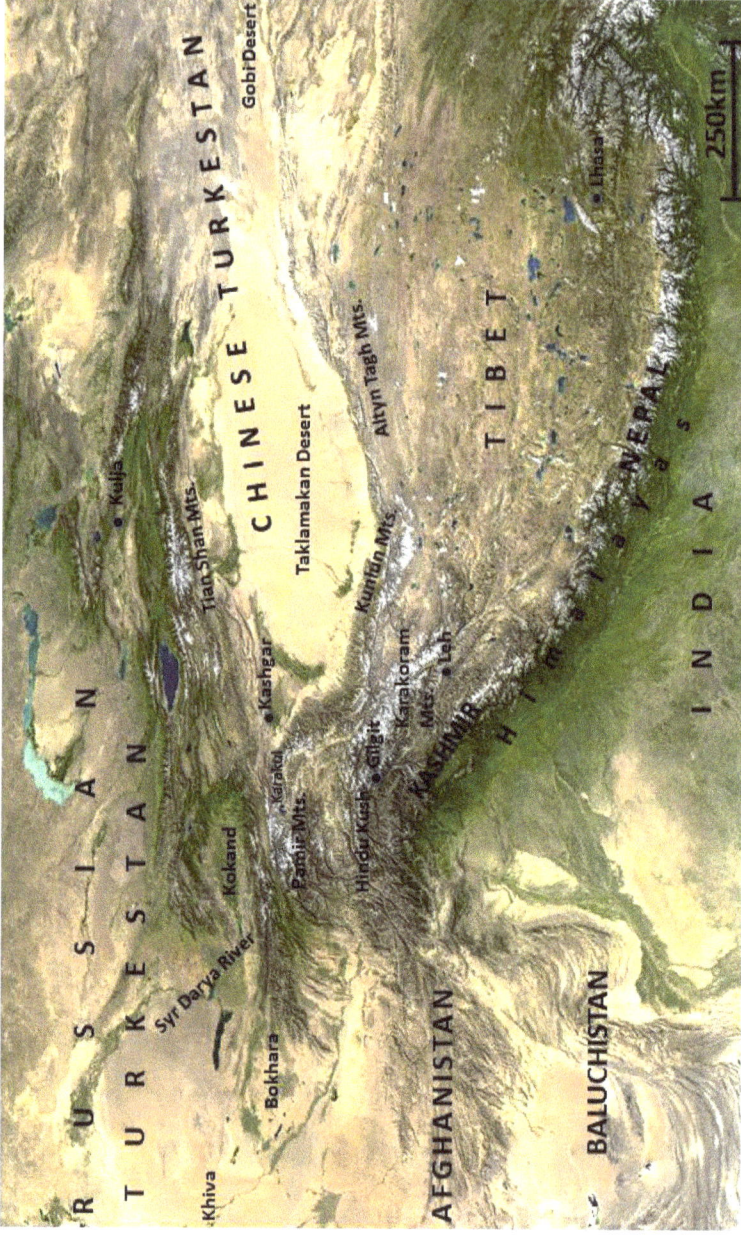

23 Map of southern Central Asia, showing localities mentioned in the text. Background image derived from NASA, Visible Earth, Blue Marble: land surface, shallow water, and shaded topography.

24 Arctic Terns, Shetland.

25 Coal Tits from *A History of the Birds of Europe*; the bird shown in the lower right depicts the British Coal Tit named by Richard Sharpe and Henry Dresser.

26 Black Woodpecker at nest hole, Finland.

27 Heinrich Gätke, from Henry Dresser's album of correspondents.

28 Saker Falcon by J. Wolf, from *A History of the Birds of Europe*.

29 Chicks of Little Stint (left), Temminck's Stint (centre) and Dunlin (right), from *A History of the Birds of Europe*.

30 Hume's Warbler as Yellow-browed Warbler, from *A History of the Birds of Europe*, based on birds sent to Henry Dresser by Edwin Brooks.

31 Immature Spotted Eagle, India.

32 *Road in Upper Norwood* by Camille Pissarro (1871). This appears to show Henry Dresser's house (St Margaret's), to the right of the gap in the row of houses.

8 Making *The Birds of Europe*

This chapter explores the range of work and practices involved in producing *A History of the Birds of Europe*, one of the most ambitious bird books of the late nineteenth century. *A History of the Birds of Europe* was a subscription book, a fairly common publishing model for large books in the nineteenth century. Subscribers would receive parts of the book periodically, and these would be rebound into volumes on completion of the project. The money that authors received for subscriptions for early parts paid for the production of subsequent parts; publishing by subscription involved a large outlay of funds in the early days. Early in 1870, Dresser and Sharpe sent out a prospectus containing examples of the articles on birds (the letterpress, as Dresser called it) and plates, in order to attract subscriptions. Their connections with the members and Fellows of several of the natural history societies – the BOU, Zoological Society of London and the Linnean – were particularly helpful in gaining subscribers (being a member of such networks was practically a prerequisite for success in publishing by subscription). The prospectus met with a good deal of interest and the partners secured enough subscribers to pay for one-third of the first issue as early as October 1870.[1]

The book was to be produced to the highest quality, in large quarto size, with illustrations from the leading bird artists. Articles and plates were to be issued monthly, covering about eight to ten species in each part. One year's worth of twelve parts cost £6 6s and the total cost of a copy was projected to be £37 16s, equivalent to over £3,000 today.[2] At the outset, the book was expected to cover 600 species, and to be completed in 1877.

The book (according to its preface) was published 'under the address of' the Zoological Society of London, 'by special permission', and printed by Taylor and Francis of Fleet Street, a firm with strong associations with many scientific subjects. The first part was issued in March 1871; by the time the first twelve parts were issued, in August 1872, they had 237 subscribers, including the King of Italy, Duke of Edinburgh and Maharajah Duleep Singh, twenty-two dukes, earls, countesses and other nobles, and just over half of the sixty-six BOU members (Lord Lilford took two copies). A list of subscribers was included in part 12, aimed to attract new subscribers. Most subscribers were in Britain, with smaller numbers in Europe, and a handful in India, North America and Gabon. Most were individuals, but there were also a small number of institutions such as the Zoological Society of London, museums in Europe and Calcutta, universities and libraries. Twenty copies were printed on thinner paper and without plates, and were given to those who had contributed information.[3]

The project required a massive amount of administration: managing the payment for individual parts, dealing with the postage of parts, and directions for binding, to say nothing of gathering information and writing the text. A large letterbook for 1871–73 (now in Manchester Museum) includes letters from 167 correspondents, including the British Museum, Smithsonian Institution, professors, a Belgian baron and four English lords.[4]

Methods

Each part of the book consisted of articles and plates on individual species of different types of bird. The first part included the European Roller, Red-footed Falcon, Marsh Sandpiper, Pine Bunting, Crested Tit, Woodchat Shrike, Eurasian Teal and Baikal Teal, very different birds. This meant that each set was very incomplete until the book was nearing completion (when it would be arranged taxonomically), encouraging subscribers to 'stay the distance', and making it impossible for subscribers to take the parts for popular groups of bird (such as birds of prey, owls, gamebirds, ducks and waders) and neglect less popular birds.

The articles were a combination of new and existing information distilled into pithy articles (generally four to eight pages in length, but up to fourteen pages). In the early parts of the book, Dresser and Sharpe both wrote the text. They requested the loan of specimens and solicited unpublished information from other collectors and travelling naturalists. The two partners investigated the latest published (and unpublished) literature, making use of the libraries at the Zoological Society of London, in their own homes and those of their friends. In 1870, Dresser also assisted Sharpe in producing the section on birds for the *Zoological Record*, an annual publication that gave references and short reviews of new literature on animals (Sharpe and Dresser, 1870b). Dresser prepared reviews of publications in other European languages, which also provided the two partners with information for *A History of the Birds of Europe*.[5] Dresser was particularly well placed to do this, with his knowledge of European languages, and he would translate information himself where he could. Through the 1870s, he also arranged for articles by Russian explorers to be translated into English in order to assist with *A History of the Birds of Europe*. Dresser sent drafts and proofs of articles to Alfred Newton for his comments, and generally followed his mentor's advice wholeheartedly.[6]

Each article began with the different scientific names that had been applied to a species in publications; Sharpe listed these names in date order to determine the oldest name, which had to be applied to the species under the rules of scientific nomenclature (on which more later). Detailed descriptions of the appearance of the male, female and juvenile of each species were given (in both Latin and English), mostly based on specimens in Dresser's collection. The geographical range and habitat were given, based on literature reviews and specimens in collections. Articles were brought to life with anecdotes on encounters with the birds from travellers' accounts, together with descriptions of species' habitat and habits. Finally, a list of the specimens that had been examined for the production of the article was given. Most of the specimens examined came from the collections of a handful of the leading BOU members, notably Lilford, Tristram, Gurney (senior), Saunders, Sclater, Godman and Salvin (and, especially, Dresser himself). They possessed collections that, when combined, were far superior to the collections of the British Museum itself at the time, demonstrating how ornithology in Britain still lay in the hands of private – and wealthy – individuals.

The plates

The plates were prepared with the same care as the text. John Gerrard Keulemans (1842–1912), a young Dutch artist who had done the illustrations for Sharpe's book on kingfishers (*A Monograph of the Alcedinidae*, 1868–71), did most of the work. He had replaced Joseph Wolf as the premier scientific bird illustrator in Britain.[7] Keulemans could take a bird study skin and produce an illustration of how he imagined the bird when it was alive (see for example plate 22). Joseph Wolf criticised the work of scientific illustrators such as Keulemans: 'There must be nothing but a map of the animal, and in a side view' (see Palmer, 1895: 57), although Keulemans undoubtedly had to follow the directions of his employers. The illustrations mostly showed a fairly restrained portrait of the bird, often in side view. Several individual birds were sometimes included in illustrations, to show the appearance in front and rear view. Male, female and young were often illustrated together. Male birds almost always dominated illustrations, at least partly because they were usually more distinctive in appearance. Interestingly, in those few species where the female is more brightly patterned than the male, the female occupies the more prominent position in the plate, with the drabber males on the sidelines.[8] In most cases, each plate consisted of a single species. For some smaller species (such as tits and warblers), two or more similar species could fit on a single plate. Many species were illustrated for the first time. Although most of the illustrations were based on study skins (mostly specimens in Dresser's own collection), there were some exceptions: the illustrations of the Black Stork and Tawny Eagle were based on birds that were living in London Zoo and Antwerp Zoo respectively, for example. The illustrations were scientifically accurate, and the scale at which they were figured was clearly marked on the plates.

The plates were prepared by lithography. Usually, an artist would prepare an illustration and copy this onto a slab of very fine-grained limestone (lithographic stone) with a special greasy crayon. The stone was then wetted with water and inked with an oil-based black ink, which was only 'accepted' by the greasy crayon lines. The inked stone was then placed in a printing press to produce the black lines of plates. The dried plates were then hand-coloured, mostly by young women who copied a plate that had been coloured by the artist. For *A History of the Birds of Europe*, Keulemans drew directly onto the lithographic stone himself, to reduce costs. The lithographs were printed by leading London-based firms (Hanhart and Hanhart, Walter and Cohn, and Mintern Brothers). The colouring of the plates was entrusted to the firm of Smith, Elder and Co. and William Hart, an artist who worked for John Gould.

As well as plates, a number of black and white engravings were included in *A History* to show particularly interesting anatomical features, for example the feet of Willow Ptarmigan (which shed part of their claws in summer) and the skull of Tengmalm's Owl with its asymmetrical ears.

Early success of *A History*

The early parts of the book met with considerable success: Alfred Wallace, Editor of the journal *Nature*, wrote how the book 'will supply a great want, since it will give in a convenient form and at a moderate price, a really good coloured figure with a full and accurate description and history of every European bird' (Wallace, 1871). In a second review, Wallace recommended the book to both general readers and amateurs (in the original sense of the word, as a lover of the subject), believing that it would 'take a high position as a scientific work, and at the same time to popularise the delightful branch of natural history of which it treats' (Wallace, 1872). The book attracted more subscribers, including some of the great museums of Europe, as well as book dealers, who could be sure of a good investment (see Table 8.1).

Dissolution of the partnership with Sharpe

Progress with *A History* was soon delayed due to the quantity of material that had to be analysed, leading to an apology to subscribers in July 1871 (appearing in the endpapers): 'The authors ... make it a point of honour to examine personally, wherever practicable, all disputed points; and as a thorough examination sometimes involves the dispatch of specimens from the Continent, a slight delay necessarily occurs.' Both partners struggled to work on the book alongside their other commitments. Dresser was working full time in the iron trade at Cannon Street during the day, then working on the book till midnight in the evenings.[9] Sharpe resigned from his position as librarian of the Zoological Society of London late in 1871 'in order to be able to devote more time to several important works on ornithology which he had in progress' (Scherren, 1905: 152). He gained an appointment at the British Museum following the death of the bird curator there, in May 1872.

Table 8.1 Subscribers to *A History of the Birds of Europe*

		August 1872	December 1873	January 1875	March 1876	May 1877
Total number of subscribers		237	268	277	287	301
Total number of copies		245	276	302	331	339
Naturalists		60	65	67	67	69
Aristocrats	UK	23	22	26	26	26
	European	3	5	5	5	5
Book dealers	UK	16 (20)	18 (22)	20 (36)	21 (54)	22 (49)
	Other	1 (4)	5 (8)	8 (16)	8 (18)	8 (18)
Museum workers	UK	3	3	3	3	3
	Other	8	9	9	9	8
Institutions	UK	2	3	3	3	3
	Europe	8	8	7	8	9
	India	1	1	1	1	1
	America	1	1	1	2	2
Private individuals	UK	155	167	195	201	211
	Europe and Turkey	6	8	8	7	7
	India	6	6	6	6	6
	Africa (including Malta)	3	3	2	2	2
	China and Japan			1	2	2
	USA and New Brunswick	2	2	1	1	1

Information taken from subscription lists issued with parts of *A History of the Birds of Europe*. Numbers in parentheses are the number of copies subscribed to.

141

Dresser became increasingly desperate at Sharpe's lack of progress.[10] As the cracks began to show, there was a dispute about the ownership of the birds that Dresser and Sharpe had been accumulating, alluded to in a letter from Dresser to Newton:

> As for our joint collection I called it so though really it has belonged to me for some time as of course Sharpe could not afford to spend money on it. He will have to get rid of his African collection [of birds] and says he will let the Museum have it. It will be a pity to see it buried there.[11]

Sharpe was not permitted to collect birds under the terms of his contract with the British Museum (a sensible rule that reduced the likelihood of conflicts of interest), so he sold his collection of African birds to the Museum (Sharpe, 1906: 477–81).

By the time Sharpe began work at the museum, in September 1872 (Gunther, 1975: 371–2), it was clear that the Dresser–Sharpe partnership could not continue as before. Sharpe wrote to Dresser to withdraw from the partnership:

> the happy combination of the labours of field and cabinet naturalists which we talked of in our prospectus is an impracticability.... I work and live for science only and if I never gained money by the work it would not affect me much, my choice between a rich and a poor life having been made five years ago.... In a word I will not be dictated to by my subscribers. With you however it is different I own. You can doubtless acquire sufficient reputation without such a scrupulous examination of every specimen which you know to exist, and considering that Bree is about to start a rival work I think you are right from your own point of view.[12]

Sharpe offered to continue to help Dresser with the synonymy (list of scientific names) in each article and relations took a turn for the better.[13] This was not to last, however, as Sharpe had a change of heart and proposed to continue as they had originally done.[14] Arthur Hay (Viscount Walden) encouraged Dresser to get clear of the partnership and offered to assist him with the synonymy of scientific names for each species.[15] Lord Lilford again offered to help financially with the project, but Dresser apparently did not take up the offer (Trevor-Battye, 1903: 254). Dresser was also overworked in the City as his partner Antonio Brady was often away, leaving him in charge of the London office.

An agreement to end the partnership was reached in December 1872. Sharpe settled a debt of £99 in specimen cabinets, bird specimens and £60 cash.[16] There were several disputes about payments for parts of *A History* and Sharpe's wife refused to surrender the account book for *A History* to Dresser except in her husband's presence.[17] The partnership was publicly terminated in June 1873 with a memorandum in part 18 of *A History*, giving Sharpe's appointment at the British Museum as the reason. The memorandum explained that Dresser intended to complete the project himself, that Sharpe would assist where necessary, and that Arthur Hay would help with the synonymy. Dresser appeared as the sole author after part 17 (although Sharpe did not contribute after part 13).

The deterioration of Sharpe and Dresser's relationship did not end there. Sharpe complained (many years later) that, at the time of the dissolution of the partnership, he was 'subjected to much abuse for "deserting him" [Dresser]'.[18] He was greatly embittered against Dresser, feelings that he would harbour for over thirty years (to be returned to in later chapters).

Dresser's work on *A History*

After finishing a day's work in the iron trade in Cannon Street, Dresser would return home to work on *A History* in the evenings throughout the 1870s and early 1880s. He worked to produce short pithy articles for *A History*, write articles on specific topics for the *Ibis* and report his discoveries at the Zoological Society of London, often illustrated with exhibitions of specimens, to say nothing of receiving (and returning) specimens to examine for the book, and handling subscriptions. Dresser had to critically assess previous work, separating the good from the bad. This meant that he put himself in the position of judge and jury, and many people consequently felt the sting of criticism in *A History*.

One of the most complicated tasks was sorting out which species were indeed separate from one another. When a naturalist encountered a bird they had not come across before, they gave it a new scientific name that they published in a journal or a book along with a description. Many supposedly 'new' species had in fact been described already and some were described many times over, especially if the species had a wide distribution across several continents or showed local variations. This confusion was brought under control in the latter part of the nineteenth century by the likes of Dresser and his contemporaries.[19]

In September 1873, Dresser spent three weeks in Germany with William Blanford to study specimens in museums.[20] They took type specimens from other collectors with them, to work out which species had been described more than once (Blanford and Dresser, 1874; Dresser and Blanford, 1874).

Once Dresser had worked out which supposed 'species' were deserving of the rank, he had to work out what scientific name to use for them. For each species, the oldest scientific name that had been given to that species was the one that had to be used: the 'law of priority'. Any other (younger) scientific names were thrown out as synonyms for the older name. Dresser examined species descriptions in obscure journals and books in a wide range of languages, and tried to make the best of vague species descriptions (often without illustrations). His investigations meant that the scientific names of many well-known species of birds had to be changed, as he discovered older names for them in increasingly obscure publications, causing great consternation among his colleagues. He 'moved' the scientific name from one species to another, causing particular upset. For example, he transferred the scientific name of the Black-throated Chat (*Saxicola stapazina*) to the Black-eared Chat (both are now treated as forms of the Black-eared Wheatear, *Oenanthe hispanica*). He also transferred the scientific name of the Common Tern (*Sterna hirundo*) to the Arctic Tern (see plate 24).

A particularly challenging investigation involved European and Indian eagles. Edwin Brooks (his biography has been covered in chapter 7) had challenged European ornithologists to give good descriptions of the various supposed species (Brooks, 1868). It turned out that the same scientific name was being applied to different species by different authors, and each species was being called a variety of names. The confusion was so great that at one point Brooks suggested that the eagles were better referred to by their common names, as their scientific names were as good as meaningless (Brooks, 1876: 500). There was a flurry of activity through the 1870s to resolve this mess and many of the leading BOU members, including Dresser, wrote articles on the eagles in question.[21] Brooks sent many (probably dozens) of eagle skins from India for Dresser to compare with European specimens in museums. Dresser also examined eagle skins from the leading British collectors, nearly ninety specimens in all, and sent the artist Archibald Thorburn to Antwerp Zoo to examine and illustrate a living eagle of a supposedly new species from Romania. Dresser collaborated with Arthur Hay, William Blanford and John Henry Gurney senior, and presented his findings at a meeting of the Zoological Society of London, exhibiting many eagle skins (Dresser, 1873). The mess was gradually ironed out, but there were many arguments and errors along the way.

For Dresser and his contemporaries, the species was the fundamental unit of classification, but no one had a good definition of what constituted a species. Exactly 'how' different two birds had to be from one another to deserve recognition was a matter of opinion. Ornithologists recognised that many species showed variation within a species, but their opinions differed on how to deal with this variation. American and German ornithologists began naming local varieties with a third part (trinomial) to their scientific name after the usual genus and species, a movement known as 'trinomialism'. Dresser was a staunch opponent of this movement as an 'objectionable retrograde step' which over-complicated matters (*A History of the Birds of Europe*, vol. 1, p. xix). This would become a key issue in years to come.

Work on *A History* led to the description of a number of new species, both in the book itself and in journals. These included (among others) a new lark from Central Asia received from Robert Swinhoe, a supposedly new pipit collected by Henry Seebohm and John Harvie-Brown in Siberia in 1875 (named *Anthus seebohmi*) (see plate 57), a new sandpiper from China (acquired from Robert Swinhoe) and a new snowcock collected by Charles Danford in Turkey (see appendix 2 and McGhie, 2011). In terms of species named in journals, Sharpe named the American Eider as a separate species, *Somateria dresseri*, after 'his excellent partner in the "Birds of Europe"' (Sharpe,

1871: 52). The two partners separated the British form of the Coal Tit and a southern European woodpecker, naming them *Parus britannicus* and *Picus lilfordi* respectively (see plate 25), and the Long-tailed Tit from southern Europe was split off as a separate species (Sharpe and Dresser, 1871b,c). Great advances were also made in understanding the range and movements of various species, based on collections of birds from different areas; an article on the distribution and variation of the Swallow, one of the most famous migratory species, was an early example (Sharpe and Dresser, 1870a). Thomas Robson provided Dresser with specimens of the White-winged Lark, which is usually found on the Central Asian steppes, that had wandered to Turkey in 1871, as well as Rustic and Pine Buntings, which are usually found much further north (Kirwan *et al.*, 2008).

Through the time of the production of *A History*, Dresser was phenomenally busy assessing records of rare birds that were found in Britain, as the risk of fraud (to inflate the value of specimens) or incorrect identifications hung over many records (see also Seebohm, 1893: 1). In the book, he wrote of the Cedar Waxwing (an American bird):

> [the Cedar Waxwing] is said to have occurred in Great Britain, but, so far as I can gather, without any valid reason. I know of more than one instance where a specimen of *Ampelis garrulus* [Bohemian Waxwing] has been killed and sent to a local taxidermist to preserve, and he has sent back in place of the bird received a mounted specimen of *Ampelis cedrorum* [Cedar Waxwing], which has consequently done duty as a British-killed example. Errors of this nature have doubtless caused the latter species to be recorded as a straggler to Great Britain. (Dresser, 1880; rebound as *A History of the Birds of Europe*, vol. 3, p. 427)

Another example concerned a supposed Lesser Kestrel (a European bird) that John Gould had acquired, supposedly from Britain; the bird turned out to be an American Kestrel, a completely different species. Dresser wrote to Alfred Newton: 'I expect it is a dealer's hoax like many other recorded specimens of rare birds'.[22] The most notorious example was the Black Woodpecker, a spectacular bird found throughout much of Europe and Asia that had supposedly occurred in Britain (see plate 26). Edmund Harting included the species as a British bird in several articles (e.g. Harting, 1865, 1872). John Henry Gurney junior developed a specialism in investigating records of birds that were supposed to have occurred in Britain, later writing: 'it is by no means in the interest of science that fictitious records should be perpetuated and copied from one book into another.... The truth, the whole truth, and nothing but the truth, is what we wish to arrive at, and with regard to British Birds that is not always easy' (Gurney, 1876: 251–2). Gurney analysed the various British records of the Black Woodpecker for Dresser and Sharpe, who published his results in the article on the species in *A History* in 1871. None of the records stood up to scrutiny, so the bird could no longer lay any claim to 'Britishness': some of the specimens were traced back to Leadenhall Market and had arrived there with gamebirds from Scandinavia. Gurney continued his investigations through the 1870s, evaluating British 'records' of such species as the Harlequin Duck (which is found in North America and Iceland), Spotted Sandpiper (from America), Eurasian Eagle-Owl and Great White Egret (both from mainland Europe) (Gurney, 1876; McGhie, 2012).

Towards the end of the project, probably in 1880, Dresser spent two weeks on Heligoland, a tiny island just off the German coast, where Heinrich Gätke (a sportsman-collector) was undertaking pioneering studies on bird migration (see plate 27). The island was a magnet to migrating birds, many of which were drawn to the lighthouse at night. Great rarities occurred there regularly, including vagrants from Siberia and North America, adding new species to the 'European' bird fauna (e.g. Seebohm, 1880b: 249–61; Gätke, 1895; Berthold, 2001).

Completion of *A History*

Through the 1870s, progress with the book continued to be slower than Dresser – and presumably his subscribers – would have liked. He delayed the production of articles on certain species while he waited for the latest information to return from expeditions. Keulemans' progress was a constant cause for concern, leading to frequent complaints from Dresser to Alfred Newton.[23] Keulemans was in great demand and was working on the illustrations for many other notable books at the same time as *A History* was being produced, for example Buller's *History of the Birds of New Zealand* (thirty-six plates), Rowley's *Ornithological Miscellany* (108 plates), Elliot's monographs on pheasants (eighty-one plates) and hornbills (fifty-seven plates), the Marshall brothers' monograph on barbets (seventy-three plates), Sclater's monograph on jacamars and puffbirds (fifty-five plates) and Shelley's great monograph on sunbirds (121 plates) (see Sitwell *et al.*, 1990, for these and further references). In 1877 Dresser engaged another artist, Edward Neale (1833–1904), to take on some of the work.[24] Later that year, Joseph Wolf provided Dresser with sketches of some birds of prey (harriers), which helped somewhat.[25] Wolf provided illustrations for fifteen plates for *A History*, mainly birds of prey (his favourite and best-known subjects) and seabirds (Palmer, 1895: 324) (see plate 28). Neale was not nearly as accomplished as either Wolf or Keulemans; however, Dresser had little choice but to use him as Keulemans was so dilatory. Neale illustrated twenty-eight plates for the book, while the remaining 600 were executed by Keulemans.[26] The hand-colouring of plates was delayed due to foggy weather in London: the task of hand-colouring the plates was done only when there was good light, in order to match the colours properly. Dresser intended to remedy this by getting the work done through the summer.[27] The lasting impression from Dresser's correspondence is of continual hard work, made more difficult during 1874–76 when he was periodically laid up with neuralgia at various times, which slowed progress with the book. Dresser was, of course, also busy during the day working in the iron trade:

I have been so terribly hard run that an express train is nothing to it and I should not wonder if, should there be a transmigration of humans into anything else, I should turn into a sort of patent perpetual motion coffee grinder and go on grinding for ever. However I am thankful to say that by perpetual pegging away at Keulemans I have got all the plates but four done, and am now working clear.[28]

The quality of the plates contributed enormously to the success of the book (see, for example, plate 29), leading to another highly complimentary review from Alfred Wallace in *Nature*:

A work like the present, so beautifully and artistically illustrated, and of which only a limited number of copies is printed, is sure to become scarce and to rise considerably in value. Lovers of nature and of art may therefore be reminded, that in becoming subscribers they are not only obtaining a valuable and most interesting book, but are at the same time making a profitable investment. (Wallace, 1875)

At roughly the same time Philip Sclater described *A History* as 'by far the most exhaustive account of the European Avifauna yet attempted' in his presidential address at the British Association for the Advancement of Science in 1875 (Sclater, 1875: 376; Sclater, 1876: 89). Sclater commented that the book's large size made it difficult to use; Dresser published *A List of European Birds* in 1881 as a kind of handlist, answering Sclater's demand fairly well.

The introduction, issued near the completion of the book, included a long essay on the biogeography of the region (written in terms of the work of Alfred Wallace and Philip Sclater), along with discussion on migration, another hot topic at the time. Perhaps most interestingly, Dresser included discussion of the ways in which birds evolve into localised forms in response to local climates and in isolation. Such open discussion of evolution and divergence was uncommon for the time, beyond the works of Alfred Wallace and Henry Seebohm, discussed in further chapters.

In the end, the book covered 624 species, a significant increase on the numbers included in previous works on the subject (admittedly Dresser also included the birds of the Azores, Canary Islands and Madeira, North Africa and western Asia, pushing the number up further). *A History of the Birds of Europe* included 586 species as occurring in Europe and adjacent South-West Asia, or 159 species more than were covered by Gould in the same area in his *Birds of Europe*. The book ran to eighty-four parts; the last parts containing plates and text on species were issued in December 1880. Part 83-4, consisting of the list of subscribers, index, references and introduction, were issued sometime in 1882 (although dated Christmas 1881) (McGhie, 2011). The book was bound in morocco leather with gold tooling when completed and made a magnificent addition to any library, in eight large, sumptuous volumes. The birds were illustrated in 633 plates; as 339 copies were subscribed to, this meant that almost a quarter of a million plates had to be individually coloured for the book.[29] A complete copy cost £52 10s, equivalent to nearly £5,000 today.[30] The project had taken thirteen years to complete (from start to finish), so long that many of Dresser's collaborators died before it was complete (including Edward Alston, Robert Swinhoe, Arthur Hay and John Gould). Dresser dedicated the book to Lord Lilford, President of the BOU, 'in recognition of the assistance he has invariably afforded … and as a token of my appreciation of his many acts of sincere friendship'.

When *A History* was completed, Dresser gave his own assessment of the project in the introduction:

> And now that the work is completed I must ask my readers to look on the result with lenient eyes. There is much that I should alter and add were I able to rewrite many of the articles; and no one can be more keenly aware of the many imperfections and shortcomings than myself. I have often wished that I had more time to compare specimens; but had I not pressed forward as much as possible it would have been impossible to complete the work in any thing like the time expended on it; and when I state that it has been entirely written before breakfast and after evening meal, during the few hours I could spare from the arduous duties of a city life, often working late into the night, and again recommencing my labours soon after daybreak, I trust I may claim some little indulgence. At the same time it has been a labour of love to me, and one which, though so heavy, I am, now that it draws to a close, reluctant to relinquish; but I hope, if I am spared and retain the health and strength I have hitherto enjoyed to so great a degree, to continue my labours, and to bring out, after having collected sufficient material, a work on the birds of the Eastern Palaearctic Region, excluding of course those species which occur there of which I have treated in the present work.

Conclusion

Dresser – ambitious, wealthy and immensely energetic – brought out one of the largest and most important of the late-nineteenth-century encyclopaedias on birds, a combination of rigorous information and the finest of scientific illustrations. The book secured his place in the canon of ornithology and helped with his steady climb within scientific society, covered in the following chapter. 'European birds' – taking in information on their occurrence in northern Asia and North America – was established as 'his' subject. Dresser asked Joseph Wolf to develop a motif for the title pages (the first mention of this comes from April 1880), showing an anthropomorphised Eurasian Eagle-Owl holding a lantern, shining the light for small birds which flit around the owl, standing inside a large 'D' for Dresser (see figure 8.1).[31] The motif of the wise owl – a clear allusion to Dresser himself – displays a spectacular lack of modesty on his part. There is more than a passing similarity to another motif of Wolf's that he had been working on since 1877. This showed another anthropomorphised Eurasian Eagle-Owl as a wise professor pondering the question as to which came first, the owl or the egg (the caption translates as 'highly learned makes the fool') (see figure 8.2). Wolf's biographer, Alfred Palmer, claimed that this was meant as a parody of stuffy intellectuals (Palmer, 1895: 180). Whether or not Dresser was aware of Wolf's subtext is unknown, but given that the two

8.1 Title page from *A History of the Birds of Europe*, showing Henry Dresser's monogram including a wise owl.

A LECTURE ON EMBRYOLOGY.

8.2 'A lecture on embryology' by J. Wolf; the text on the lectern translates as 'Highly learned makes a fool'.

were close collaborators it seems likely that he was. Palmer later wrote that he regretted that Wolf was responsible for the design on the title pages for *A History* (Palmer, 1895: 115).

In the previous two chapters, we met with a disparate group of individuals who, for one reason or another, collected birds and eggs from distant parts of the world. In this chapter we have seen how one person (largely) brought their information and results together, synthesising their individual experiences and encounters with birds into one great whole. Collections of specimens were central to this work, as the rawest of materials on which to make decisions about where each species existed in the world, and which were worthy of recognition as separate species. The production of *A History of the Birds of Europe* involved the detailed examination of 10,637 bird specimens (mostly by Dresser, although some would have been examined by Sharpe when he was working on the project). Almost half of these (4,385 specimens) came from Dresser's own collection, but he also examined birds from 102 other collections. His close friends' collections provided the majority of specimens for examination: 1,079 specimens from Howard Saunders' collection, 928 from Henry Tristram, 407 from Arthur Hay, 394 from John Henry Gurney junior, 383 from Osbert Salvin and Frederick Godman, 255 from Robert Swinhoe and 241 from Edmund Harting, for example. On the other hand, only a relatively small number of the specimens examined came from museum collections (1,046 specimens in total, 10 per cent of all specimens), including 475 birds in the British Museum (4 per cent of all birds examined). These figures amply demonstrate the extent to which private collectors still dominated ornithology, especially in Britain, during the mid-nineteenth century.

Notes

1 Smithsonian Institution Archives (SIA hereafter), RU2021, letter from Henry Dresser (HED hereafter) to George A. Boardman (GAB hereafter), 22 October 1870.

2 The relative value of the investment was calculated as the purchasing power, using www.measuringworth.com, accessed 11 March 2017.

3 SIA, RU2021, letter from HED to GAB, 15 June 1871.

4 Manchester Museum (MM hereafter), ZDH/1/3.

5 Cambridge University Library (CUL hereafter), MS.Add.9839.1D.317, F142, letter from HED to AN, 11 February 1871.

6 See, for example, CUL, MS.Add.9839.1D.320, F188, letter from HED to AN, 28 February 1871, and MS.Add.9839.1D.325, I104, letter from HED to AN, 7 May 1871.

7 See, especially, Anker (1938), Jackson (1975), Dance (1978), Skipworth (1979), Keulemans (1982), Lambourne (1990) and Sitwell *et al.* (1990) on bird illustration and its leading artists.

8 As shown in the plates of the Red-necked Phalarope (vol. 7, plate 537) and Grey Phalarope (vol. 7, plate 538). In the plate of the Dotterel (vol. 7, plate 526), it is a female in breeding plumage that is illustrated, rather than a male.

9 National Museums Scotland Library (NMSL hereafter), GB 587, JHB 15/239, letter from HED to John A. Harvie-Brown (JAHB hereafter), 9 September 1871.

10 CUL, MS.Add.9839.1D.348, F908, letter from HED to AN, 14 July 1872. See also CUL, MS.Add.9839.1D.350, F697, letter from HED to AN, 21 July 1872.

11 CUL, MS.Add.9839.1D.351, F707, letter from HED to AN, 26 July 1872.

12 Letter from Richard B. Sharpe (RBS hereafter) to HED, 26 September 1872, quoted in McGill University Library, Rare Books and Special Collections, Blacker-Wood Autograph Letter Collection (MUL, B-W hereafter), letter from HED to RBS, 29 October 1878.

13 CUL, MS.Add.9839.1D.358, F940, letter from HED to AN, 27 September 1872.

14 CUL, MS.Add.9839.1D.359, F941, letter from HED to AN, 2 October 1872.

15 CUL, MS.Add.9839.1D.363, G027, letter from HED to AN, 25 November 1872.

16 MUL, B-W, letter from HED to RBS, 11 March 1874; the letter includes a duplicate of the settlement between HED and RBS.

17 MUL, B-W, letter from HED to RBS, 11 June 1873.

18 Natural History Museum Library and Archives, London, DF231/2/457 Department of Zoology: Vertebrate Section, Reports to Trustees and other Official Documents, 1896–1909, duplicate letter from RBS to E. Ray Lankester, 18 April 1902; NHM, DF231/2/469–70, Department of Zoology: Official documents, copies of reports to Trustees by RBS, with supporting letters, lists and notes, 1899–1903, letter and accompanying report from RBS to E. Ray Lankester, 18 April 1902.

19 Analyses of taxonomic activity have focused on the dates when currently recognised species were first published and the publication dates of new species' names, rather than analyses of when the modern 'picture' of species lists was developed. See Johnson (2005) and 'Historical patterns of avian taxonomy', www.zoonomen.net/avtax/hist/histfr.html, accessed 8 April 2017.

20 CUL, MS.Add.9839.1D.366, G160, letter from HED to AN, 5 August 1873.

21 BOU members who wrote about the eagles included John Henry Gurney senior, Howard Saunders, Henry Elwes, William Blanford and RBS.

22   CUL, MS.Add.9839.1D.351, F707, letter from HED to AN, 26 July 1872.
23   CUL, MS.Add.9839.1D.316, F123, letter from HED to AN, 4 February 1871.
24   NMSL, GB 587, JHB 15/240, letter from HED to JAHB, 12 August 1877.
25   NMSL, GB 587, JHB 15/240, letter from HED to JAHB, 8 December 1877.
26   NMSL, GB 587, JHB 15/240, letter from HED to JAHB, 28 April 1878.
27   NMSL, GB 587, JHB 15/240, letter from HED to JAHB, 13 February 1876.
28   NMSL, GB 587, JHB 15/240, letter from HED to JAHB, 7 January 1879.
29   Some subscribers dropped out before the project was completed; a total of 332 subscribers took the book at some time, involving parts of 377 copies of the book.
30   The relative value was calculated as the purchasing power, using www.measuring worth.com, accessed 11 March 2017.
31   Elliott Coues used a similar motif for himself, with a bespectacled owl reading his 'Key' of 1872; see the cover of Coues (1874).

9 A central figure: society life in the 1870s

This chapter explores Dresser's life in scientific society through the 1870s, when he was working on *A History of the Birds of Europe*. He enjoyed an increasingly international reputation: by 1872 he was a member of the Imperial Society of Naturalists of Moscow and of the German Ornithological Society, and a corresponding member of the Boston Society of Natural History. He had become part of the machinery involved in the production of scientific knowledge, which involved field collectors, experts, elite (and elitist) societies, publications and publishers, and a readership.

Dresser and his close colleagues in the BOU were the great 'swells' of ornithology (a term used by Louis Mandelli – see Pinn, 1985: 12): wealthy men with a high social standing. They collaborated by supporting each other with their individual projects; at the same time, they worked independently on 'their' projects, as scientific writing was largely a single-handed affair. When Dresser was producing *A History of the Birds of Europe*, his closest associates were working on their own book projects. Alfred Newton was revising Yarrell's *A History of British Birds*; Ernest Shelley produced *A Handbook of the Birds of Egypt* (1872) and gathered African birds as a basis for more books; and John Gould, who wrote his most famous books alone (with illustrations produced by his team), was issuing *The Birds of Asia* (1849–83), *The Birds of Great Britain* (1862–73) and *The Birds of New Guinea* (1875–88), as well as monographs on various groups of birds.[1] In terms of partnerships, Edgar Layard (1824–1900, a British diplomat who spent time in South Africa) and Richard Sharpe produced *The Birds of South Africa* (1875–84), and Osbert Salvin and Frederick Godman began a great work on the natural history of Central America, the *Biologia Centrali-Americana*. This was issued in 215 parts (rebound as sixty-three volumes) between 1879 and 1915. Incredibly, 19,263 new species of plants and animals were made known to science (see Godman and Salvin, 1918: 4).

Collaborating sometimes required careful negotiation: as early as 1871, Henry Tristram told Dresser and Sharpe they would 'tread on his toes' if they wrote about Great Grey Shrikes, as he intended to do so (they did so in any case – see Sharpe and Dresser, 1871a).[2] 'Lesser' figures were sometimes reminded of their position in publications: Carl Craemers (whom Dresser knew through his business) once wrote to John Harvie-Brown: 'I once saw a pair [of Choughs] and shot one but did not preserve it, see Dresser's Birds of Europe on the Chough, I am understood there to be the collector <u>in Mr. D's service</u>!!'[3]

Dresser continued to be very active in London natural history society. He was, of course, closely involved with the BOU, as it consisted of his closest circle of friends. He was also a regular attendant at the fortnightly meetings of the Zoological Society of London; he exhibited many specimens to the audience, including new species and particularly rare specimens he had recently acquired for his collection, such as the rare Labrador Falcons he received from H. F. Möschler (Dresser, 1875b). In 1872, Alfred Newton encouraged him to join the Zoological Club, an elite band of twenty-five members of the Zoological Society of London.[4] This did not go smoothly, as Dresser was black-balled by another member. He wrote to Newton:

> I did not know until I read your letter that I had (for the first time in my life) been black-balled. Of course it is no small annoyance to me…. I think I know who is my 'friend' though how I can have trod on his toes I cannot imagine. I shall certainly <u>not</u> steer clear of them in future.[5]

Dresser found out that he had been blackballed by someone he considered a friend and 'not unconnected with a sporting paper'. This was almost certainly Edmund Harting, who was both a member of the Zoological Club and the natural history editor for *The Field*, a newspaper devoted to field sports (thus answering well as a 'sporting paper').[6] Dresser had been critical of Harting in *A History of the Birds of Europe* regarding the occurrence of the Black Woodpecker in Britain (discussed in chapter 8). Dresser was elected into the select club the following year and remained a member until 1906. Members met for dinner

9.1 Photograph of the grave of Thomas Jerdon, provided to subscribers who paid for the tombstone (detail of photograph from Henry Dresser's album of correspondents).

at the Pall Mall Restaurant on Cockspur Street, occasionally dining on an unusual animal (Zoological Club, 1933).[7]

In 1874, Dresser and fifteen other ornithologists established a subscription fund for a tombstone in Norwood Cemetery for Thomas Jerdon, famous for his work on the birds of India; Dresser was evidently the secretary for the fund. Each subscriber was provided with a large photograph of the tombstone (see figure 9.1). Dresser had risen far enough in society to be able to make polite demands of John Gould – whom the young Dresser had been so desperate to impress – to settle his subscription.[8] Dresser also acted as a go-between for the grander ornithologists, notably Alfred Newton.[9] Lord Lilford occasionally used Dresser to pass on financial assistance to other ornithologists with their book projects anonymously (Drewitt, 1900: 234).

The 'naming debate'

While he was working on *A History*, Dresser gave birds their scientific names based on the Strickland Code of scientific naming, which had been formally adopted by the British Association for the Advancement of Science in 1843 (after being discussed at its 1842 meeting). This contained some important principles: the law of priority (that the first published scientific name for a species had to be used); the adoption of Linnaeus's binomial system, which gives each species a unique two-part scientific name consisting of a genus and species (*Turdus merula* for the Blackbird, for example); and that Linnaeus's twelfth edition of his *Systema Naturae* (1766) was the starting point of scientific nomenclature, so no names could pre-date 1766 (Farber, 1982; Melville, 1995; Rookmaaker, 2011). Dresser overturned a number of long-accepted names as he found older names for the same species in obscure publications. In this, he was a purist, as were Alfred Newton and William Blanford, while Philip Sclater, Osbert Salvin, Frederick Godman and Henry Seebohm favoured a more judicious approach, to avoid having to change long-established names for birds (Blanford, 1874; Salvin, 1874; Sclater, 1874: 173).[10] Dresser would also put forward complex, rather speculative arguments on what bird had been referred to by what name, leading Sclater to write:

> It cannot be too often repeated that the term to be adopted as the permanent designation of a species should not be a subject of conjecture, or even of disputed evidence, but the first term that is certainly applicable to it. Were this wholesome rule adhered to more strictly, we should cease to be perplexed by such startling changes in our ordinary nomenclature as have lately been suggested to us from several quarters. (Sclater, 1877)

Through this debate we see how Dresser had become confident and skilled enough to propose significant changes to ornithology, and how he did not hold back from disagreeing with his contemporaries.

The BOU and 'The Den'

After Dresser parted company with Sharpe on *A History of the Birds of Europe*, he needed ready access to the library at the Zoological Society of London to make progress with the book. He rented a four-storey townhouse at 6 Tenterden Street, close to the Society's offices in Hanover Square, and moved in in February 1873.[11] Dresser shared the house with his closest ornithological friends: Lord Lilford occupied the ground floor; above him were Osbert Salvin and Frederick Godman; Dresser was on the second floor, and above him were Henry Elwes, Thomas Buckley (see figure 9.2) and Ernest Shelley (Manson-Bahr, 1959: 60). R. H. Porter, a publisher and bookseller, also used the address, and published many of the books written by the house's occupants; his wife acted as housekeeper to the ornithologists. The house was initially known as the 'Ibidaeum' but soon became known as 'The Den'. The best descriptions of The Den are to be found in Lord Lilford's biography (by his sister):

> In 1873 my brother took two rooms on the ground-floor at No. 6 Tenterden Street, Hanover Square, which was then the headquarters of the British Ornithologists' Union. Here he stayed on and off when in town, and '*The Den*,' as it was familiarly called, became known to all his friends. Visitors dropped in during the evening, whist was played, pipes were smoked, and a running talk was kept up on the passing topics of the day, and the never-passing topics connected with zoology. Ladies who were not afraid of a somewhat smoky atmosphere looked in on their way to balls and parties, and were sure of a friendly welcome. (Drewitt, 1900: 89)

BUCKLEY, THOMAS EDWARD, B.A., F.Z.S.

9.2 Thomas Buckley, from Henry Dresser's album of correspondents.

The Den was, in effect, a private club: a bachelor pad for a group of wealthy gentlemen-collector-naturalists and their invitees and, as a near contemporary wrote, 'without Clubs what would bachelor-life in London be?' (Smith, 1910: 158). Lilford was the eldest resident, at forty-one years of age; Elwes and Buckley were the youngest, at only twenty-eight years of age. The Den served as the first 'official' office of the BOU; the annual BOU meeting – the most important event in the Union's calendar – was held there, followed by dinner at the Grosvenor Hotel in Bond Street; no doubt Dresser was a regular attendant. The Den served as the office of the BOU until 1889 (Mountfort, 1959: 13).

Dresser kept his bird skin collections at The Den, as did Lord Lilford, Shelley, Godman and Salvin; the wooden cabinets for all of these collections followed an ingenious design by Salvin.[12] Lord Lilford kept his prized group of eggs of the Great Auk – for which he had paid a king's ransom – at The Den, along with a mounted Great Auk and skeleton, and some live animals:

> amongst them was a half-grown Boa, which on one occasion escaped from its box and was lost for some weeks, only to be eventually found during the 'spring cleaning' coiled up behind some of the books. Lilford used to allow this animal to crawl over him, and on one occasion the creature having coiled itself round his body commenced to squeeze him most unpleasantly; he, however, with some little difficulty freed himself from its embrace and at once put it in a box and sent it straight to the Zoological Gardens. (Evans, 1909b: 87)

Visitors to The Den

Visitors poured through The Den after the fortnightly meetings at the Zoological Society of London. The collections belonging to Dresser and the other occupants were studied by ornithologists, underlining their scientific importance (Gurney, 1876: 238; Seebohm, 1877a,c,d). The Russian Nikolai Severtzov (introduced in chapter 7) spent several weeks with Dresser at The Den in 1875.[13] Severtzov brought the bird specimens he had collected in Central Asia with him, to work out if any species were new. Dresser stored Severtzov's specimens in his collection cabinets.[14] This led to an incident that was retold many years later by Joseph Wolf's biographer, Alfred Palmer. John Gould visited The Den and tried to obtain skins from Severtzov, flattering him that he 'was a naturalist greater than Cuvier or Linnaeus' and that 'it was in the interests of science necessary that he should borrow and examine them [his bird skins]'. Gould's attempt was foiled as Dresser had locked the cabinets and taken the key with him to his office in town. Severtzov told Dresser: 'That minute his face change. He go straight down the stairs, and at every step he say – <u>Damn</u> Mr. G!' (Palmer, 1895: 74–6). 'Mr. G' (whose name had been altered, presumably by Palmer) was Dresser, as revealed by a letter to Alfred Newton: 'I have now secured all the birds he [Severtzov] can possibly spare and Gould has not had any innings at all, tho' he tried it on, but when he came to talk Severtzoff over he found I had locked up the birds and <u>taken the key</u> to the City with me.'[15] It is more than likely that it was Dresser who told Palmer the story.

Severtzov was greatly interested in Darwin's ideas on evolution and natural selection. Dresser arranged for Severtzov to meet the great naturalist at Down House in Kent, on Monday 12 September (see figure 9.3).[16] Darwin made a great impression on Severtzov (Severtzova, 1946: 61). Darwin was less impressed and wrote to his son George: 'an awful Russian bore has been here and has tired me'.[17] Darwin's daughter Henrietta wrote in similar vein:

> A Russian ornithologist forced his way in accompanied by an English ditto [Henry Dresser] as keeper.... The Russian was the most awful of all the foreigners that I've ever seen in this house. He was a perfectly unintelligible unwashed savage. Hideously ugly with a voice to match and his face all gashed with the marks of where other savages had tried to cut his head off.[18]

Following the visit, Severtzov wrote to inform Darwin that he had left a series of specimens of Asiatic thrushes with Dresser that (Severtzov believed) demonstrated the emergence of one species from another. Severtzov was keen that Darwin should include this in his publications, but Darwin did not make use of the information.[19]

Robert Swinhoe visited The Den in June 1875 after retiring from the consular service in China.[20] He and Dresser began work on a book to be called *The Birds of China* and issued a prospectus to potential subscribers.[21] Philip Sclater referred to the project in his presidential address to the British

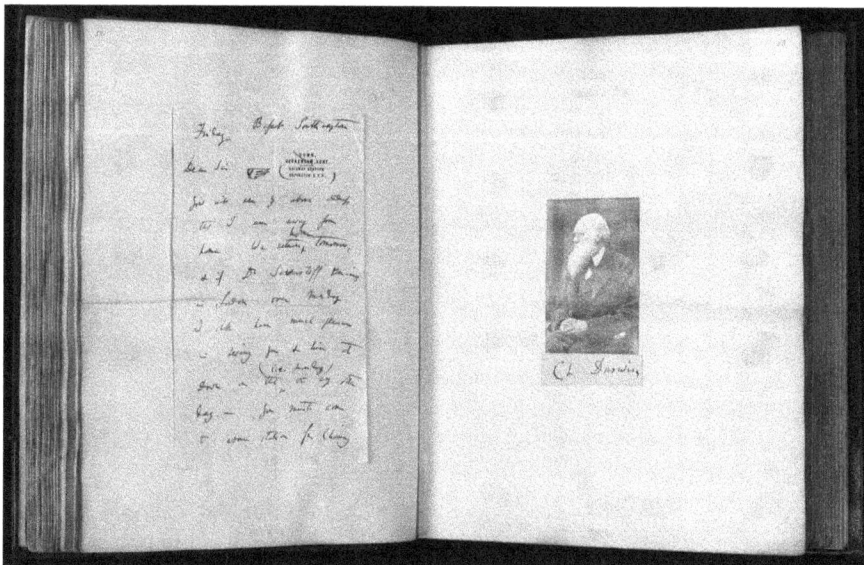

9.3 Letter and photographic portrait from Charles Darwin to Henry Dresser, from Henry Dresser's album of correspondents.

Association for the Advancement for Science in 1875, insofar as Swinhoe was preparing a book on Chinese ornithology 'for which he has secured the co-operation of one of our most competent naturalists' (Sclater, 1875: 380; Sclater, 1876: 99). Swinhoe wanted to issue the book at once but Dresser wanted to wait until they had at least eighty subscribers.[22] The project never got off the ground as Dresser found that he did not have time to rewrite the text from Swinhoe, and it was put to one side in February 1876.[23] Swinhoe died not long afterwards, in October 1877, bringing an end to the project.

Thomas Brewer, an American ornithologist who made a tour of European collections in 1876, visited 'a few of the private collections of natural history in which England abounds'. His published account mentions collections belonging to Howard Saunders, Henry Tristram and Osbert Salvin (Salvin's collections of Central American birds, eggs and insects were kept at The Den). Brewer described Dresser's egg collection as 'the most complete collection of the eggs of the birds of Europe probably in existence. It is admirably arranged, containing many fine suites of the least common kinds, and very many species not in any other collection' (Brewer, 1877: 481). The result of such visits was that Dresser fell behind with *A History* and had to work on the book from five in the morning before going to work in the City.[24]

Telegrams and letters arrived at The Den from far and wide, bringing news of English travellers who were far overseas. In 1875, Henry Feilden – naturalist to the British Arctic Expedition – sent a telegram to The Den telling his friends that he was quite safe and well on Ellesmere Island, north-west of Greenland (see figure 9.4).[25] Two years later Dresser received news from Henry Seebohm (introduced in chapter 6) who was travelling in Siberia at the time. Dresser passed the news on to a friend:

> Have you not heard of Seebohm's shipwreck. The *Thames* [the ship] is a total loss at the mouth of the Yenesei but all hands are safe. As yet the news are only by telegraph but ere long there ought to be a letter with full particulars. I hope he has saved some if not all his collection.[26]

Seebohm gave a lecture on his expedition to Siberia at the Zoological Society of London in 1877, which Dresser described to John Harvie-Brown:

> Seebohm had a grand night of it at the Zoo and had quite an ovation. He has got quite a nice lot of things [specimens]. Of course there was a big smoke and long drink at our Den afterwards and all hands turned up – even little Bowdler [Sharpe's middle name was Bowdler and he was sometimes referred to as Bowdler Sharpe] who seldom comes to the Zoo now but affects the Linnean [Society].[27]

Seebohm was an excellent public speaker, a trait that he shared with his father, who had travelled widely as a Quaker minister. Shortly afterwards, in early 1878, Seebohm took over Godman's and Salvin's rooms at The Den and stored his collection there.[28]

Dresser and Seebohm were matched in their business interests as well as in ornithology: Seebohm was a partner in a steelworks in Sheffield, comparing

9.4 Henry Feilden, from Henry
Dresser's album of correspondents.

9.5 Henry Seebohm, from Henry
Dresser's album of correspondents.

well with Dresser's involvement with the Bowling Iron Company (discussed in
chapter 3). Seebohm's firm (Seebohm and Dieckstahl) had business premises
alongside Dresser in Cannon Street and Dresser's youngest brother, Arthur,
actually worked for Seebohm during 1876–79.[29] Dresser helped Seebohm
establish himself in ornithological society and Seebohm gave Dresser some
of the best specimens from his expeditions to Siberia, showing that the two
were on good terms. However, their relationship was strained from the time
of Seebohm's earliest publications, when Seebohm wrote critically of Dresser
(Seebohm, 1877c). The two would fall out in spectacular fashion in the following decade (see figure 9.5).

Henry Dresser and colonial ornithologists

In order to produce *A History of the Birds of Europe*, Dresser relied on
information and specimens from many field collectors and travellers. They
needed Dresser to publicise their discoveries, whether at the meetings of the

Zoological Society of London or in his publications. There was self-interest on both sides, and their alliances could be uneasy ones. Dresser's relationship with Edwin Brooks serves as an example of this. Brooks, who worked on the Etawah railway (now in Uttar Pradesh), was a first-rate ornithologist and knew the leading ornithologists in India (all of whom were British) (chapter 7). How they became acquainted is unknown, but by 1871 Brooks was writing to Dresser fortnightly.

Through the 1870s, Dresser arranged for Brooks' articles to be published in Britain; he exhibited his specimens at the Zoological Society of London (Dresser, 1872), consulted other ornithologists for him, checked information at the Society's library and specimens in the British Museum. He received many consignments of Brooks' bird skins and eggs at the London docks, and dispersed them to other ornithologists as gifts and by sale. In return, Brooks provided Dresser with large amounts of novel and largely reliable information, as well as specimens to help with *A History of the Birds of Europe* (common, rare and previously unknown to science).

Brooks made an expedition to Cashmere (Kashmir) in 1871, in search of the nests of a particular species of small warbler (Brooks, 1872) (see plate 30).[30] He hoped to cover the cost of the expedition by selling the bird's eggs to British collectors.[31] Brooks complained to Dresser when Henry Tristram advertised the warblers' eggs for sale in the *Ibis*: 'The railway board will haul me over the coals for dealing in eggs!!'[32] Brooks sent thousands of bird skins to England, including those of large eagles and cranes, via the Suez Canal.[33] He gradually became fed up with giving specimens away: 'I am so out of pocket sending these big boxes home to say nothing of the expense of getting the specimens here. People think these things cost us nothing but they are greatly mistaken.... I must give it up for I am robbing my family with my expensive amusements.'[34]

Dresser tried to place Brooks' articles in the *Ibis* or, if rejected, presented them at the Zoological Society of London or included the information in *A History*.[35] Few made it into the *Ibis* as the Editors did not always accept Brooks' supposedly new species: 'I sent lots of papers these last 3 years [to the *Ibis*] till I am sick of doing so as they are never made any use of.'[36] In the end, he sent most of his articles to the *Proceedings of the Asiatic Society of Bengal* or to Allan Hume's journal *Stray Feathers*.[37]

Brooks was strongly opposed to evolution and rejected notions that a species could consist of localised races (subspecies): 'My friend Dr. Jerdon laboured hard one day to convince me that we were descended from monkeys.... I think, in regard to "subspecies," we should not go beyond facts. Speculation that one species is derived from another is hurtful to our science' (Brooks, 1884: 236–7). He held fast to the older practice of giving a two-part scientific name for each different type of bird, no matter how small the difference between it and other species.[38] Brooks criticised 'those stupid Darwinites',[39] and the 'warped intellects' of Darwin and Wallace: 'I refuse the evolution theory as pure speculation'.[40]

The most heated disagreements between Brooks and the 'swell naturalists' in England were about eagles (discussed in chapter 8). Brooks was a keen observer of these birds: he showed that two groups of species (marsh eagles and hunting eagles) could be separated by the shape of their nostrils (Anderson, 1872: 685–6; Gurney, 1872)! Dresser sent Brooks two eagle skins, a Lesser Spotted Eagle from Danzig and a Spotted Eagle from the Volga, to compare with Indian eagles.[41] Brooks was adamant that the Lesser Spotted Eagle was identical to the Indian Spotted Eagle: 'It is mature <u>Aquila Hastata</u> [Indian Spotted Eagle] upon this I would bet 100£. Anderson would bet his life he says upon either birds being what I have told you…. So much for sending me a shabby bird. Oh Mr. Dresser the two old eagles were two capital jokes.'[42] Brooks proposed to name the specimen from the Volga as a new species but Dresser, Salvin and Newton quashed the article.[43] Dresser (and his collaborators) often disagreed with Brooks' views, to the latter's growing frustration: 'I know the true eagles of this country without boasting better than any man and can tell them at any stage … so don't you hint I don't know our Indian eagles…. I could soon send you a cart load of eagles spotted and otherwise but cannot afford the carriage.'[44] When Brooks visited The Den in June 1876, Dresser expected to have a 'grand battle over the eagles'.[45] Brooks was ultimately proven correct on many counts; he showed that the bird that was known as the 'Tawny Eagle' of India was an entirely different bird from the true Tawny Eagle of Africa (and which also occurred in India). The Indian bird became known as Brooks's Eagle before being confirmed as a rare colour variation of the Spotted Eagle (see plate 31).[46]

Brooks wrote in an argumentative, 'chaffy' style (as he himself called it) and once told Dresser that his grumbling was only intended to be good-natured.[47]

> I know you have but very little time to write to me and I can always make allowances for you, but make an effort sometime and consider what exiles we are here and how any little variety in the way of an English letter, especially an ornithological one, is greatly appreciated.[48]

Brooks' letters to Dresser reveal some of the complexity and the unequal balance of power that existed between the great 'swells' and those who worked at great distances from Western intellectual structures.

Richard Sharpe and the British Museum

While Dresser and his friends enjoyed the high life at The Den, significant developments were underway at the British Museum. Albert Günther, a specialist on fish and reptiles, took on the running of the bird collection in 1872 after the previous curator, George Gray, died. Almost immediately, he set to work on a great catalogue of the bird collection, using the same model as his *Catalogue of Fishes in the British Museum*.[49] The project was to be more than just a catalogue of the Museum's own collection, but to be a comprehensive encyclopaedia of the birds of the world. Richard Sharpe was taken on to assist with the project. Osbert Salvin had been recommended for the position by Alfred Newton, but Günther wanted an assistant who would work steadily for him, rather than a wealthy dilettante (Gunther, 1975: 372). Günther worked Sharpe extremely hard on the *Catalogue of the Birds in the British Museum* (usually known simply as the *Catalogue of Birds*), although he found Sharpe 'too ready to flit from one work [book] to another for steady and methodical work' (quoted in Gunther, 1975: 375). Sharpe wrote the first four volumes of the *Catalogue* during 1874–79, undertaking most of the work in the evenings (Gunther, 1975: 375).

Sharpe's entry to the Museum coincided with a period of rapid expansion (Knox and Walters, 1992). The American Thomas Brewer had visited the British Museum during his tour of European collections in 1876:

> The writer [Brewer] will say that all he did or could see was very disappointing.... It is, however, but justice to say that, since the zoölogical portion of the museum has been under the charge of Mr. Sharpe, a systematic rearrangement has been begun, and, so far as it has proceeded, is a great improvement.... The contrast between the Museum of London and that of Newcastle-upon-Tyne, or that of Liverpool, cannot but be painful to the national pride of a true English naturalist. (Brewer, 1877: 479–80)

Through the 1870s, the natural history collections were separated from the rest of the British Museum collections, to be housed in a new institution, the British Museum (Natural History) (the present Natural History Museum in South Kensington). Collections flooded in from the Empire and colonies; others were acquired with the power of imperial wealth. Sharpe worked single-mindedly to develop the collection, donating many specimens at his own expense. He used the *Catalogue of Birds* as a vehicle to enhance the collection, with great success.[50] When he first entered the Museum, in 1872, the birds of prey skins were kept in a few wooden boxes on shelves in the cramped bird stores; eleven years later (when they moved to South Kensington) they took up 108 of the same boxes, each three feet long by one and a half feet wide and requiring two men to lift them (Sharpe, 1906: 87). Among the most notable acquisitions during the 1870s were Alfred Wallace's collection of 2,474 birds from the Malay Archipelago and Sharpe's own collection of 3,583 African birds and kingfishers (Sharpe, 1906: 481). Sharpe's acquisitiveness even threatened to slow the production of the *Catalogue* (Gunther, 1975: 376).

While Sharpe was making his mark at the British Museum, he was a rare visitor to The Den. His relationship with Dresser was strained to say the least and he felt that Dresser's friends 'subjected him to much abuse' (see chapter 8); there is also Dresser's patronising reference to 'little Bowdler' quoted above. Sharpe did make an occasional appearance at The Den, having tea there with Dresser and Robert Swinhoe on one occasion, at Swinhoe's request.[51] When Count Tommaso Salvadori visited London from Turin to study bird collections, Dresser wrote: 'Bowdler has him [Salvadori] in tow and will of course keep him away from that evil neighbourhood about The Den for fear of ornithological contamination'.[52]

Dresser's work with bird conservation

Through the 1870s, Dresser served as the Secretary to the British Association for the Advancement of Science 'Close-Time Committee' (discussed in chapter 4), set up in 1868 to explore the desirability (and practicality) of protecting birds through the breeding season. He organised the Committee's meetings at The Den,[53] drafted Bills to go before Parliament and presented the Committee's progress at the annual BA meeting in most years. His name occasionally appeared in *The Times* as Secretary to the Close-Time Committee: he commented on bird-catching and served as an expert witness in a court case on cruelty to swans (although both he and Newton were keen to distance themselves from the anti-cruelty movement).[54]

At the 1870 BA meeting in Liverpool, Dresser reported on the success of the Sea-birds Preservation Act, implemented in the previous year (Close-Time Committee, 1871). He also reported that wildfowl (ducks and geese) were declining and, being important for food and sport, this was a potentially serious matter. He proposed, on behalf of the Committee, that wildfowl receive protection in the same way as had been achieved for seabirds in order to increase their numbers.

In 1872 the Committee prepared a 'Bill for the Protection of Wild Fowl'. Dresser mustered support among the politicians, with a view to including their names on the Bill.[55] The Bill was brought into Parliament in February, but was altered drastically by politicians, being extended to all wild birds. The members of the Close-Time Committee were very unhappy with this direction, fearing that protection for all birds (many of which did not require protection) would dilute the protection of those that were in need of action; penalties for killing rare birds were not high enough to deter their persecutors (Wollaston, 1921: 140–3). Dresser and two other Close-Time Committee members gave evidence to a Select Committee, but to no avail, and the Bill was passed as the Wild Birds Protection Act in August 1872.

At the 1873 BA meeting in Bradford the Committee tabled some key aims for bird protection in Britain: unrealistic legislation would quickly be repealed; only birds in need of protection should receive it; birds of prey could not be effectively protected (although their numbers were low as a result of persecution); a law protecting wildfowl could be successful if penalties were greater than the birds' market value; and wildfowl were in decline as a result of over-hunting and loss of habitat to agriculture. A Bill was put before the House of Lords in 1874 for the protection of wild birds during the breeding season, but this bore no relation to the recommendations of the Close-Time Committee or the House of Commons Select Committee. The Close-Time

Committee reported to the 1874 BA meeting (in Belfast) that the only way to prevent declines in bird numbers was to protect adult birds during the breeding season; no mention was made of the effects of egg collecting (Close-Time Committee, 1875).

The original intentions of the Committee were realised in 1876, when a Bill for the Preservation of Wild Fowl was enacted as law, with an incredible majority (337 in favour, thirteen against). This established a close season between February and mid-July, when wildfowl could not be killed. In 1878, Dresser gave evidence to an enquiry into the Scottish herring fishery, demonstrating that seabirds caused less damage than had been alleged (see Close-Time Committee, 1879). In 1880, an MP introduced a Bill into Parliament to amend the bird protection laws. Dresser and some other Close-Time Committee members reworked the Bill as it went through its three readings. Lord Lilford and Lord Walsingham, both keen ornithologists, were also involved in the House of Lords. The Wild Birds Protection Act of 1880 established a close season for all wild birds between 1 March and 1 August, meaning that they could not be killed while they were breeding. Importantly, heavier penalties were exacted for destroying rare birds; explanatory amendments were made in 1881 and the Act remained unmodified for fourteen years.

Through his membership of the Close-Time Committee, Dresser had helped lay the foundations of nature conservation in Britain (see also Wollaston, 1921: 136–59; Barclay-Smith, 1959; Evans, 1992; Bircham, 2007: 314–17). One commentator has noted: 'Bird protectionists ... may claim to have been the earliest, and, at least until very recently, the most energetic and successful arm of the world conservation movement' (Nicholson, 1970: 197).

Settling down

In 1878, Dresser's life took a major change, as he married Eleanor Walmisley Hodgson (1854–1937): 'The young lady (Miss Nellie Hodgson) is an old friend and her tastes run tolerably together with mine so I shall go on at ornithology much the same as usual, only I can't be quite as often at The Den.'[56] She was the daughter of a solicitor, Charles Hodgson, from fashionable Belgravia. The couple were separated by a rather large age gap, with Dresser aged thirty-nine and Eleanor twenty-three. Dresser's bachelor years, spent smoking in the evenings at The Den with his fellow bachelor ornithologists, came to an end. He still kept his collections at The Den and used it as an address for his ornithological activities. They married on 7 March 1878 at St George's Church in Hanover Square, with the ceremony performed by Henry Tristram. Eleanor took the name Walmisley Dresser from the time of her marriage. After honeymooning on the Isle of Wight, the newlyweds moved back to The Firs with Dresser's parents in Upper Norwood: 'I am in full swing in my old bird room. I work two evenings per week in town and the rest here and have more than half my books here and find that my wife can and does help me a good deal.'[57] Henry Dresser and Eleanor moved into a luxurious detached villa, St Margaret's, only a few doors down South Norwood Hill, in April 1878 (see plate 32). Their first child, Henry Joseph, was born there in 1879 and a daughter, Brenda Eleanor, was born the following year.[58]

Notes

1   Gould produced some of the most famous and sumptuous books on birds, for example *The Birds of Europe* (1832–37) and *The Birds of Australia* (1840–48),

2   Cambridge University Library (CUL hereafter), MS.Add.9839.1D.303, E506, letter from Henry Dresser (HED hereafter) to Alfred Newton (AN hereafter), 17 February 1870.

3   National Museums Scotland Library (NMSL hereafter), GB 587, JHB, note in the John A. Harvie-Brown (JAHB hereafter) archive, accompanying a list of birds sent to Harvie-Brown in April 1876.

4   CUL, MS.Add.9839.1D.346, F849, letter from HED to AN, 16 May 1872.

5   CUL, MS.Add.9839.1D.361, F995, letter from HED to AN, 8 November 1872.

6   CUL, MS.Add.9839.1D.362, G006, letter from HED to AN, 13 November 1872.

7   Sclater, Newton, Salvin, Godman, Harting, Blanford, Elwes, Alfred Wallace and Swinhoe were all members during the 1870s. Richard Bowdler Sharpe (RBS hereafter) was a member during 1892–93.

8   Natural History Museum Library and Archives, London (NHM hereafter), Z MSS GOU A/3, letter from HED to John Gould, 17 October 1874.

9   CUL, MS.Add.9839.1D, correspondence from HED to AN, contains many offers of assistance.

10 Salvin told HED that his treatment of one scientific name was 'pure pedantry'; CUL, MS.Add.9839.1D.437, H395, letter from HED to AN, 4 April 1877.

11 CUL, MS.Add.9839.1D.365, G252, letter from HED to AN, 1 March 1873.

12 Manchester Museum (MM hereafter), ZDH/8/2/2, letter from HED to W. E. Hoyle, 26 April 1900. Drawers fitted into slots in the cabinet carcasses; they were made in multiples of a fixed depth, meaning that deep drawers could be placed between shallower drawers as required. Drawers had wooden lids to protect the bird skins from insects and airborne soot.

13 CUL, MS.Add.9839.1D.411, I203, letter from HED to AN, 7 September 1875.

14 CUL, MS.Add.9839.1D.407, I118a, letter from HED to AN, 15 August 1875; CUL, MS.Add.9839.1D.410, I197, letter from HED to AN, 1 September 1875.

15 CUL, MS.Add.9839.1D.411, I203, letter from HED to AN, 7 September 1875.

16 Charles Darwin's letter to HED is now in John Rylands Library, Manchester. It was sent from Southampton, 9 September 1875. Bound in English Manuscript 1404. See also CUL, MS.Add.9839.1D.412, I209, letter from HED to AN, 12 September 1875.

17 CUL, DAR 210.1: 47, Darwin Papers, letter from Charles Darwin to George Howard Darwin, 13 September 1875.

18 CUL, DAR 251: 1612, Darwin Papers, letter from Henrietta Litchfield to Leonard Darwin, 14 September 1875.

19 Letter from Nikolai Severtzov to Charles Darwin, 25 September [1875], quoted in Burkhardt and Secord (2015: 374–6).

20 McGill University Library, Rare Books and Special Collections, Blacker-Wood Autograph Letter Collection, letter from HED to RBS, 22 June 1875.

21 CUL, MS.Add.9839.1D.404, A805, letter from HED to AN, 22 July 1875.

22 CUL, MS.Add.9839.1D.414, I235, letter from HED to AN, 29 September 1875.

23 NMSL, GB 587, JHB 15/240, letter from HED to JAHB, 13 February 1876.

24 NMSL, GB 587, JHB 15/240, letter from HED to JAHB, 1 July 1876.

25 CUL, MS.Add.9839.1D.411, I203, letter from HED to AN, 7 September 1875.

26 NMSL, GB 587, JHB 15/240, letter from HED to JAHB, 2 September 1877.

27 NMSL, GB 587, JHB 15/240, letter from HED to JAHB, 8 December 1877.

28 NMSL, GB 587, JHB 15/240, letter from HED to JAHB, 3 February 1878.

29 Information on Arthur Dresser's career comes from A. R. Dresser, *Travels in New Brunswick, Canada and Manitoba*, National Archives of Canada, Ottawa, Microfilm A-1536. Details of his time at Cannon Street are also taken from *Kelly's London Directory* for 1877–80.

30 Called the Yellow-browed Warbler (*Reguloides superciliosus*) at the time, now split into Hume's Warbler (Hume's Leaf Warbler) (*Phylloscopus humei*) and, farther north, the Yellow-browed Warbler (*Phylloscopus inornatus*).

31 MM, ZDH/1/3, letter from W. E. Brooks (WEB hereafter) (Saidahad, Cashmere) to HED, 21 June 1871; MM, ZDH/1/3, letter from WEB (Etawah) to HED, 6 September 1871.

32 MM, ZDH/1/3, letter from WEB to HED, 1 December 1871.

33 MM, ZDH/1/3, letter from WEB to HED, 6 March 1872.

34 MM, ZDH/1/3, letter from WEB to HED, 4 June 1872.

35 See MM, ZDH/1/3, letter from WEB to HED, 23 October 1871.

36 MM, ZDH/1/3, letter from WEB to HED, 1 December 1871.

37 MM, ZDH/1/3, letters from WEB to HED, 1 December 1871, 11 January and 21 February 1872. See also Pittie (2011).

38 MM, ZDH/1/3, letter from WEB to HED, 31 October 1871.

39 MM, ZDH/1/3, letter from WEB to HED, 6 September and 23 October 1871. See also MM, ZDH/1/3, letter from WEB to HED, 1 December 1871.

40 MM, ZDH/1/3, letter from WEB to HED, 5 April 1872.

41 MM, ZDH/1/3, letter from WEB to HED, 19 December 1871.

42 MM, ZDH/1/3, letter from WEB to HED, 19 December 1871.

43 NMSL, GB 587, JHB 15/239, letter from HED to JAHB, 18 April 1872.

44 MM, ZDH/1/3, letter from WEB to HED, 24 May 1872.

45 NMSL, GB 587, JHB 15/240, letter from HED to JAHB, 7 June 1876.

46 MM, ZDH/1/3, letter from WEB to HED, 13 August 1872. See also Brooks (1873a–c, 1875).

47 MM, ZDH/1/3, letter from WEB to HED, 5 April 1872.

48 MM, ZDH/1/3, letter from WEB to HED, 19 January 1872.

49 Eight volumes written over a thirteen-year period and published 1859–70.

50 Günther's prefatory notes to the various volumes of the *Catalogue* are a good source of information on the Museum's spectacular growth.

51 McGill University Library, Rare Books and Special Collections, Blacker-Wood Autograph Letter Collection, letter from HED to RBS, 22 June 1875.

52 CUL, MS.Add.9839.1D.448, H867, letter from HED to AN, 21 September 1877.

53 NHM, DF200/11/146, Department of Zoology: Keeper of Zoology's Correspondence and Files 1785–1997, letter from HED to A. Günther, 27 April 1877; NHM, DF200/13/197, Department of Zoology: Keeper of Zoology's Correspondence and Files 1785–1997, letter from HED to A. Günther, 15 May 1878.

54 H. E. Dresser, 'Bird-catchers', *The Times*, 12 March 1874, p. 10, col. e; 'Swan-nicking: at the Petty Sessions at Slough', *The Times*, 31 January 1878, p. 6, col. e.

55 CUL, MS.Add.9839.1D.341, F553, letter from HED to AN, 12 February 1872.
56 NMSL, GB 587, JHB 15/240, letter from HED to JAHB, 3 February 1878.
57 NMSL, GB 587, JHB 15/240, letter from HED to JAHB, 28 April 1878.
58 Their children used the double-barrelled Walmisley-Dresser throughout their lives.

33 Hairy-fronted Muntjac, illustration of a specimen given (alive) to London Zoo by Henry Dresser, and named as a new species by Philip Sclater.

34 Arabian Green Bee-eater, from *A Monograph of the Meropidae*.

35 Red-bearded Bee-eater, from *A Monograph of the Meropidae*.

36 Short-legged Ground Roller, from *A Monograph of the Coraciidae*.

37 'Eastern Shore-lark', from *A History of the Birds of Europe*.

38 Greenshank, from *A History of the Birds of Europe*.

39 Black-eared Wheatear *Oenanthe hispanica* as *Saxicola stapazina*, from *A History of the Birds of Europe*.

40 Syrian Serin *Serinus syriacus* as Tristram's Serin *Serinus canonicus*, from *A History of the Birds of Europe*.

41 British Marsh Tit *Poecile palustris dresseri*, Norfolk.

42 *Coracias weigalli*, from *A Monograph of the Coraciidae.*

43 *Coracias mosambicus*, from *A Monograph of the Coraciidae.*

44 The extinct Canary Islands Oystercatcher (as African Black Oystercatcher), from *Supplement to A History of the Birds of Europe*, based on a bird in Henry Dresser's collection.

SOFT-PLUMAGED PETREL.
ÆSTRELATA MOLLIS.

45 The rare Desertas Petrel (as Soft-plumaged Petrel), from *Supplement to A History of the Birds of Europe*, based on a bird in Henry Dresser's collection.

46 Pallas's Warbler, Burnham Overy Dunes, Norfolk, October 2010.

47 'Labrador Falcon' (dark-phase Gyr Falcon) painted by J. Wolf in 1875, based on a specimen in Henry Dresser's collection (shown in plate 1, see also plate 19).

10 The 1880s: the rise of rivalry

This chapter explores Dresser's activities through the 1880s, a decade of great change for him and his family. His father died of a stroke on 28 January 1881 at The Firs aged seventy-seven (his estate amounted to under £3,000). After this time, Dresser's mother lived with two of her daughters close to The Firs. Dresser and Eleanor moved to Farnborough, Kent, in 1883, to a newly built mansion amid orchards that Dresser named Topclyffe Grange.[1] His childhood home, Farnborough Lodge, lay less than two kilometres away across open fields. Neighbours included Sir John Lubbock (later Lord Avebury); Charles Darwin's home lay a kilometre further away. Dresser commuted daily into the City by train from nearby Orpington. He and Eleanor had their youngest child, Phyllis Caroline Eeles, at Topclyffe Grange in 1884. Dresser's health caused him problems throughout the 1880s, with sciatica affecting one of his legs in 1882–83,[2] and severe stomach pain at various times.[3] His family was hit with sadness again at the end of the decade: his mother died of bronchitis, on Christmas Eve 1889, aged eighty-three.

Life in scientific society

The 1880s were typified by growing rivalry among ornithologists and institutions. A cartoon depicting a meeting of the Zoological Society of London appeared in *Punch* in 1885, showing serious-looking naturalists examining an assortment of zoological curios (see figure 10.1).[4] Dresser looks decidedly downbeat, perhaps because this was a time of illness. The Zoological Society of London, like other natural history societies, had a long-standing history of competition and argument among members. Charles Darwin had complained many years earlier of the zoologists' 'mean, quarrelsome spirit', having seen the speakers 'snarling at each other in a manner anything but like that of gentlemen'.[5] Naturalists could be forceful with one another in print too, and arguments and disagreements found their way into natural history journals. The 'brother Ibises' of the BOU appear to have been especially quarrelsome to judge from the *Ibis*. Most arguments were about which kinds of birds deserved to be recognised as separate species ('good species'), what scientific names these species should be known by and which species had occurred in Britain.

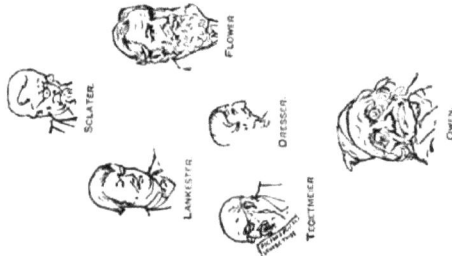

10.1 'Meeting of the Zoological Society at Hanover Square' by H. Furniss, from Scherren (1905).

The BOU and Zoological Society of London continued to be dominated by Dresser and his closest colleagues. Newton, Salvin, Godman and Sclater held the most eminent positions, but Dresser was not far behind them, becoming a member of the Council of the Society in 1882 (a role he held during 1882–88 and 1889–95),[6] and Secretary and Treasurer of the BOU during 1882–89.[7] He was elected an honorary member of the American Ornithologists' Union on its establishment in 1883 and to the Athenaeum, one of the most select London clubs, in March 1885 with support from Alfred Newton (see Cowell, 1975).[8] He often lunched at the Club, where Salvin and Sclater were also members.[9]

Dresser was a frequent contributor at meetings of the Zoological Society of London, exhibiting many specimens of rare birds that had been found in Britain, and unusual birds and eggs received from southern Europe and Central Asia, including new species.[10] Other naturalists made their own pronouncements on specimens Dresser received from overseas. Philip Sclater announced the discovery of a new species of deer based on a specimen that Dresser had received alive from Ningpo, China, and donated to London Zoo. The deer had been sent over by a Scottish businessman, Alexander Michie, who associated with Robert Swinhoe (Sclater, 1885) (see plate 33). A pickled hermit crab from Saint Helena that Dresser donated to the British Museum (Natural History) was also announced as a new species at the Society's meetings (Miers, 1881: 275–6).[11]

Leading zoologists also used Dresser's publications for their own articles and books. Ernest Shelley named a supposedly new species of bee-eater *Merops dresseri*, after 'his friend … in acknowledgment of the valuable services rendered to ornithology by his large work on the Birds of Europe' (Shelley, 1882: 304), although Dresser subsequently showed that this 'new' bee-eater belonged to a species that was already known (Dresser, 1883a). John Gould included information from *A History* in his fabulous *Birds of Asia*,[12] and Philip Sclater and Alfred Wallace included information from Dresser and from *A History of the Birds of Europe* in their work on biogeography (Sclater, 1875, 1876; Wallace, 1876, 1880). Others borrowed birds from Dresser's collection in researching their own books (e.g. Meyer, 1887).

On 16 June 1887, the Zoological Society of London held a special meeting and garden party at London Zoo to celebrate Queen Victoria's Golden Jubilee. The Maharajah of Kuch-Behar (who had given a Great Indian Rhinoceros to the Zoo) was awarded the Society's silver medal and the President (William Flower) gave a speech on progress in zoology. Among the participants were: the Queen of Hawaii; the Thakore Sahib of Limbdi; the Maharajah of Bhurtpore; a variety of earls and lords; Sir Joseph Hooker and Thomas Henry Huxley; and Council members of the Society, including Dresser. The event was reported at length in *The Times*, which listed the leading attendants. Through his position in scientific society, Dresser found himself listed among queens, maharajahs and all manner of aristocracy; in the report, he was mistakenly referred to as an FRS (Fellow of the Royal

Society), the leading scientific achievement in Britain.[13] Dresser did stand as a candidate for the Fellowship of the Royal Society during 1889–90, when his thirteen nominators included Philip Sclater, Osbert Salvin, Frederick Godman, Lord Walsingham, William Flower (Director of the British Museum (Natural History)) and William Blanford, among others; however, he was not successful in gaining entry to the Society (discussed further in the following chapter).[14]

Unlike most of his peers, Dresser's ornithology was confined to his 'spare' time outside of business. One reviewer of *A History of the Birds of Europe* noted: 'It is the more remarkable as being not the work of a professed naturalist entirely devoting himself to the subject, but of a gentleman engaged in business all day' (Anon., 1882: 460). Precisely what was meant by a 'professed naturalist' was not made clear but a distinction was beginning to emerge between those who spent all of their time on ornithology and those who could spend only their leisure time on birds. Even the individuals involved held conflicting views of what amateurs, professionals and scientists were (see Lucier, 2009; Lewis, 2012). There are two fundamental approaches to the study of professionals and professionalisation. Under the so-called 'trait' approach, a profession is defined as a full-time occupation, characterised by formal training, organised practices and some kind of certification. Conversely, the 'power' approach emphasises the 'ongoing quest to maintain status, privilege, and monopolistic domination' (see Barrow, 1998: 4). Of these, the power approach can be applied more readily to the likes of Dresser and his peers, at a time when there were few institutions or paid individuals.

Encyclopaedias, books and collections

Dresser worked on four encyclopaedic projects on birds through the 1880s. He acquired large numbers of birds from poorly known parts of Europe to keep his 'main' collection of the 'birds of Europe' up to date. To take one example, the birds of the Atlantic islands – the Azores, Canary Islands and Madeira – were comparatively little known. Dresser bought hundreds of bird skins and eggs from Ramón Gómez, a pharmacist and natural history collector from Tenerife who visited England in 1889 to sell his collections. These included the rarest of birds, such as two skins of the Canary Islands Oystercatcher, now extinct and known from only a handful of specimens (see plate 44).[15] Henry Tristram (whose travels have been explored in chapter 7) visited the Canary Islands with his friend Edmund Meade-Waldo (1855–1934), a noted early conservationist and falconer (although perhaps best known for having seen a sea-serpent off the coast of Brazil in 1905), in the late 1880s and early 1890s. They discovered new birds, including the endemic Canary Islands Stonechat, and distinctive kinds of Chaffinch and African Blue Tit on the various islands, naming some as new species. They gave specimens of most of the birds they encountered, including some type specimens of their new birds, to Dresser.

Dresser worked on a monograph on the bee-eaters (colourful kingfisher-like birds) of the world. From 1882, he built a collection of 200 bee-eater skins, including almost all known species. He visited the Muséum National d'Histoire Naturelle in Paris to study its specimens and was offered some duplicate birds from a box by the curator. Dresser was amazed to discover that the box contained five specimens of the Arabian Green Bee-eater (*Merops viridis cyanophrys*), considered to be a separate species and previously known from only two specimens (see plate 34). The curator gave Dresser two of the five specimens. The bee-eaters book was issued during 1884–86, with exceptionally fine illustrations by Keulemans based on Dresser's specimens (see plate 35). Dresser dedicated the book to the late Arthur Hay, Marquess of Tweeddale, from 'his sincere admirer'.[16] In the book, Dresser gave an account of his 'discovery' of the Arabian Green Bee-eaters in the Muséum National d'Histoire Naturelle, which was rather critical of the museum staff.[17] Dresser's book was favourably reviewed in the *Ibis*, although the reviewer predicted (rightly) that his account of Arabian Green Bee-eaters would cause offence to the staff of the Paris Muséum (Anon., 1885a).

As early as 1883 (while he was already hard at work on the bee-eaters book), Dresser began work on another monograph, on rollers (another group of colourful birds). He built a collection of 124 roller skins, including many rare specimens that cost £5–£6 each.[18] Alfred Newton put Dresser in touch with his correspondents in Madagascar to obtain specimens of rare and unusual species (the ground rollers and Cuckoo Roller) that occurred there.[19] One of these was the Rev. James Wills (1836–98), of the London Missionary Society. Wills was based at Antananarivo during the 1880s and '90s, where he collected birds, moths and rocks that he sold to museums, including the Smithsonian and the British Museum (Natural History) (Oberholser, 1900). Wills provided Dresser with a variety of specimens, including the extraordinary ground rollers (see also Wills, 1893) (see plate 36). John Kirk (1832–1922), the Consul-General of Zanzibar and one-time travelling companion of David Livingstone, provided Dresser with Cuckoo Roller specimens from the Comoros (see Shelley, 1879).

Dresser began work on a great book on the birds of northern and eastern Asia to accompany *A History of the Birds of Europe*, writing to Robert Ridgway (at the Smithsonian) in 1884: 'I have for long been quietly but steadily collecting N. Asiatic birds and have a fair collection, which I trust to increase largely before I begin to write on the subject.'[20] This idea was no flash in the pan and two years later he was again asking Ridgway for specimens towards this project,[21] although it never got off the ground.[22]

Dresser helped Lord Lilford with his *Coloured Figures of the Birds of the British Islands* (issued during 1885–98) by lending bird skins to the artist, Archibald Thorburn, to assist with the illustrations for the plates.[23] A rumour circulated that Lilford had produced the book as he found Dresser's *Birds of Europe* too bulky to take on his Mediterranean cruises, although this story has been dismissed by one commentator (Radclyffe, 1994).

Work on the BOU *List of British Birds*

At the annual BOU meeting in 1878, a committee had been set up to 'draw up a list of British birds, according to the most approved principles of modern nomenclature', to iron out the differences of opinion between ornithologists on birds' scientific names (Anon., 1878b). The List Committee was also to assess records of rare birds that had supposedly occurred (and often been shot) in Britain, as no two authorities were in agreement on this point (see Bircham, 2007; Knox, 2007). The committee consisted of Salvin and Sclater (joint Editors of the *Ibis*), Godman, Dresser, Newton (who soon dropped out), Seebohm and Henry Wharton. Howard Saunders and Richard Sharpe joined the group in 1879.

Sclater, Salvin and (particularly) Seebohm criticised various changes that Newton and Dresser had made to birds' scientific names (Salvin, 1875; Sclater, 1879; Seebohm, 1879a,b, 1880a). Seebohm complained of Newton's devotedness to adopting the oldest scientific name used for a species (the law of priority) and criticised Dresser, who, he wrote, had 'outeroded Herod' in following Newton's changes to scientific names unfalteringly (Seebohm, 1879c: 428–9). He criticised Dresser's transferral of the scientific name of one species to another, and called on ornithologists to protest 'against the revolutionary attempts of Messrs. Newton, Sharpe, and Dresser to corrupt the ornithological morality of the present age' (Seebohm, 1879c: 436–7). Elsewhere, he wrote 'I have done my best to <u>cure</u> some of the confusion caused by the ill-judged attempts of Messrs. Newton, Sharpe, and Dresser' (Seebohm, 1883a: 121; see also Seebohm, 1880b: 297). Seebohm wanted to ignore the strict rule of priority to avoid changing long-established scientific names (a return to 'the good old times', as he put it); he preferred the principle of '*plurimorum auctorum*' (latterly '*auctorum plurimorum*') – 'of many authors' – so that where 'many authors' used a scientific name, it should continue to be used in favour of any obscure older name for the same species (Seebohm, 1881: xi–xii). This principle met with particular criticism: Stejneger (1891) wrote how it was 'intended to strike terror to the hearts of those authors who believe in an inflexible law of priority' and criticised Seebohm for using his own principle when it suited him. The BOU List Committee had a difficult job in reaching agreement and met seventy-one times to complete the task! Most of the many changes that Dresser had made to birds' scientific names were accepted in the final list, published in 1883 (British Ornithologists' Union, 1883).

The List Committee was also tasked with producing a list of birds that had occurred in Britain, as there were numerous improbable or dubious records of rare birds. Many of these were backed up with preserved specimens, but birds could be relabelled and passed off as more interesting than they actually were. Dresser investigated many records of rare birds in Britain through the 1880s, inspecting specimens in museums and private collections (e.g. Dresser, 1885b). He exhibited specimens of rare vagrant species at meetings of the Zoological

Society of London (most of which had been shot in Britain), usually on behalf of people who lived far from London and continuing in the 'what's hit is history and what's missed is mystery' vein (e.g. Dresser, 1885a).[24] The most extraordinary investigation concerned a Blue-tailed Bee-eater – a colourful species from South and South-East Asia – allegedly shot on a slag-heap by the side of the River Tees in 1862 (Mather, 1986: 411–12) and listed by John Hancock in *A Catalogue of Birds of Northumberland and Durham* (1874). Dresser exhibited the bird at the Zoological Society of London in 1883, when he pointed out how unlikely it was that the bird had occurred in Britain (Dresser, 1883b). He scored it from his copy of Hancock's book with the note 'a mistake' (McGhie, 2012).[25]

Lumps, splits and three-part names

During the late nineteenth century, naturalists worked without any clear or agreed definition of what constituted a species, yet worked tirelessly to separate 'good' species from 'bad' ones. Where one species 'ended' and another 'began' was a matter of great disagreement (Stresemann, 1975; Haffer, 1992; Melville, 1995; Knox, 2007). From the late 1870s, the way that leading figures such as Dresser created (split) and merged (lumped) species was criticised with increasing frequency (e.g. Clifton, 1879). Dresser followed the traditional practice of giving a unique two-part scientific name to each distinct type of bird (species) and ignored variation within each supposed species. Henry Seebohm, on the other hand, gave a third part (called a trinomial) to scientific names to recognise each local variation, or subspecies, an approach that was gaining popularity in America (Seebohm, 1879a).

The scientific names for birds in America were as confused as they were in Britain (there were three different lists of names in use), so a group of ornithologists founded the American Ornithologists' Union (AOU) in 1883 to draw up an agreed list. Like the BOU, the AOU was a closed shop, with twenty-one founding members. Elliott Coues, one of the foremost members, in 1883 wrote of the group: 'We are everything and everybody in this country worth anything in ornith[ology]'.[26] A subcommittee of the AOU drew up its Code for naming birds, with some important differences from the British Strickland Code. The Americans took the tenth edition of Linnaeus's *Systema Naturae* (1758) as the starting point for scientific nomenclature, rather than the twelfth edition (1766), and trinomials were to be widely adopted (see Sclater, 1896; Barrow, 1998: 86–7). The Americans adopted 'trinomialism' as a cause, sometimes claiming it as an invention rather than an innovation (Coues, 1884: 244). Dresser wrote to Newton about the Americans' approach: 'What do you think of the new movement as regards trinomial nomenclature. I can't agree with it at all.'[27] Coues visited London in July 1884 to promote the AOU's scientific naming practices at a conference at the British Museum (Natural History) (Anon., 1884). Sharpe, Blanford, Saunders and Harting all spoke against trinomials (Dresser was not present), while Seebohm spoke in their favour. In the end, the 'binomialists' triumphed in Britain, if only because they were too conservative to adopt the innovation. Dresser did give scientific names to some subspecies, but only in two parts (genus and species), with the word 'subspecies' as a prefix, a confusing approach. Leonhard Stejneger, in the United States, criticised Dresser for the subjective way that varieties were or were not given their own scientific names (Stejneger, 1887).

The AOU Code and check-list of birds was issued in 1886, complete with subspecies and trinomials. Its adoption of the tenth edition of Linnaeus's *Systema Naturae* (the first to use a two-part binomial nomenclature) instead of the twelfth (the last that Linnaeus worked on) resulted in many changes to scientific names (American Ornithologists' Union, 1886). Richard Sharpe reviewed the AOU Code, the BOU's *List of British Birds* (1883) and Seebohm's innovations in scientific naming in *Nature*, taking the opportunity to dismiss his former partner as no more than an 'indefatigable compiler' rather than a taxonomist (Sharpe, 1886: 169).

In addition to the British Strickland Code and the AOU Code, the Americans had established the 'Dall Code' (1878) for plants and animals, and the Sociéte Zoologique de France (1881) and Congrès Géologique International (1882) both issued their own sets of rules (for naming animals, and plants and animals respectively). The need for some kind of international agreement on scientific names was becoming increasingly clear to cut through the confusion. To meet this need, the International Zoological Congress was established to settle matters of general importance, such as scientific naming, and to meet every three years. At the first meeting, in Paris in 1889, some rules for scientific naming were presented and adopted by the Congress, including the controversial trinomial scientific names (Melville, 1995: 17). This would be the start of a long period of dispute.

The two Henrys

During the 1880s, Henry Dresser and Henry Seebohm had a spectacular public spat. They had been on good terms in the early 1870s, sharing lodgings at Tenterden Street (see chapter 9). Seebohm developed a special affectation for clearing up the 'blunders' (as he called them) of his contemporaries; Dresser became his favourite target (Seebohm, 1877b: 128–9; Seebohm, 1882a: 104; Lloyd, 1985: 117).[28] There was a dispute over specimens: Seebohm bought Robert Swinhoe's collection in 1879, two years after Swinhoe died (see Anon., 1878a,c). He offered to purchase specimens missing from the collection 'in the interests of science' (Anon., 1879). This request did not meet with much success, as Swinhoe specimens are to be found in a number of collections, including Dresser's.

Dresser was a confirmed evolutionist, accepting that species had arisen over time from other species, yet he and his contemporaries were largely silent on the subject of evolution. They used scientific naming as an end in itself, for description, classifying and cataloguing. Henry Seebohm, on the other hand, was particularly outspoken on the subjects of nomenclature and evolution (Seebohm, 1879a: 19–20; Seebohm, 1882a: 292–7). He believed that local variations deserved their own scientific name, so that they could be used to investigate evolution (Seebohm, 1882b: 425–8; Seebohm, 1882c). Seebohm was just as outspoken on the views of his contemporaries: 'They [modern writers] profess to believe in the theory of the development of species, but they never dream of looking at birds from an evolutionary point of view. In their hearts they still cling to the old-fashioned notion of special creations' (Seebohm, 1882a: 292). The ultra-radical Seebohm wrote of 'Dr. Dry-as-dust and Professor Red-tape' (a probable allusion to Dresser and Alfred Newton) and their views from 'the pre-Darwinian dark ages of ornithology'. Seebohm suggested they should 'be exiled to Siberia for a summer to learn to harmonise their system of nomenclature with the facts of nature' (Seebohm, 1882a: 297).

Seebohm issued the first volume of *A History of British Birds* in 1883, with criticisms of Dresser on almost every page (mostly in footnotes). Seebohm also chose to ignore the scientific names for birds included in the BOU's *List of British Birds*.[29] He concluded the first volume: 'If I have criticised the work of any of my fellow ornithologists too severely, I ask their pardon, and hope that they will pay me back in my own coin, by correcting my blunders with an unsparing hand' (Seebohm, 1883b: xxi). A number of years later, Leonhard Stejneger wrote: 'Mr. Seebohm … has never handled his colleagues with gloves, and he himself would be the first one to resent any attempt at establishing a mutual admiration society' (Stejneger, 1891: 101). In the second volume (1884) it was the BOU's *List of British Birds* (and the committee that produced it) that was in for criticism. Seebohm gave the book 'The gold medal for the best example of slipshod literature … with errors on almost every page.… The volume is well worth purchasing as a literary curiosity' (Seebohm, 1884c: 234–5). It is worth recalling that Seebohm had been on the committee that produced the *List*.

Dresser complained to Newton about Seebohm's remarks on the BOU *List*,[30] and took Seebohm to task in a letter in the *Ibis* on the way he had treated one group of birds, the Shore Larks (see plate 37):[31]

> He [Seebohm] prefaces his remarks (which are written in the spirit that commonly pervades his criticisms of the authors on whose labours his own book is based) as follows: – 'Dresser, in his "Birds of Europe," has so confused the synonymy of the Asiatic species and races of Shore-Larks that I have had some considerable difficulty in disentangling the skein;' … no unbiased ornithologist can for a moment doubt that Mr. Seebohm has made an egregious 'blunder' in uniting these two birds, as will easily be seen.…
>
> I am so averse to entering upon controversy that I should have followed my practice of leaving unnoticed the hostile comments of Mr. Seebohm, were it not that in the interests of science it would be wrong for me to allow an error of this

kind to pass unchallenged. Surely a writer, who apparently affects the character of an ornithological critic with a special vocation to point out and supply the shortcomings of all his predecessors, should be a little more careful, and should compare specimens before committing himself to print. (Dresser, 1884: 117–18)

Seebohm retorted in the next issue of the *Ibis* with more criticism of Dresser (Seebohm, 1884a). Things went quiet after this, at least in print, until Seebohm issued the last part of his *History*, in 1885, with extraordinary criticism of Dresser. *A History of the Birds of Europe* was put down as 'the compilation of a writer whose personal knowledge of birds appears to be very small, though his opportunities of acquiring it have been large' (Seebohm, 1885: xvi). The articles in *A History of the Birds of Europe* were in for similar criticism:

> the writer of the extraordinary article in question was absolutely ignorant of every thing connected with the Greenshank [see plate 38], except the information which a series of skins might afford.... What a misfortune it is that the writer of this remarkable article could not renew his youth like the Greenshank, and reappear as a young author in first plumage! He would probably be careful not to dip his quills into the ink until he had either worked out the subject for himself or taken the trouble to read up some author who had done so.... It is scarcely possible to imagine any person so ignorant of his subject voluntarily undertaking the task of teaching others. (Seebohm, 1885: 154)

Seebohm criticised the way Dresser and Sharpe ('two juvenile ornithologists') 'made a desperate effort to achieve notoriety by introducing novelties into nomenclature' by altering scientific names, and congratulated them (sarcastically) for the way that they 'heroically sacrificed their reputation for common sense and sound judgement for the good of the science they loved' (Seebohm, 1885: 284). Dresser wrote to Newton: 'He [Seebohm] takes good care to abuse everyone all round – evidently so as to cast an extra halo around himself posing as the "only Jones" in ornithology'.[32]

Dresser took Seebohm to task again in the *Ibis* in 1886, after Seebohm separated the Wren from St Kilda as a new species in the *Zoologist*, a rather unusual place to publish such an important discovery as a new species of British bird (Seebohm, 1884b; Dresser, 1886). Seebohm asked Dresser to send him his specimens of Wrens, to check what he had written in the *Ibis*;[33] Dresser took his specimens to Seebohm's house, where the two argued over them, much to Dresser's amusement (described below).[34] Shortly afterwards, Seebohm wrote another critical article on the old chestnut of Dresser's alteration of the name of the Black-eared Wheatear many years before (Seebohm, 1886) (see plate 39). Seventy years later, one commentator wrote of 'some of the astonishing rudeness that was allowed to appear in print [in the *Ibis*]. Dresser and Seebohm, for example, attacked each other in the pages of the "*Ibis*" with a violence not met with today' (Moreau, 1959: 22). Henry Seebohm appears to have been the main aggressor in the dispute; his motives are unclear, but self-advancement surely played a part.

The two Henrys were closely matched in terms of interests – both ornithological and commercial – but they had widely differing outlooks. Both were involved in metal production, but for separate (probably competing) firms. Dresser was allied to Alfred Newton and conservatism, both in the sense of precision, careful fact-gathering and avoiding leaps of interpretation, and in his political leanings. Seebohm, on the other hand, was radical, both politically and scientifically: he came from a Quaker family with strong liberal and reformist associations, and he also had a propensity to make bold leaps of interpretation where birds were concerned. To give evidence of how birds and politics were intertwined, when Dresser and Seebohm argued about Wren specimens, Seebohm suggested that birds should be measured 'by the eye' rather than with measuring equipment. Dresser wrote to Newton that this was 'such a joke – he told me that he should now feel compelled to vote with the Conservatives. I reminded him what an ultra Radical he used to be and he said that he had not altered but that the present political parties had so altered their platform that it was they and not he that had altered.'[35] To say who was 'right' in their approach depends on the viewer's outlook: Dresser was certainly more 'mainstream' in his adherence to British scientific traditions, while Seebohm's views are, in some ways, more in keeping with modern approaches to science (see also Davison, 2013; this subject is returned to in chapter 15).

The rise of the British Museum (Natural History)

The British Museum (Natural History), or BM(NH) for short, had become a respected and trusted research institution, worthy of accepting type specimens that would be preserved in perpetuity. Increasingly, it dominated the ornithological scene in Britain, especially after the completion of the new museum building in 1881 (Günther, 1888: vi; Johnson, 2005: 182). Alfred Newton, meanwhile, was extremely successful in developing the collections of the Museum of Zoology at the University of Cambridge, and gave no birds or eggs to the BM(NH) after 1859 (Sharpe, 1906). Osbert Salvin was the first Strickland Curator of Ornithology at the University of Cambridge, during 1874–82; in spite of this association with Cambridge, he gave most of his collections – held jointly with Frederick Godman – to the BM(NH) in the 1880s. When Salvin gave up the Cambridge post, Mrs Strickland consulted Dresser on a suitable replacement. Dresser wrote to Alfred Newton, 'I would like the berth well enough for myself, but can't afford to take it.'[36]

The BM(NH)'s collection developed very rapidly through the 1880s, notably with the acquisition of Allan Hume's collection of Indian birds and eggs, which added 59,612 bird skins and 15,965 eggs in 1884 and practically doubled the size of the bird collection. Sharpe wrote of it, justifiably, as 'one of the most splendid donations ever made to the nation, and added to the Museum' (Sharpe, 1906: 390–3; see also Sharpe, 1885). Almost equal in size and scientific value was Godman and Salvin's collection, with 3,191 eggs presented in 1884 and 52,120 birds the following year (they were still being catalogued in 1898). The Museum's collection trebled in size between 1870 or so and 1888 (Günther, 1888: v), and continued to increase in spite of the 'continual weeding-out of all absolute duplicates' (Flower, 1898: vii). It is interesting to note that even the great Hume collection was not intended to be kept in its entirety: Günther (the Keeper of Zoology at the Museum) estimated that 25,000 specimens (out of a collection of almost 60,000 skins) would be added to the permanent collection after 'the elimination of duplicates' (Günther, 1888: vi).

As the BM(NH) collection expanded, Sharpe – overburdened with work – persuaded his superiors to engage specialists to write volumes of the *Catalogue of the Birds in the British Museum*. Seebohm, Sclater and Hans Gadow (who replaced Salvin at the University of Cambridge Museum of Zoology) wrote volumes of the *Catalogue* in the 1880s, and each gave some of their collections to the Museum. Sclater's collection of South American birds (8,803 skins and 442 type specimens) went to the Museum in 1884 (in instalments until 1890). Ernest Shelley passed his collection, which mostly consisted of African birds, to the Museum in 1889, around the time he was writing a volume for the *Catalogue*. The Museum also purchased many of the collections that were flooding into England from around the world (see Sharpe, 1906). It is quite possible that Sharpe invited particular people to write volumes of the *Catalogue* in order to obtain their collections for the Museum. Most of the prominent and often wealthy collectors did not part with their collections for free. Sclater and Shelley and, later, Edward Hargitt and Howard Saunders sold their collections to the Museum. Shelley felt obliged to do so as he had a family to keep, and complained that the Museum was slow in paying for the collection in the hope that he would die and it would get the collection free as a bequest (see Johnson, 2005: 184)! Many of the main donors had effectively finished with their collections, having published articles and books on them under their own names; Allan Hume was a notable exception, but his great book project on the birds of India had been derailed when his papers went missing (see chapter 7). Some evidence for this line of argument comes from the case of Henry Seebohm's donations to the Museum: although he had written the *Catalogue of Birds* volume on warblers and thrushes, he kept his collection of these birds as he intended to produce a monograph on thrushes (completed by Sharpe following Seebohm's death).[37] Godman and Salvin formed collections of petrels and pigeons with a view to producing monographs on these groups of birds. They never found time to do this, so they donated their collections to the BM(NH) instead, petrels in 1888 and pigeons the following year (Sharpe, 1906: 367, 368). Salvin wrote about his old petrel specimens in a volume of the *Catalogue of Birds*, while another expert wrote about his pigeons. Many years later, Godman restudied his old petrel specimens when he produced a *Monograph of the Petrels* (1907–10).

Dresser and museums

Dresser had next to no involvement with the BM(NH) or with the *Catalogue of Birds*. He was thanked for assistance by several authors of volumes of the *Catalogue*, but by no means all.[38] *A History of the Birds of Europe* was used as a source of descriptions of many species of birds in the various volumes of the *Catalogue*, showing its continuing importance as a source of reference of basic, standardised information.

While many of Dresser's contemporaries were parting with their collections to the BM(NH), he was adding to his own collection to assist with his various book projects, discussed above. Nevertheless, by the 1880s he was beginning to part with significant chunks of his collection. He sold 700 bird skins, including his North American collection (572 skins), to the Royal Scottish Museum (RSM) in Edinburgh in 1889 for £70; this collection included many specimens Dresser had obtained from his fellow Ibises.

The politics of naming

Ornithologists often named new species after other people, and this was (and is) generally considered to represent a very great honour (e.g. Mearns and Mearns, 1988: xix–xx; Mearns and Mearns, 1992; Boelens and Watkins, 2003: 11–16). Although many names were given as compliments, such as *Picus lilfordi* (by Sharpe and Dresser) and *Serinus canonicus* (by Dresser to honour Henry Tristram) (see plate 40), others were given for less magnanimous reasons. Leonhard Stejneger named the British Marsh Tit after Dresser in 1887, although he contradicted Dresser's work on the species in *A History of the Birds of Europe*. Charles Dixon, who collected the St Kilda Wrens that Seebohm described as new, took Dresser's criticism of the Wren's status personally. After some extraordinary criticism of Dresser, he wrote how specimens of the British Marsh Tit 'now go down to posterity (wrapped in an ill-concealed sarcasm), thanks to the acumen of Dr. Stejneger, under the name of *Parus palustris dresseri*!', although there is no evidence that Stejneger meant the name as anything but a compliment to Dresser (Dixon, 1893: 152–3) (see plate 41). The finest example of this kind of polite insult comes from Henry Seebohm, who suggested that the black-throated variety of the Black-eared Wheatear should be named *Saxicola dresseri* 'in commemoration of his [Dresser's] ineffectual attempt to rectify the nomenclature of the genus' (Seebohm, 1886: 195). Those who worked with competing ornithologists had to find their own ways of maintaining working relations with each of them: Ladislas Taczanowski (curator of Warsaw Zoological Museum) provided both Dresser and Seebohm with birds. He named two new kinds of South American Brush Finch in 1883, one each for Dresser and Seebohm.

A trip abroad

In May 1889, Dresser and his friend Colonel Hanbury Barclay spent three weeks in Spain, cruising along the Guadalquivir River between Seville and the sea. They travelled by steam launch, accompanied by a skipper who doubled up as the cook, a lad and a guide. As they travelled, they paid locals (some of whom made a living from collecting gulls' eggs) to show them around the countryside in search of birds. They encountered many interesting ones, including rare Spanish Imperial Eagles, Greater Flamingos and Great Bustards, and had to dodge fierce Spanish bulls from time to time. The trip resulted in a collection of over 500 eggs and some bird skins; two tiny wader chicks (Collared Pratincoles) were taken alive, but they died on the train journey to England (see Dresser, 1890a).

Conclusion

Through the 1880s, ornithology continued to drift away from being the pursuit of men of means who worked singlehandedly on their own book projects, and towards becoming standardised, institutionalised and professionalised. Collections that had been in private hands were passed to museums, notably the BM(NH). Museum collections exert a kind of gravitational pull on one another: large collections attract smaller ones from donors, while small collections exert less of a pull. Dresser stood apart from the BM(NH) but was still a force to be reckoned with, occupying a prominent position in scientific society. In addition to his roles in the BOU and the Zoological Society of London, he was President of the Yorkshire Naturalists' Union during 1889–90.[39] His presidential address at its annual meeting in Hull consisted of a very long account of his collecting trip to Spain with Hanbury Barclay.[40] Nevertheless, Dresser's relations with the ornithological staff at the BM(NH) meant he was increasingly isolated from the Museum and its scientific ambit, a trend that would continue in the following decade.

Notes

1   Cambridge University Library (CUL hereafter), MS.Add.9839.1D.465, J960, letter from Henry Dresser (HED hereafter) to Alfred Newton (AN hereafter), 5 September 1882. The house lay in five acres of orchards, as recorded in Bromley Local Studies Centre, Bromley tithe and manor records, reference 1217/12. The house was demolished in the mid-1930s; the location is preserved in the name Topcliffe Drive.

2   CUL, MS.Add.9839.1D.465, J960, letter from HED to AN, 5 September 1882; CUL, MS.Add.9839.1D.467, H099, letter from HED to AN, 7 May 1883; CUL, MS.Add.9839.1D.468, K208, letter from HED to AN, 12 October 1883.

3   CUL, MS.Add.9839.1D.481, K925, letter from HED to AN, 30 July 1885.

4   The Zoological Society of London moved offices from 11 Hanover Square to 3 Hanover Square in 1883.

5   Letter to J. S. Henslow, 30–31 October 1836, quoted in Burkhardt and Smith (1985: 514).

6   Information on HED's membership of the Council of the Zoological Society of London was provided by M. Palmer, Zoological Society of London, in 2005.

7   HED was elected as a Fellow of the Linnean Society on 4 November 1880, nominated by Howard Saunders, Osbert Salvin, Edmund Harting, Philip Sclater, Frederick Godman and other prominent naturalists (*Proceedings of the Linnean Society of London*, 1880–81: 1; Gina Douglas, pers. comm., 29 November 2006). He served on the Society's Council for a year from 24 May 1883; see the *Proceedings of the Linnean Society* 1882–83: 10 and 1883–84: 14.

8   CUL, MS.Add.9839.1D.479, K603, letter from HED to AN, 24 March 1885.

9   CUL, MS.Add.9839.1D.485, L011, letter from HED to AN, 11 January 1886.

10 See the list of Dresser's publications for full references.

11 Other presentations involving HED included an antler of a Mexican deer (collected in Texas) (Brooke, 1879: 919), and the head of a rare female Moose, with antlers, HED bought in New Brunswick in 1859 (Alston, 1880: 298).

12 Gould wrote of 'my friend Mr. Dresser' in the article on the 'Cinereous Bullfinch' in vol. 5 of *The Birds of Asia*.

13 'The Zoological Society', *The Times*, 18 June 1887, p. 8, col. e.

14 Information on Dresser's candidature of the Royal Society in 1889–90 comes from the *Proceedings of the Royal Society* and information provided by Laura Outterside (Royal Society) in 2017.

15 Lord Lilford was critical of Gómez for collecting large numbers of rare Fuerteventuran birds (Trevor-Battye, 1903: 261–2). The Canary Islands Oystercatcher was not distinguished from the more widespread African Oystercatcher (then called the African Black Oystercatcher) at the time (see Valledor de Lozoya, 2013).

16 CUL, MS.Add.9839.1D.475, K435, letter from HED to AN, 5 August 1884. HED had Keulemans' paintings bound in an album, which still exists in private hands.

17 The circumstances by which Dresser discovered and acquired the rare Arabian Bee-eaters are also described in Natural History Museum Library and Archives, London, DF231/2/457, Department of Zoology: Vertebrate Section, Reports to Trustees and other Official Documents 1896–1909, letter from Richard B. Sharpe to E. Ray Lankester, 18 April 1902.

18 MM, ZDH/8/1/1a–2, letter from HED to Sydney Hickson, 5 April 1899.

19 CUL, MS.Add.9839.1D.474, K379, letter from HED to AN, 26 March 1884; CUL, MS.Add.9839.1D.479, K603, letter from HED to AN, 24 March 1885; CUL, MS.Add.9839.1D.488, L031, letter from HED to AN, 22 January 1886; CUL, MS.Add.9839.1D.501, L502, letter from HED to AN, 18 September 1887.

20 McGill University Library, Rare Books and Special Collections, Blacker-Wood Autograph Letter Collection (MUL, B-W hereafter), letter from HED to Robert Ridgway, 12 February 1884.

21 MUL, B-W, letter from HED to Robert Ridgway, 2 October 1886.

22 CUL, MS.Add.9839.1D.485, L011, letter from HED to AN, 11 January 1886.

23 HED noted which specimens Thorburn illustrated for Lilford in his catalogue and on the specimens' labels.

24 CUL, MS.Add.9839.1D.478, L955, letter from HED to AN, 4 December 1884.

25 This book is housed in the collections of the University of Manchester Library, Manchester.

26 MUL, B-W, letter from Elliott Coues to Joel Allen, 4 November 1883. See also Lewis (2012: xiii).

27 CUL, MS.Add.9839.1D.475, K435, letter from HED to AN, 5 August 1884.

28 Seebohm was also critical of Captain Joseph Wiggins, with whom he travelled in Siberia. He wrote how, being an Englishman, the Captain had an 'unlimited capacity to blunder', and wrote openly of the Captain's manner and arguments with the crew, which Seebohm put down to the Captain's insistence on having a teetotal ship (Seebohm, 1882a: 161–3). Wiggins' biographer wrote how Seebohm's criticism of Wiggins for events outside his control was 'a little ungracious ... but naturalists are never satisfied!' (Johnson, 1907: 152–3).

29 Anon. (1883: 114) wrote critically of Seebohm's use of disputed scientific names.

30 CUL, MS.Add.9839.1D.469, K232, letter from HED to AN, 3 December 1883.

31 HED drafted a letter 'on the spur of the moment' and sent it to AN for his comments before sending the revised letter to the *Ibis*: CUL, MS.Add.9839.1D.470, K241, letter from HED to AN, 11 December 1883. Newton agreed with HED, and added his own comments; see CUL, MS.Add.9839.1D.471, K244, letter from HED to AN, 13 December 1883.

32 CUL, MS.Add.9839.1D.483, K956, letter from HED to AN, 25 August 1885.

33 CUL, MS.Add.9839.1D.490, L052, letter from HED to AN, 1 February 1886.

34 CUL, MS.Add.9839.1D.491, L057, letter from HED to AN, 3 February 1886.

35 CUL, MS.Add.9839.1D.491, L057, letter from HED to AN, 3 February 1886.

36 CUL, MS.Add.9839.1D.466, K081, letter from HED to AN, 17 March 1883.

37 Thus refuting Johnson's (2005: 184) claim that authors of volumes were obliged to pass their collections to the Museum.

38 Dresser's help was acknowledged in the volumes on birds of prey (Sharpe, 1874: viii); babblers (Sharpe, 1881: viii); warblers and thrushes (Seebohm, 1881: viii); tits and shrikes (Gadow, 1883: viii); woodpeckers (Hargitt, 1890: x); and gamebirds (Ogilvie-Grant, 1893: ix).

39 See *The Naturalist*, January 1889, no. 162, p. 11. HED attended its bank-holiday excursion to Teesdale in 1889; see *The Naturalist*, September 1889, no. 170, pp. 279–88.

40 See *The Naturalist*, June 1890, no. 179, p. 176.

11 The 1890s: the continuing rise of the British Museum (Natural History)

Henry Dresser was fifty-two in 1890; through the decade, he and Eleanor lived a comfortable life in Farnborough. They suffered a great tragedy when their nine-year-old daughter Phyllis died of mumps in 1893. This had a tremendous impact on both Henry and Eleanor: she often spent time away from home in London on charitable work ('mission work'), while he would take trips away (to Lilford Hall in 1895 for example) so as not to be left alone at Topclyffe Grange.[1] The Bowling Iron Company, for which Henry Dresser was the sole London agent, was in growing financial difficulties by the 1890s, due to the diminution of its coal and iron ore mines (the coke-producing 'Better Beds' coal seams were worked out by 1896), and competition, both from the neighbouring ironworks at Low Moor and from iron- and steelworks in north-east England and the East Midlands. The firm went into liquidation in 1898 (see Long, 1968: 177; Richardson, 1976: 73). Dresser gave up its agency in 1893 but remained at 110 Cannon Street and worked as an agent for the British Mannesmann Tube Company, which produced seamless steel pipes at its works in Landore, close to Swansea (which Dresser occasionally visited).[2] Dresser also served as the official liquidator of the Irwin's Peak Mining Company when it was wound up in 1894; Irwin's Peak (now called Torreys Peak), in the Rocky Mountains in Colorado, was the site of short-lived silver mines in the 1880s.[3]

Book projects

Dresser completed his book on rollers in 1893, dedicating it to Alfred Newton (Dresser, 1893b). He also published several papers on these birds: he named a specimen of the Racket-tailed Roller in Henry Tristram's collection as a new species, *Coracias weigalli* (see plate 42) (Dresser, 1890b). He also split off the southern race of the African Purple Roller as a new species, naming it *Coracias mosambicus* (see plate 43) (Dresser, 1890c).[4] Richard Sharpe wrote about rollers and bee-eaters in volume 17 of the *Catalogue of the Birds in the British Museum*, published in 1892. He wrote that he had planned to name the bird that Dresser had named *Coracias mosambicus* as a separate species (to be called *Coracias olivaceiceps*) in the *Catalogue*, and refused to accept Dresser's name for the bird (Sharpe, 1892: 25).[5] Sharpe also rejected *Coracias weigalli* as nothing more than a young individual of the well-known Racket-tailed Roller (Sharpe, 1892: 23). Dresser was proven correct when Tristram received a second specimen from no less a figure than the Bishop of Zanzibar (Tristram, 1894; see also Clancey, 1969).

A more ambitious book was the *Supplement to A History of the Birds of Europe*, begun in earnest in 1890.[6] Dresser worked hard to obtain specimens towards the book, including downy chicks, skins and eggs of rare and newly described species, and birds from western Asia.[7] Henry Tristram and Edmund Meade-Waldo continued to provide him with rare birds from the Canary Islands; Padre Ernst Schmitz (1845–1922), a German Catholic priest based in Madeira, also provided him with specimens of rare endemic seabirds and forest birds at this time (see plate 45).[8]

As he worked on the *Supplement*, Dresser investigated several particularly difficult groups of species, including shrikes, dippers and long-tailed tits. These studies resulted in a series of articles and exhibitions at the Zoological Society of London, where he expanded upon and corrected the views he had presented in *A History of the Birds of Europe* (e.g. Dresser, 1892). The *Supplement to the Birds of Europe* was issued in 1895–96, in nine parts; the preface states that it was published 'by special permission at the Zoological Society of London'. Dresser dedicated the book to his late daughter Phyllis, from 'her sorrowing father'. The *Supplement* included the birds of western Central Asia and Persia – enabling Dresser to include many species encountered by his Russian collaborators – as well as new birds recently received from the Atlantic islands. The book covered 114 species, including twenty-two rare vagrants, thirty-four species found in the eastwards extension of the area, twenty-six birds that had been given full species status and fourteen species discovered since the publication of *A History of the Birds of Europe* (1871–82). Dresser wrote how he felt 'compelled' to recognise distinct geographical races of birds as separate species, demonstrating how he modified his naming practices to some extent to march with the times (see, especially, Dresser, 1895–96: 168, 174). The plates were illustrated by Keulemans, although some were based on illustrations by Archibald Thorburn (including two based on living birds in Lord Lilford's aviary) and Joseph Wolf. Many species were illustrated for the first time (see plates 44 and 45). The *Supplement* was very favourably reviewed (Anon., 1897a).

The long-held dream of issuing a *Birds of the Eastern Palaearctic* fell by the wayside as the days of sumptuous subscriber volumes were drawing to a close. Instead, Dresser began work on a compact manual on Palaearctic birds to include the latest information on every bird species found in Europe and northern Asia.

Collecting records of rare birds

Dresser and some of his contemporaries, notably Henry Seebohm and John Henry Gurney junior, continued to assess reports of rare birds to keep some kind of 'British list', and to separate fact from fiction and fraud (McGhie, 2012). Seebohm (1893: 1–2) noted that most erroneous records of 'British' birds were down to misidentifications, people importing dead birds into Britain (deliberately or accidentally) and foreign birds escaping from captivity. Gurney wrote how, on investigation, the collectors of rare birds were often reported to be dead when he tried to trace them: 'an inconvenient circumstance which naturally casts some doubt on the marvellous statements attributed to them' (Gurney, 1890: 129). Oliver Aplin (1890) also wrote of his suspicions that rare northern birds could be passed off as British (i.e. as having been killed in Britain) by collectors or taxidermists, as their frozen corpses could be bought in wildfowl markets and made into perfectly good bird skins.

Dresser's articles and exhibitions on unusual occurrences of birds in Britain included the announcement of the first Pallas's Warbler (which breeds in Siberia and winters in South-East Asia), shot in Norfolk in 1896 (see plate 46), a record of an Osprey shot in Dorset in 1897, and notes on some Daurian Partridges (from Central Asia) he saw in Leadenhall Market, rolled in paper and frozen in a barrel for transportation to Britain (Dresser, 1897a,b, 1898). In an appendix to the *Supplement* (pp. 417–26), he wrote how 'the greatest caution' was needed in assessing records of foreign waterfowl as they were often kept in captivity; he rejected records relating to species commonly kept as cagebirds, including a Cedar Waxwing (an American bird) that he thought had been 'either "changed at nurse" by the bird-stuffer or had escaped from confinement'. 'Changed at nurse' meant that the taxidermist had swapped the specimen for another, whether accidentally or deliberately (see also chapter 8).[9]

The most scandalous case of foreign birds being passed off as British was the 'Hastings Rarities' affair: large numbers of rare birds (mostly as dead corpses) were reported as having come from a small area of Kent and East Sussex centred on Hastings (hence the name), including many 'firsts' for Britain. The affair reached its peak in the 1890s and early 1900s, when ornithological journals regularly reported the discovery of yet more rare species in the vicinity of Hastings. Most of the records involved a taxidermist, George Bristow of St Leonards-on-Sea (near Hastings), who is generally credited as being involved in the fraud (whether single-handedly or with associates).

Ornithologists were incredulous at the 'discoveries' around Hastings by 1915, but there is one intriguing suggestion that they were suspicious from as early as 1890, when Dresser wrote to John Harvie-Brown:

> I send as it may amuse you a communication I have not received 'thankfully'. I wrote a very civil reply and carefully evaded any opinion and I am now to have some further new? species to report on. All of course only inhabit that part of Kent – and I am receiving a letter of four sides every day.[10]

It seems likely that Dresser was referring to having a 'new' bird to add to his *Supplement*, either a new species or a bird not normally found in Britain.[11] The 'Hastings Rarities' affair was brought to light through a statistical analysis of records of rare birds, revealing the unlikelihood of so many rare birds occurring in so small an area. This was put down to fraud, with the blame being directed towards Bristow (Nelder, 1962; Nicholson and Ferguson-Lees, 1962, 1971; Nicholson *et al.*, 1969). Most authorities accept Bristow's guilt, although this is not unanimous (e.g. Harrison, 1968, 1971; Morris, 2010: 274–7). Intriguingly, Hastings was also associated with the infamous Piltdown Man hoax of 1908–15.

Relationship with Walter Rothschild

Walter Rothschild (1868–1937) (see figure 11.1) was the eldest son of an immensely wealthy banker; his wealth was accompanied by an almost insatiable desire for specimens of rare birds, mammals and butterflies. Rothschild was given a museum on the family estate at Tring (Hertfordshire) as a twenty-first birthday present. He needed a curator so Albert Günther recommended a young German, Ernst Hartert (1859–1933) (see figure 11.2), who was working with Sharpe on a volume of the *Catalogue of the Birds in the British Museum*. Rothschild and Hartert met for the first time in Dresser's London office in 1892; their partnership became an extraordinary success (Rothschild, 2008: 129). Rothschild's collection developed rapidly as he bought up collections and financed collecting trips around the world; some others, notably Alfred Newton, were envious of his buying power (Rothschild, 2008: 75–6).

11.1 Walter Rothschild, from Henry Dresser's album of correspondents.

11.2 Ernst and Claudia Hartert, from Henry Dresser's album of correspondents.

Dresser arranged a good number of exchanges and acquisitions for Rothschild.[12] These included the skull of an extinct Steller's Sea Cow from the Smithsonian,[13] birds from Mikhail Menzbier in Moscow, and from Morocco, the Canary Islands[14] and Madagascar.[15] Dresser purchased an extremely rare mounted hybrid Black and Hazel Grouse for £15 and sold it on to Rothschild for the same price.[16] In return, Rothschild provided Dresser with access to rare birds in his collection to help with his *Manual of Palaearctic Birds*. Dresser wrote: 'I would sooner borrow from Tring than go to the Brit. Museum', presumably because of his difficult relationship with Sharpe.[17] The interplay of 'you scratch my back and I'll scratch yours' is apparent through their correspondence; the two men had a great deal in common beyond their interest in birds and collections, as they both worked in business in the City.

Correspondence between Dresser and Rothschild's curator, Hartert, mixed discussions on birds with family matters. Dresser sought Rothschild's support in finding his son a commission in the army after he failed to get into Sandhurst.[18] Dresser asked for Rothschild's help reluctantly, writing to Hartert: 'I am shy of troubling him [Rothschild] for I know how much he will be pestered by people asking favours'.[19] Rothschild wrote a note of recommendation for Henry junior, who subsequently enlisted with the Royal Buckinghamshire Yeomanry.[20]

Position in scientific society

Dresser was a member of the Council of the Zoological Society during 1889–95 and a frequent contributor to its meetings. He also helped with the running of the BOU and arranged the election of several members, including Rothschild and Hartert (Sclater, 1909: 55).[21] Many naturalists visited Dresser's home to consult him and study his collection, including Hugh Popham (returned from a trip to the Yenisei in Siberia in 1897),[22] Joseph Wolf, who spent Christmas with the Dressers in 1892,[23] and Otto Finsch, who visited from Bremen in 1895.[24] Dresser took care to name his visitors and their impressions of his collection in his correspondence to other ornithologists: 'I have just had Mr Pleske of St Petersburg staying with me in order to go through my collection and am glad to say he was very pleased with it and says it is the best private Palaearctic collection he knows of'.[25]

In 1893–96, Alfred Newton (with others) issued his famous *Dictionary of Birds*, full of praise for Dresser and the *Birds of Europe*: 'European ornithologists have been all but unanimously grateful to Mr. Dresser for the way in which he brought this enormous labour to a successful end' (pp. 41–2); the book represented 'an enormous forward stride' (p. 344). Henry Seebohm fared less well (e.g. pp. 345, 734) and was criticised for making 'downright errors and wild conjectures', with his 'misuse of language and absence of reasoning power' (p. 44).

Dresser was a candidate for the Fellowship of the Royal Society in 1892 and 1896–97; in 1892 he had the same supporters as he had had in 1889 (see previous chapter), with additional supporters including Sir John Lubbock (afterwards Lord Avebury) and Henry Tristram (rather surprisingly, Alfred Newton was not among Dresser's nominators, although he was clearly a strong admirer of Dresser's work, as outlined above, and the two continued to correspond through the 1890s). Some of his ornithological acquaintances were up for election around the same time, including Henry Seebohm, Howard Saunders and Henry Lansdell. None of these were elected: the days of independent gentlemen-naturalists were passing in favour of those who worked in institutions in paid positions. Henry Elwes was an exception to this trend: he became a Fellow in 1897.[26]

The Society for the Protection of Birds was formed in 1891 by the union of two societies, one in Didsbury (Manchester) and the other in Croydon. These had been set up by middle-class women to protest at the killing of birds for the millinery trade: feathers from over 34,500 birds-of-paradise and 86,000 egrets were sold in London in 1898 alone, for example.[27] The Society campaigned by distributing leaflets and had some influence as it was supported by a number of aristocrats and leading ornithologists (Samstag, 1988: 40). Dresser played a leading role in the Society as an ornithologist of note.[28] He edited its Educational Series of pamphlets on the distribution, identification and legal protection of various birds – mostly species under persecution from gamekeepers and collectors – and joined the Society's Council in 1899 (Dresser, 1896–97; Anon., 1910a).

Richard Meinertzhagen provides an interesting first-hand account of his early meetings with Dresser in the 1890s, giving some insight into Dresser's relationships with other ornithologists. Meinertzhagen's comments should be treated with caution, as many of his 'recollections' have been shown to be fabrications.[29] He wrote that Sharpe introduced him and his brother to Dresser, whom he described as 'a curious character, without much humour, inclined to be secretive and jealous…. Dresser had correspondents all over the world and was probably better known to foreign ornithologists than any other British ornithologist.' Meinertzhagen and his brother met Dresser again in 1894, along with Seebohm, Saunders and Sclater: 'Dresser looked sad and serious, taking little part in the conversation, though Saunders tried to cheer him up, chaffing him about his "Birds of Europe", but Henry Eeles would have none of it and was determined to sulk'. Meinertzhagen wrote that Saunders and Seebohm could be free with their criticism of Dresser (Meinertzhagen, 1959: 49–50).

The British Ornithologists' Club

Richard Sharpe established the British Ornithologists' Club (BOC) in 1892, part of the ongoing rise of the BM(NH). The Club was set up so BOU members could meet monthly (from October to June) to display their latest acquisitions, and so descriptions of new species could be published quickly in the Club's *Bulletin*. Almost half of the 200 British BOU members had joined within a year of the Club's establishment. Members met on the third Wednesday of the month for a dinner, talks and exhibitions of bird skins, eggs and bones (Manson-Bahr, 1951, 1952; Snow, 1992). Meetings were held in London, at the Mona Hotel in Covent Garden and the Restaurant Frascati on Oxford Street. Sharpe and his assistant William Ogilvie-Grant dominated the earliest meetings with accounts of specimens in the BM(NH). This situation quickly changed as Hartert and Rothschild (who began attending in 1892 and 1893 respectively) exhibited extraordinary numbers of rare birds that Rothschild had acquired. Some of these were described as new subspecies and were given the controversial trinomial names (e.g. Hartert, 1892).

Dresser attended BOC meetings during 1892–93 and after 1896, exhibiting his new species there rather than at the Zoological Society of London or reporting them in the *Ibis*. These included a new long-tailed tit from Macedonia (Salvadori and Dresser, 1893), and a new warbler from Palawan (in the Philippines) exhibited on behalf of the collector-explorer John Whitehead (Dresser, 1893a).[30] One person who was absent from BOC meetings was Alfred Newton, who referred to its meetings by 'a most uncomplimentary name' (Wollaston, 1921: 70). Philip Manson-Bahr recollected that Newton 'who had constituted himself the doyen of British ornithology, kept aloof and was not himself a member of the Club, but always a severe critic of its personnel and policy' (Manson-Bahr, 1951: 2).

The *Catalogue of Birds* was completed in 1898, despite the continual expansion of the BM(NH) collection. Sharpe had written the entries for almost half of the 11,548 species covered,[31] based on his examination of almost half a million bird skins, equating to an average of fifty specimens a day – every day – since he joined the Museum in 1872 (Sharpe, 1898: 63)! When the final volume was published, the BOC devoted a volume of the *Bulletin* as an index of the *Catalogue*. Philip Sclater penned a verse (originally in Latin) 'to commemorate the names of the eleven Authors of the Catalogue of Birds' (see Sharpe, 1906: 85; Snow, 1992: 2–3). Could there be any better evidence of the rise of the BM(NH): a verse commemorating the authors of its *Catalogue of Birds* and who contributed to the growth of its collection, printed in the journal of the most elite club of British ornithologists, edited by the Museum's bird curator? Alfred Newton, meanwhile, had 'a deep-rooted dislike' of the 'B.M. Cats' (according to his biographer Alexander Wollaston), and he wrote to Henry Tristram: 'you are easily pleased if you can find delight in B.M. Cats; as a *whole* a more useless litter was never kitted' (original emphasis retained, see Wollaston, 1921: 235 and footnote pp. 235–6).

Membership of the BOC grew fairly rapidly, to include some younger enthusiasts who would rise to become important figures in the world of ornithology. Among these were Norman Ticehurst (1873–1969, who first attended the BOC in 1896) and Harry Witherby (1873–1943, who attended from 1897). Witherby came from a London-based publishing family; he began producing small books on birds from his early twenties and wrote articles on birds in the popular science magazine *Knowledge* (published by his family), including regular reports on rare birds seen (and often shot) in Britain (Knox, 2007: 609–15; Ogilvie *et al.*, 2007: 3–5; Mussell, 2009: 335–6).

More disputes over scientific naming

There were ongoing disputes over how birds, and other animals, should be given their scientific names. The German Ornithological Society drew up a set of rules during 1889–91, as did the German Zoological Society in 1894 (ahead of publishing a vast encyclopaedia on every known species of animal, *Das Tierreich*, giving three-part names as the norm) (Stresemann, 1975: 250–2). The set of rules for naming animals that had been adopted by the 1889 International Zoological Congress was published after the second Congress, of 1892, held in Moscow. At the third International Zoological Congress, in Leiden in 1895, a group of five zoologists was set up as an 'International Commission on Scientific Nomenclature', to draw up a code for scientific naming that was to be presented at the next Congress. Philip Sclater was one of the five commissioners; he subsequently spoke on the differences between the British Strickland Code and the German Zoological Society's Code at a meeting of the Zoological Society of London in 1896 (Sclater, 1896), sticking fast to the 'British' traditions: that the twelfth edition of Linnaeus's *Systema Naturae* should be the starting point for scientific nomenclature (rather than the tenth, as advocated by the American Ornithologists' Union, the International Zoological Congress and the Germans); he criticised the acceptance of erroneously spelt scientific names (advocated by the AOU), as 'no one with a pretence to a classical education is likely to submit to the causeless infliction of such barbarisms' (p. 313). Sclater was particularly critical of those, including Dresser and Newton, who strictly adhered to the law of priority, noting that it was 'the extremist and the sensationalist, who strive to astonish us by carrying out the law of priority to its "bitter end," that have caused the disgust which many of us feel at the mere mention of priority in nomenclature' (p. 315).[32] Joel Allen, a leading American trinomialist, wrote:

> Mr. Sclater concedes practically nothing…. We must say, with regret, that this looks like unwise conservatism, bordering on perversity; for the few British naturalists who still stick to the British rules can hardly expect the rest of the world to waive their better judgment in favor of insular sentiment and traditions. (Allen, 1896: 327)

Hartert, almost alone in Britain, continued to advocate for the use of trinomialism to study subspecies, which helped in the understanding of evolution. He deplored the confusion caused by naming subspecies with two-part scientific names (as Dresser would sometimes do), leading to some strongly worded comments: 'we must agree that the **scientific systematic treatment of living animals DEMANDS the recognition of subspecies**, if systematic zoology is to be more than a pastime, and if it is to take the important place in science which it ought to hold' (Hartert, 1896: 366, original emphasis).

The subject continued to be debated, particularly in Germany (Stresemann, 1975: 259–61). The International Commission on Scientific Nomenclature did not present a proposal for naming animals at the 1898 Congress in Cambridge, as the five commissioners could not agree. The Commission was expanded and told it had to present a proposal at the next Congress; Howard Saunders was one of those who joined it (Sedgwick, 1899: 55; Melville, 1995: 23). It was clear that the British ornithologists, including Dresser, were losing ground in the 'battle' over scientific nomenclature.

Dispute over specimens

Through the 1890s and the following decade, Richard Sharpe made several accusations that Dresser had obtained specimens in an underhand way from the BM(NH). When Henry Seebohm died in November 1895 he bequeathed his collection to the Museum, but some specimens were out on loan and others had been promised to other ornithologists. Dresser wrote to Ernst Hartert about some Asian Grey-backed Thrushes the month after Seebohm's death:

Seebohm had promised me a pair out of the last collection he bought and told me when I saw him only five days before his death that he was sorry that owing to his illness he had not been able to let me have them – but now that the Brit. Mus. take over his collection I don't suppose I shall ever get them.[33]

The following year, Dresser repeatedly asked Sharpe for the return of some of his thrush specimens from Siberia. These had presumably been lent to Seebohm for his projected book on thrushes and went to the BM(NH) with the rest of Seebohm's collection.[34] Things got worse after Dresser entertained Hartert, his wife and the German ornithologist Otto Kleinschmidt at his home in September 1897.[35] Sharpe appears to have alleged that Kleinschmidt told him that Dresser's collection included some bird skins (an African Tawny Eagle, some Fulmars and some white Puffins) that had belonged to Edward Hargitt, and which should be in the BM(NH) with the rest of Hargitt's collection. Sharpe must have written to Dresser to this effect, to which Dresser replied:

> Your letter of the 25th came duly to hand, and I take the earliest opportunity to reply. Who is this Mr. Kleinschmidt who claims to know more about my collection than I do, and more about poor Hargitt's than he himself did when alive? Your memory has however been playing you sad tricks, and if Hargitt's spirit (which is with the Saints I trust) were permitted to revisit this wicked world it would tell you the same, for Hargitt would never have said that I had his Puffins and Fulmars, inasmuch as I find on reference that I never borrowed them. As regards Aquila rapax [Tawny Eagle] it is somewhat remarkable that, to my knowledge Hargitt never even mentioned to me, during the seventeen years which have elapsed since I returned it, that this skin was missing, and on reference I also find nothing named about it in his letters to me.[36]

Dresser wrote to Hartert to find out what was going on,[37] and again to Sharpe:

> The Mr. Kleinschmidt I know, and who was here, has I believe only recently visited England for the first time, and never saw Hargitt, so I could never imagine that you referred to him. Nor can I understand why you point out that Hargitt never gave me, or intended to give me, his specimen of Aq. rapax, for I never put forward such an absurd claim…. There is no doubt that I have not got the bird, nor has it ever been in the house.[38]

He wrote again to Sharpe:

> Referring to your accusation that I had annexed a specimen of Aq. rapax from the Hargitt collection and your statement that Mr Kleinschmidt was your authority, I give the following translation of a statement received from him, viz 'I do not recollect to have seen Aquila rapax in Dresser's collection, nor do I recollect to have mentioned anything to Sharpe respecting any single specimen in the Dresser collection.'

Dresser concluded: 'I think that this already bears out my statement that your memory has been playing you tricks, and proves that your accusation was made without any basis whatever.'[39] Sharpe wrote to Dresser to retract the accusation and the matter was dropped, but this would not be the last time he would be suspicious of Dresser.[40]

Dresser and Sharpe were not the only ornithologists to disagree over the ownership of specimens. Joseph Wolf complained that he lent John Gould (who was well known for his acquisitiveness) a rare falcon specimen that was never returned (Palmer, 1895: 74). Louis Mandelli (the Darjeeling tea–planter introduced in chapter 7), too, complained about the actions of Allan Hume:

> Yes, Hume is a brute, in fact, I call him a swindler, as far as birds are concerned.... What else would be thought of a man who promised to help me <u>and very grand and magnificent promises they were</u> to make my collection of Indian birds as perfect as he possibly could, in order only to get out the best and the rarest things to be found up here, and then leaving me on the lurch now, as he has found out that I am no more his slave subservient to his sneaking and bland manners and hypocritical ways? ... I should say that swindler is too mild a term for such a man after having got out from me about 5000 birds and given only in return about 800, the commonest birds in India, 400 of which went down the khud [hill], as they were not worth the carriage.[41]

Ownership of specimens was a matter of the utmost importance: collections were the material on which careers and reputations were built, both as a basis for publications and as material possessions.

Parting with collections

The collections of the BM(NH) continued to swell through the acquisition of formerly private collections. Among these was Henry Seebohm's collection, including 20,299 bird skins and mounts, 230 skeletons and 16,290 eggs that were acquired after Seebohm's death in 1895 (Sharpe, 1906: 274, 470–5). Indeed, most of the leading private bird collections had passed to the BM(NH). When Henry Tristram sold his massive collection to the Liverpool Museum in 1891, a note appeared in the *Ibis*:

> It was, if not the last, almost the last, of the great undispersed private collections which were amassed by wealthy cultivators of this science in England during the past half-century or more, nearly every one of which has now become incorporated in the National Museum of Natural History at South Kensington, either by gift or by purchase. No such general collections are now being made. (Anon., 1897b: 488)

Dresser, meanwhile, kept his collections out of the hands of the BM(NH), probably owing to his poor relationship with Sharpe. He did part with some of his collections during the 1890s, again favouring the Royal Scottish

Museum (RSM) in Edinburgh: his collection of North American bird eggs (630 specimens) went to the RSM in 1899, as did some birds from Brazil and hummingbirds.

Transferring specimens to museums was a complicated affair, with a variety of social and economic factors involved in a kind of 'politics of donation'. Donors could benefit financially or by attaining a reputation as a philanthropist (especially if they gave their collections *gratis*); alternatively, they could benefit from exchanges of specimens. These subtleties are exemplified well by Dresser's transferrals to museums. To give one example, he offered Hartert a rare specimen (a Rufous-chested Plover from Chile) in 1896, in exchange for other birds that he required for his collection.[42] Hartert offered Dresser only a poor exchange of specimens, so in the end Dresser gave the bird (rather than sold it) to the RSM.[43]

As the nineteenth century drew to its close, both Dresser's and Eleanor's health continued to deteriorate. Henry junior was on the verge of leaving home for the army, and the time had come for Dresser and Eleanor to think about relocating to a smaller home. With this in mind, Dresser began to make arrangements to dispose of his entire bird collection in 1899. Alfred Newton acted as intermediary, writing to Sydney Hickson, Professor of Zoology at Owens College (now the University of Manchester):

> Dresser's collection of Palaearctic birds must be quite the largest ever formed by one man, and setting aside the British Museum, must be the largest in the world.... You no doubt know that Tristram's collection was bought for the Liverpool Museum, and I think £2000 paid for it – its value being probably £3000 or thereabouts. It seems to me that Manchester might like to follow the example of its rival city, and become possessed of Dresser's collection, which has been formed with a more definite purpose.[44]

This letter was accompanied by a letter from Dresser himself, explaining how he had 'spared neither time, trouble nor expense to make it [his collection] as complete as possible'.[45] William Hoyle (see figure 11.3), the Keeper of Manchester Museum (part of Owens College), wrote to one of the Museum's trustees to ask for their support. He argued that Dresser's eminence as an ornithologist and the fact that the collection was the basis of his famous books would give the Museum 'a high standing in the scientific world'.[46] This tactic did not meet with success, as the trustee replied that the collection would be a 'useless burden' and that Dresser should accept an offer of $25,000 from Chicago for the collection![47] Hoyle replied that he could not agree with the collection going to Chicago 'on account of the almighty dollar'.[48] He approached a local philanthropist, John Thomasson of Bolton, to see if he would meet the £2,000 price tag. Thomasson declined at first, but wrote that, if Dresser would accept £1,000, he would provide the funds, preferring his support to remain anonymous.[49] Dresser accepted the offer, which came at the same time as an alternative offer from the Liverpool Museum.[50] The transfer of the collection was accelerated as Dresser was given an offer for the

11.3 William Hoyle, from Henry Dresser's album of correspondents.

rent of Topclyffe Grange, meaning that he would have to vacate the house of his possessions and collections.[51]

The collection was transported to Manchester in June 1899 by horse-drawn cart and by train, in twenty-three large wooden cabinets.[52] Dresser sent almost every bird skin that he possessed, retaining only the Waxwings he had collected when he was a young man in Finland (these were subsequently donated to the Museum),[53] and promised to add to the collection to keep it as complete as possible.[54] A complimentary notice of the acquisition appeared in the *Ibis* (arranged by Dresser): 'the acquisition of this valuable collection is indeed a piece of singular good fortune for the Manchester Museum, and therefore for all students of ornithology in this neighbourhood … Owens College is indeed to be congratulated upon possessing so valuable a collection' (Anon., 1899b; see also Anon., 1899c,d).

With his precious collection safely out of the way, Dresser described his changing circumstances to Ernst Hartert in June 1899:

> I began to find that it was more than I could bear to be alone in charge of the servants, when these liver attacks trouble me and having received a good offer to take this place on lease, I decided to accept and have been busy getting out all last week and we shall be quite clear out on Wednesday next. My furniture, pictures, library and egg collection I shall store till I decide where to settle down, and shall probably take a house close to Hyde Park, as if we live in London Mrs. Dresser is bound to live at home and I can run out into the country whenever I

want a change. As regards my bird collection I have had an offer for it from the Manchester Museum which I have accepted for no-one but myself cares for it here and if I died suddenly it would be put up to auction and dispersed, whereas in Manchester it will be kept together, and be available for the purpose of study.... Till further notice you had better direct to me to my office 110 Cannon Street EC. for we are undecided where we shall find a house that will suit us and in the meanwhile I am staying with my sisters at Norwood for I prefer it to town in this warm weather, and my wife and Brenda are in town as my wife won't leave her mission work.[55]

Eleanor, Brenda and Henry junior travelled to Switzerland the following month for the benefit of Eleanor's health; Henry joined them there in August.[56]

Conclusion

Years were creeping up on Dresser, sixty years of age in 1898 and in poor health; in spite of this he continued to work in the City. His scientific–social circle was changing, as many of his longstanding colleagues passed away during the 1890s.[57] Ornithological society was also changing: the days when ornithology was run by a small number of men of independent means were not entirely over (as the leading societies continued to be dominated by a handful of wealthy individuals), but the writing was on the wall. The great private collections that were so distinctive of Britain had almost all passed to the BM(NH). Dresser still held a prominent and indeed pivotal position in ornithological society: a good demonstration of this comes from Alfred Palmer's biography of Joseph Wolf, for which Dresser was the source of a number of anecdotes (Palmer, 1895). Many of these painted Dresser in a favourable light, such as a story concerning the Labrador Falcon specimen in Dresser's collection that Wolf once painted (see plate 47), or which poked fun at Dresser's contemporaries and rivals (such as the anecdote concerning John Gould referred to in chapter 9). Powerful and influential as he was, Dresser was becoming increasingly anachronistic as other leading scientific figures – many of whom Dresser had known for forty years – aligned themselves with the BM(NH). He kept himself clear of the Museum but continued to gravitate towards his old friend Alfred Newton.[58]

Notes

1   Natural History Museum Library and Archives, London (NHM hereafter), TM1/6/19/136–7, Tring Museum Correspondence (TMC hereafter): Correspondence A–D, 1894, letter from Henry Dresser (HED hereafter) to Ernst Hartert (EH hereafter), 20 February 1894; NHM, TM1/12/14/380–1, TMC: Correspondence A–C, 1895, letter from HED to EH, 19 April 1895.

2 NHM, TM1/26/19/77, TMC, letter from HED to EH, 1 September 1897.

3 He is listed as secretary to the company in Skinner and Skinner (1891: 179). He is listed as the company's liquidator in the *London Gazette*, issue 26,491, 2 March 1894, p. 1316.

4 See also Cambridge University Library, MS.Add.9839.1D.480, K691, letter from HED to Alfred Newton (AN hereafter), 22 May 1885.

5 He refused to accept the name as there was no evidence that it came from Mozambique. This was an insufficient reason to justify rejecting a scientific name and HED's new name is still in use.

6 National Museums Scotland Library (NMSL hereafter), GB 587, JHB 15/240, letter from HED to John A. Harvie-Brown (JAHB hereafter), 21 October 1890.

7 HED wrote to Robert Ridgway asking to borrow specimens of downy chicks, and also wrote that he preferred to keep every bird he had illustrated for future reference. McGill University Library, Rare Books and Special Collections, Blacker-Wood Autograph Letter Collection (MUL, B-W hereafter), letter from HED to Robert Ridgway, 14 November 1892.

8 See Bannerman and Bannerman (1963–68) for biographical information on, and references to, the work of collectors who were active in Madeira, the Canary Islands and the Azores.

9 HED attributed the Blue-tailed Bee-eater referred to in chapter 10 to the same causes.

10 NMSL, GB 587, JHB 15/240, letter from HED to JAHB, 21 October 1890.

11 The pages of the *Bulletin of the British Ornithologists' Club* and the journal *British Birds* contain many references to rare birds from around Hastings, admitted as British by most ornithologists at the time.

12 NHM, TM1/1/116, TMC: Correspondence A–D, 1890–93, letter from HED to EH, 16 January 1893; NHM, TM1/62/7/65, TMC: Correspondence D–F, 1900–02, letter from HED to EH, 13 November 1902.

13 NHM, TM1/1/17/118–19, TMC: Correspondence A–D, 1890–93, letter from HED to EH, 17 February 1893, sent with a letter from F. W. True (Smithsonian Institution) to HED, 21 January 1893.

14 NHM, TM1/1/17/115, TMC: Correspondence A–D, 1890–93, letter from HED to EH, 16 January 1893.

15 NHM, TM1/12/14/379, TMC: Correspondence with HED, 1895, postcard from HED to EH, 24 January 1895; NHM, TM1/12/14/380–1, TMC: Correspondence with HED, 1895, letter from HED to EH, 19 April 1895.

16 Henry exhibited this specimen at a meeting of the Zoological Society of London in 1876 (Dresser, 1876c). NHM, TM1/1/17/129–30, TMC: Correspondence A–D, 1890–93, letter from HED to EH, 26 September 1893.

17 NHM, TM1/33/16/202–3, TMC: Correspondence C–E, 1898, letter from HED to EH, 14 January 1898.

18 NHM, TM1/33/16/202–3, TMC: Correspondence C–E, 1898, letter from HED to EH, 14 January 1898; NHM, TM1/42/8/82–3, TMC: Correspondence D–F, 1899, letter from HED to EH, 19 July 1899.

19 NHM, TM1 /33/16/187–8, TMC: Correspondence C–E, 1898, letter from HED to EH, 4 February 1898; NHM, TM1/33/16/204–5, TMC: Correspondence C–E, 1898, letter from HED to EH, 29 January 1898.

20 NHM, TM1/42/8/85, TMC: Correspondence D–F, 1899, letter from HED to EH, 30 December 1899.

21 NHM, TM1/6/19/134, TMC: Correspondence A–D, 1894, letter from HED to EH, 9 April 1894.

22 Henry assisted Popham by identifying his specimens and was repaid for his troubles in rare thrush specimens (Popham, 1897).

23 NHM, TM1/1/17/112, TMC: Correspondence with HED, 1893, letter from HED to EH, 5 January 1893.

24 NHM, TM1/12/14/378, TMC: Correspondence with HED, 1895, letter from HED to EH, 24 January 1895.

25 MUL, B-W, letter from HED to Robert Ridgway, 7 September 1892.

26 Information on the candidature of Fellows of the Royal Society comes from the *Proceedings of the Royal Society*, information provided by Laura Outterside (Royal Society) in 2017 and certificates of election of successful candidates (https://royalsociety.org/fellows, accessed 7 January 2017).

27 See A. Newton, 'The plume trade' (letter), *The Times*, 25 February 1899, p. 16, col. f.

28 HED was also a member of the Yorkshire Naturalists Union's Protection of Wild Birds' Eggs Committee (formed in 1891) through the 1890s.

29 See Cocker (1989) and Garfield (2007) for discussion on Meinertzhagen's fraudulent behaviour, and Knox (1993) and Rasmussen and Prŷs-Jones (2003) for discussion of his ornithological frauds.

30 Sharpe and Seebohm refused to recognise it as distinct; they were responsible for describing birds from Whitehead's expedition.

31 According to Günther in the preface to vol. 27, p. vi.

32 Sclater's account was followed by presentations from Flower (who preferred the twelfth edition of the *Systema Naturae* as the starting point for scientific nomenclature) and Hartert (who preferred the tenth edition). Hartert said that if the law of priority was to be used at all it had to be used consistently.

33 NHM, TM1/12/14/382–3, TMC: Correspondence with HED, 1895, letter from HED to EH, 12 December 1895.

34 NHM, DF230/1, Department of Zoology: Bird Section Correspondence, Letterbook A–Z (1896), letter from HED to Richard B. Sharpe (RBS hereafter), 10 September 1896.

35 NHM, TM1/26/19/78, TMC: Correspondence with HED, 1896–97, letter from HED to EH, 21 September 1897.

36 MUL, B-W, letter from HED to RBS, 28 November 1897.

37 NHM, TM1/26/19/79–81, TMC: Correspondence with HED, 1896–97, letter from HED to EH, 1 December 1897.

38 MUL, B-W, letter from HED to RBS, 1 December 1897.

39 MUL, B-W, letter from HED to RBS, 11 December 1897.

40 Dresser wrote to Sharpe to acknowledge Sharpe's retraction of the accusation. See MUL, B-W, letter from HED to RBS, 16 December 1897.

41 Letter from Louis Mandelli to Andrew Anderson, 28 January 1876, quoted in Pinn (1985: 28–9).

42 NHM, TM1/19/17/390–1, TMC: Correspondence C–E, 1896, letter from HED to EH, 5 January 1896.

43 NHM, TMI/19/17/392–3, TMC: Correspondence C–E, 1896, letter from HED to EH, 2 February 1896.

44 Manchester Museum (MM hereafter), ZDH/8/1/3–4, letter from AN to Sydney Hickson (SH hereafter), 7 April 1899.

45 MM, ZDH/8/1/1a–2, letter from HED to SH, 6 April 1899.

46 MM, ZAD/1/33/2, letter from William E. Hoyle (WEH hereafter) to Robert D. Darbishire (RDD hereafter), 13 April 1899.

47 MM, ZAD/1/33/3, letter from RDD to WEH, 16 April 1899.

48 MM, ZAD/1/33/4, letter from WEH to RDD, 17 April 1899.

49 MM, ZDH/8/1/5, 7, 8, 15, 18, 34, letters between WEH and John Thomasson, April–July 1899.

50 MM, ZDH/8/1/9, letter from HED to WEH, 4 May 1899.

51 MM, ZDH/8/1/10, letter from HED to WEH, 8 May 1899.

52 MM, ZDH/8/1/1a, letter from HED to WEH, 6 April 1899; MM, ZDH/8/1/16, letter from HED to WEH, 18 May 1899.

53 MM, ZDH/8/1/19, letter from HED to WEH, 19 June 1899; MM, ZDH/8/1/28, letter from HED to WEH, 26 June 1899; MM, ZDH/8/1/26, letter from WEH to HED, 26 June 1899.

54 MM, ZDH/8/1/35, letter from HED to WEH, 17 July 1899; MM, ZDH/8/1/40, letter from HED to WEH, 9 Nov 1899.

55 NHM, TM1/42/8/78–9, TMC: Correspondence with HED, 1899, letter from HED to EH, 18 June 1899.

56 NHM, TM1/42/8/82–3, TMC: Correspondence with HED, 1899, letter from HED to EH, 19 July 1899; MM, ZDH/8/1/39, letter from HED to WEH, 11 September 1899.

57 Including Henry Seebohm (1895), Henry Drummond-Hay (1896), Lord Lilford (1896), Alexander More (1898), Osbert Salvin (1898) and Joseph Wolf (1899).

58 As evidence of the continuing close relationship between HED and AN, only Philip Sclater and Lord Lilford feature more frequently in AN's preserved correspondence in the University of Cambridge (some of HED's later correspondence to AN is clearly missing from the preserved letters). See http://www.zoo. cam.ac.uk/department/library/library-docs/alfred-newton-papers.pdf, accessed 7 January 2017. See also AN's views on HED's *A History of the Birds of Europe* earlier in this chapter.

12 Working independently, 1900–5

The Dresser family – Henry, Eleanor and Brenda – moved back to London in 1900, to 28 Queensborough Terrace near Hyde Park. Henry junior was away fighting in the Second Boer War.[1] Dresser's health was still poor, with bouts of rheumatism and stomach problems (see figure 12.1).[2] He avoided spending time alone at home when he could, possibly because of his health problems.[3] In 1903, his daughter Brenda fell ill, so she and her mother journeyed to Assouan (Aswan) in Egypt to take advantage of the climate.[4] Dresser's extended family was fast diminishing: two of his brothers and two sisters passed away between 1903 and 1905.

Apart from periods of ill-health, Dresser still worked full time in the City; his work took him to the Admiralty at least once a week and he associated with

12.1 Portrait of Henry Dresser, aged around sixty.

some of the most prominent Admiralty officials.[5] In July 1901, he and Brenda visited SS *Discovery* at the London docks, ahead of the British National Ant-arctic Expedition. On board, they met Edward Wilson, the naturalist, assistant surgeon and Expedition artist (Wilson died in 1912 along with Robert Scott, on the British Antarctic Expedition's return from the South Pole).[6]

The International Zoological Congress of 1901

The fifth International Zoological Congress was held in Berlin in August 1901. British naturalists formed a rather small contingent, only forty-four of the 500 participants. These included Dresser, Alfred Newton, Richard Sharpe, Philip Sclater, Howard Saunders, Ernst Hartert and Walter Rothschild.[7] The International Commission on Scientific Nomenclature presented its long-awaited rules for scientific nomenclature. Firstly, the chair of the Commission (Charles Stiles of the Smithsonian Institution) announced that Philip Sclater and Howard Saunders, the only two British ornithologists among the group of fifteen commissioners, had resigned. Sclater's reasons are unclear but none of the recommendations he had called for had been adopted (see Sclater, 1896, discussed in the previous chapter). The Commission put forward its set of rules that included the use of three-part scientific names for subspecies, the use of the tenth edition of Linnaeus's *Systema Naturae* as the starting point for scientific nomenclature, and strict adherence to using the oldest scientific name – 'absolute priority' – for a species. The rules were accepted and adopted by the Congress (Stiles, 1902: 882–90; Melville, 1995: 25–7).[8] The British ornithologists had lost their long-running battle, and the American Ornithologists' Union and other 'trinomialists' had won the day.

At the Congress, Anton Reichenow, President of the German Ornitho-logical Society, made a plea that three-part scientific names should be used judiciously, rather than applied to all species:

> The stone has been set rolling and it rolls on. The concept of subspecies is no longer confined to the minor deviations of a geographical nature;... We must turn back! The species must remain what it was until now, the smallest indivisible unit in the system, the essence of an individual being. (Reichenow, 1902: 912–13, translated from German)[9]

Reichenow had voiced the same concerns many years earlier (see also Haffer, 1992). The ensuing discussion demonstrated the full range of view-points, but British ornithologists played little part in this, at least as reflected in the published proceedings (Matschie, 1902a).

Incredibly, no mention of the International Zoological Congress was made in the *Ibis*, although many of the leading British ornithologists had taken part in the proceedings. This was nothing new, as the previous three Congresses had gone unnoticed in the *Ibis* and the only mention of them was in a handful of articles published following the various Congresses. From this point on, the British ornithological establishment – including the ornithologists at the British Museum (Natural History) (the BM(NH)) – chose to ignore the recommendations of a body that sought international agreement among scientists, a situation that clearly could not continue forever. Only a few British ornithologists – and few British zoologists generally – attended the sixth International Zoological Congress, in Berne in 1904.

Disputes over Henry Seebohm's bird specimens

Early in 1901, William Ogilvie-Grant of the BM(NH) began borrowing birds from Dresser's former collection, now housed in Manchester Museum.[10] He asked for the loan of two specimens of a particularly rare shrike, *Lanius funereus* (now considered to be a subspecies of the Great Grey Shrike) (see plate 48), writing to the Museum's Director, William Hoyle, 'you have at least two I fancy'.[11] Hoyle could not find the birds so proposed that he would write to Dresser about them, if Ogilvie-Grant thought they should be in the Museum. Ogilvie-Grant wrote to Dresser directly himself, about two shrikes 'lent you for examination by Seebohm…. If you still have them will you kindly return them to me.'[12] Dresser replied that he had borrowed the two specimens from Seebohm and returned them many years before, but that Seebohm gave him one of the two birds in exchange and that this bird should be in Manchester Museum. Ogilvie-Grant told Hoyle he had been in touch with Dresser and wrote: 'I think if you make further search you will find it [the specimen Dresser claimed Seebohm had given him] and possibly also the other bird lent by Seebohm to Dresser just before the former's death.' Hoyle found the two shrikes and lent them to Ogilvie-Grant, who then requested the loan of skins of another bird, a kind of wheatear from North Africa (*Saxicola seebohmi*; see plate 49), for which he had 'stupidly forgot to ask', writing to Hoyle that there should be two examples in Manchester Museum.[13] One specimen was found and sent to Ogilvie-Grant.[14]

All this was a ploy hatched between Ogilvie-Grant and Sharpe to find out if some birds that were missing from Seebohm's collection when it went to the BM(NH) were in Dresser's collection in Manchester. Sharpe described the circumstances in a lengthy report (sent with a covering letter) to the Director of the BM(NH), Ray Lankester, dated 18 April 1902.[15] He wrote how he did not believe that Seebohm would have exchanged specimens with Dresser, writing 'I knew him intimately for 20 years, and the last person that he would have exchanged with would be Mr. Dresser, of whom he had a very poor opinion':

> These two specimens of <u>Lanius funereus</u> must have been acquired by him [Seebohm] with great difficulty. They were the only ones in England, and Seebohm would as soon have parted with one of them as he would have thought of jumping over the moon…. I believe that Mr. Dresser, not having returned the

specimens to Mr. Seebohm in 1895, as he says he did – but, imagining that after Mr. Seebohm's death no enquiry would be made and that the Seebohm collection would be sold, deliberately added the two specimens of <u>Lanius funereus</u> to his collection and sold them to Manchester.

Sharpe told Lankester he had obtained a specimen of *Saxicola seebohmi* from Hoyle, labelled as being the 'type'. This also bore a label from Dresser, recording that the bird had been received in exchange from Seebohm and was a 'cotype'. A cotype is a vague term meaning that the specimen was mentioned in the article where the species was first described, but was not the basis of the main description; the term is no longer used, but equates to a paratype or paralectotype. Sharpe wrote how 'two specimens were lent by Seebohm to Dresser in 1895. What has become of the second specimen we shall never probably know, but the bird sent from Manchester is no "co-type", but the <u>actual</u> type.'[16] Sharpe acknowledged that, as he had a poor relationship with Dresser, 'any opinion as to the purloining of these Seebohm specimens, might not be somewhat prejudiced', so he consulted another ornithologist who had known both men for many years – probably Howard Saunders – on what to do about the matter. Sharpe proposed that Dresser be invited to the BM(NH) to be given the opportunity to present evidence to Lankester that Seebohm had given him the specimens in question. Sharpe concluded: 'Meanwhile the same strict supervision will be exercised over him [Dresser], if he visits the Bird-room.'

Even more revealing is the covering letter that accompanied the report. The letter outlines the background to Sharpe and Dresser's partnership on *A History of the Birds of Europe*, as well as Sharpe's career and progression at the BM(NH). It demonstrates both the basis and the extent of Sharpe's ill-feeling towards Dresser. It tells how Alfred Newton suggested that Sharpe enter into partnership with Dresser on *A History of the Birds of Europe*, and that Sharpe considered this to be the 'great mistake' of his life. Sharpe stood to earn £2,500 from the book but Albert Günther and his superior, John Gray, induced him to join the Museum 'at a great pecuniary sacrifice … the attraction being that I should write the *Catalogue of Birds*… I gave up the money interest, for the sake of the science, being assured that promotion in the service of the Museum would be by merit alone!' Sharpe wrote how 'great pressure' was brought on him to give up the *Birds of Europe* in order to devote all of his time to the *Catalogue of Birds*, presumably by Günther, and he was 'assured that the Trustees would not fail to recognise any sacrifice made on behalf of the Museum'. He wrote how, after the passage of thirty years, he had 'never had the slightest acknowledgement from the Museum':

> At the time I was subjected to much abuse from Mr. Dresser's friends for 'deserting him', although as a matter of fact, I gave over to him the whole of the stock of the <u>Birds of Europe</u> and he was free to continue the work, and, although he says that he made nothing out of the work, my opinion is that he cannot have made less than £5000.

Sharpe, who had ten daughters and struggled financially, held a poorly paid position at the Museum for twenty-three years until he received a promotion as a result of pressure from ornithologists (Gunther, 1975: 368, 372–8, 383; Jackson, 1994).

Sharpe wrote how Dresser behaved with 'the grossest ingratitude' and he had planned to denounce Dresser in *Nature* and the *Athenaeum* (a British literary magazine), but refrained from doing so on advice from Howard Saunders, who was known to both Sharpe and Dresser. He noted how Dresser was always placed under close supervision when he studied the BM(NH) collections:

> It must be clearly understood that I have never detected Mr. Dresser in attempting to take any specimens from our Museum, but, from a knowledge of the man's character, I have always felt that he was the sort of person who, if he could annex a valuable specimen without fear of discovery, would not hesitate to do so.

Sharpe's complaints appear to come down to four points: that he lost out financially from the *Birds of Europe* as a result of entering the Museum; that his advancement at the Museum was slower than he had been given reason to expect; that he felt he was mistreated by Dresser and his friends; and (possibly as a consequence of his ill-feeling towards Dresser) that he was suspicious Dresser would remove specimens from the Museum.

Sharpe wrote to explain the situation to William Hoyle in Manchester, repeating a good deal of his criticism of Dresser. He wrote how he imagined that the label attached to the *Saxicola seebohmi* specimen had been produced by Dresser after Hoyle drew his attention to the specimen in his collection. Sharpe asked for the return of the three birds, and suggested that their presence in Manchester should be put down to 'some mistake…. We willingly adopt the view that the specimens were sent in error, and desire to avoid letting the matter come before the Trustees [of the BM(NH)].'[17] Hoyle accepted that Sharpe had made a good case for the three specimens to be restored to the Seebohm collection, but absolutely refuted Sharpe's allegation about the label with the *Saxicola seebohmi* specimen.[18] Sharpe asked Hoyle to look for some specimens missing from Edward Hargitt's collection 'for curiosity's sake', including the Tawny Eagle he had accused Dresser of having in the previous decade: 'I am not going to claim them! But I should like to know if they are there…. We all thought that the Dresser label [attached to the leg of this specimen] on the *Saxicola seebohmi* was quite new. I withdraw all I said, and this only shows that one can be a little too clever sometimes.'[19] Hoyle spoke with the Museum Committee about the matter, and the three birds were duly sent to Sharpe. The Tawny Eagle and other birds missing from Hargitt's collection were not found in Manchester Museum, nor is there any evidence that they had ever been in Dresser's collection.[20]

Sharpe's assertion that Seebohm never gave or exchanged specimens with Dresser is incorrect: Seebohm provided Dresser with specimens from his Siberian expeditions (mentioned in *A History of the Birds of Europe*) and a specimen of the St Kilda Wren Seebohm named as a new species (see Dresser, 1886). Sharpe clearly believed that some of the birds that Dresser possessed should have been in the BM(NH), just as Dresser believed that he had made prior arrangements to acquire some of the birds that went to the Museum. Dresser saw Seebohm only five days before his death, and told Hartert that Seebohm had apologised to him for not having arranged the transfer of two rare thrushes.[21] It seems unlikely that Dresser would knowingly misappropriate specimens and then place them in a public museum for all to see. However, the motivations and intentions of collectors who can no longer defend themselves or explain their actions make for a moot argument.

The *Manual of Palaearctic Birds*

Dresser continued to work on his *Manual of Palaearctic Birds*, covering the same area as the *Birds of Europe*, as well as Asia south to Afghanistan, the Himalayas, Tibet, China, Korea and Japan. He produced brief accounts incorporating the most recent discoveries, covering the birds' distribution and habitat, and wrote pithy descriptions of their different plumages, nests and eggs. Descriptions of birds were mostly taken from specimens in Dresser's own collection. He also consulted specimens belonging to Walter Rothschild and the BM(NH) for those species he did not possess, although this involved some difficulty, as he outlined to Ernst Hartert in 1902: 'It is so very kind of Mr. Rothschild to lend them [bird skins], and indeed I don't know what I should do without the assistance you have so unfailingly rendered me, for at the Brit. Museum I can get no assistance that they have the power to withhold.'[22]

The *Manual* was issued in two parts, in 1902 and 1903, in royal octavo size at a cost of 25s. It was published at the office of the Society for the Protection of Birds in Hanover Square. The book was intended to be of practical use for field naturalists and travellers, but could not be called pocket-sized. It provided information on 1,219 species (including a few 'subspecies') within 950 pages, a triumph in condensing and distilling information. Dresser stuck to his practice of avoiding subspecies where possible:

> The endless manufacture of subspecies … often based on very trifling differences of tint, seems calculated rather to puzzle and discourage than to assist the beginner…. For this reason, besides being in principle a binomialist, I have declined the recognition of such so-called 'subspecies,' as those who have described them have so little confidence in as to need the aid of trinomials. (p. iv)

Dresser did include some entries on 'subspecies' in groups in which he had a particular interest, including dippers, long-tailed tits and shrikes. These subspecies were listed after the 'main' species, and each was given its own unique two-part (rather than three-part) scientific name, for example 'subsp. *Cinclus pyrenaicus*'. In the case of dippers, Dresser wrote that the various local forms were best treated as subspecies, and that he believed these had descended from one parent species, demonstrating his acceptance of the evolution of species over time (p. 25).

A measure of the exciting times in which Dresser was writing his *Manual* can be gauged by just how little was known about many birds. The nest and eggs of just over 200 of the 1,219 bird species and subspecies covered in the *Manual* were listed as being completely unknown. Among these were such fascinating species as the Spoon-billed Sandpiper, Ross's Gull and the Knot.[23] Other birds whose breeding grounds and nests remained to be discovered included a tremendous variety of East Asian birds, from sea eagles to mysterious ocean-going storm petrels.

Dresser's and Hartert's manuals on birds

Dresser's *Manual* was very well received. A reviewer, possibly Philip Sclater, wrote in the *Ibis*: 'It … will be much appreciated by all students of Palaearctic birds, amongst whom may probably be included nearly all the members of our Union. We therefore strongly recommend it to their notice.' The reviewer commended Dresser's conservative approach to recognising and including subspecies. They also took time to criticise 'splitters' – namely Elliott Coues and the American Ornithologists' Union, and Ernst Hartert at Rothschild's museum in Tring – who made wide use of subspecies and trinomials (Anon., 1903a). The *Manual* also received high praise in the *Zoologist* (Anon., 1903c) and *The Field*, where it was described as 'one of the most important contributions to zoology which has appeared for some time' (Anon., 1903b); across the Atlantic, Joel Allen wrote that Dresser was 'entitled to great credit for having placed before the public such a concise and excellent manual' (Allen, 1903).

While Dresser's *Manual* was well suited to the old guard among the BOU, the first parts of Hartert's *Die Vögel der Paläarktischen Fauna* (1903–22) were not so well met. A reviewer in the *Ibis*, almost certainly Sclater, compared Hartert's work directly to Dresser's *Manual*, which they admired for its 'steadfast adherence to the old-fashioned binomial system of nomenclature'. The reviewer attacked Hartert's three-part 'monstrosities', for example *Pica pica pica* (Magpie), as 'almost ridiculous'. The reviewer complained that Hartert asked readers to virtually give up the binomial system of scientific naming:

> Here we most decidedly decline to follow him…. We shall, no doubt, be stigmatized by some of our friends as 'fossils' and 'antediluvians'; but we believe that the great majority of sober-minded ornithologists, in spite of the efforts of the new school, will stick to the binomial system. (Anon., 1904a: 293)

Hartert responded with 'some anticriticisms' to 'what was apparently meant for a review' of his book:

The 'Editors' compare my treatment of species and subspecies with that of Mr. Dresser, whom they 'praised for his steadfast adherence to the old-fashioned binomial system of nomenclature,' and with whom they agree because 'even he recognises subspecies in certain cases.'

He argued that Dresser's selective approach to subspecies was the weakest point of the book:

It is true that Mr. Dresser ends his book with the sentence: 'Subspecies described under trinomial titles I have not considered it necessary to be included'; but is that a scientific method? (Hartert, 1904: 544)

He wrote that ornithologists were no longer content with this 'old system', and that advances in science meant that subspecies were of great interest and required recognition 'and neither Mr. Dresser nor the Editors of "The Ibis" will be able to stop the progress in that direction, whatever they may do' (p. 545). Hartert took Sharpe, Sclater and Dresser to task on many points; his 'anticriticisms' were practically the last thing he wrote in the *Ibis* while Sclater was Editor. Harry Witherby subsequently wrote to the *Ibis*: 'those who cling to two names, and two names only, should describe and name nothing but distinct species and leave geographical races entirely alone' (Witherby, 1905a). This was a sensible approach, but it went unheeded.[24]

Walter Rothschild compared Dresser's and Hartert's manuals at a meeting of the BOC, noting how Dresser had 'dealt with his subject throughout on ancient lines'. Rothschild was concerned that Dresser's work was being characterised as for the use of field naturalists, while Hartert's was for museum workers only. He pointed out how Dresser's avoidance of subspecies and trinomials had led him to 'two most ludicrous errors' as 'proof of how far preconceived prejudices can lead even a veteran ornithologist to make rash statements' (Rothschild, 1904: 88). On the other side of the Atlantic, Joel Allen reviewed Hartert's book in the *Auk*: 'although we have a recent popular manual on the birds of the same region, the present work is to be most heartily welcomed as an exposition of the subject from a technically up-to-date standpoint' (Allen, 1904). The 'recent popular manual' clearly refers to Dresser's *Manual of Palaearctic Birds*, which Allen himself had praised the year before.

Confusion reigned in British ornithology, so that ornithologists named subspecies in different ways in the same journals.[25] Throughout the discussion on the subject, Dresser was portrayed as the embodiment of tradition both by the old guard of the BOU and by Hartert and Rothschild. Whatever disagreements he had caused within the BOU through his naming practices in the past, these were as nothing compared with the radical innovations of Hartert and Rothschild. They, on the other hand, criticised Dresser for sticking to tradition. Dresser was effectively the figurehead for 'old-fashioned' British ornithology, even over Sharpe of the BM(NH).

Dresser begins a book on bird eggs

While the *Manual* was being published, Dresser began an enormous project on the eggs of the birds of Europe and Asia. He returned to his old obsession of egg collecting and sought out eggs of rare and newly discovered species for the book. He attended many auctions at Stevens' at this time, seeking rare specimens among the private collections that were being sold off there. He also negotiated with many museums for specimens, including even the BM(NH) in 1902, despite his ongoing disagreements with the Museum's bird curators. Dresser made several trips to the Continent to study eggs in museums and private collections to research the book. The first trip was to Hungary, Croatia and Bosnia in May 1902. Henry junior, home on sick leave from the Indian colonial service, accompanied his father. Dresser visited Budapest Museum and travelled on to a large private hunting estate at Bogyiszló, eighty kilometres south of Budapest. Dresser was interested in observing birds on their breeding grounds and acquiring eggs, and later wrote 'I never shoot a bird during the breeding season' (Dresser, 1903). The party travelled to Agram (Zagreb), accepting an invitation from the Viceroy of Croatia to collect eggs there. They then travelled on to Bosnia, where Dresser travelled with Othmar Reiser, the curator of Sarajevo Museum. They visited colonies of Pygmy Cormorants, herons, egrets and terns at the Hutovo Blato, a wonderful marsh close to the Adriatic, and collected many eggs. Returning to Budapest, Dresser met his old friend Frederick Selous, a famous army officer, big game hunter and egg collector. Dresser wrote up his experiences as 'Birdsnesting in Lower Hungary, Bosnia, etc.', published in *The Field* (Dresser, 1903).

Publications on birds and eggs

In 1900, Dresser had seen a colour photograph of a bowl of fruit produced by the 'three-colour process', invented by Carl Hentschel.[26] Here at last was a method that looked like it could reproduce the subtle colouring, markings and textures of bird eggs, which were notoriously difficult to replicate (as can be seen by the generally poor illustrations of eggs in previous books on the subject). Dresser collaborated with the firm André and Sleigh, a Hertfordshire-based printer, to produce coloured plates of eggs using the process, initially for articles and latterly as the basis of a great book on European bird eggs. He presented the process at a meeting of the BOC in April 1901, with illustrations of eggs and reproductions of pictures (Dresser, 1901a). He had a watercolour of a Labrador Falcon by Joseph Wolf reproduced using the process, and advertised copies of this for sale in the *Zoologist* (Dresser, 1901b).

Dresser's use of the three-colour process was novel enough to warrant a lengthy article co-authored by Dresser in *Nature*, the leading scientific weekly magazine. The method involved photographing eggs, life-size, with three filters that corresponded to the colours red, green and blue on long exposures of two hours for each filter. The three colour-filtered negatives were then used to produce acid-etched printing blocks made of copper. Each plate had to pass through three presses (which printed the colour for the relevant printing block), meaning that the plates had to be aligned perfectly to avoid blurring of the image (Dresser and Trueman Wood, 1902; Wall, 1925).

Work towards the 'eggs book' led to an extraordinary series of exhibitions and announcements at BOC meetings and a series of papers on 'new' eggs in the *Ibis*, beginning in 1901 and running through the decade. The first article followed a common model, with an article by Dresser on the eggs of some rare Siberian thrushes illustrated by a plate produced by the three-colour process (see plate 50) (Dresser, 1901c). This was followed by an article by the field collector, Hugh Popham, describing the trip on which he collected the eggs (Popham, 1901). Other papers in the series covered little-known buntings, finches, shrikes and other birds from Siberia and Central Asia.

As well as his publications on eggs, Dresser described a new Copper Pheasant from Japan (Dresser, 1902). He collected Red Grouse from throughout Britain on behalf of Otto Kleinschmidt in 1901, requesting specimens in a letter in *The Field* (Dresser, 1901d), so that he could study regional variation in the birds. Dresser made the birds into study skins to send to Kleinschmidt, while their bodies were made into soup.[27] Kleinschmidt named the Hebridean Red Grouse *Tetrao dresseri* in return for Dresser's assistance (see Witherby, 1923).

Collaboration with Russian ornithologists

Dresser's collaboration with Russian ornithologists on his eggs book was a particularly distinctive feature of his work. He did not speak or write Russian but corresponded with Russian ornithologists in English, German and French.[28] He knew the leading museum curators very well, but it was two independent ornithologists who contributed the most to the great egg project, Nikolai Zarudny (1859–1919) and Sergei Buturlin (1872–1938). Zarudny was Ukrainian and taught in military schools during 1879–1919 (see figure 12.2). He undertook five collecting expeditions in western and western-central Asia between 1884 and 1892 (Zarudny, 1888, 1896) and four to Persia (in 1896, 1898, 1900–1 and 1903–4), often with official support. His 1900–1 expedition covered 4,500 kilometres and amassed 3,140 birds and about 50,000 insects (Ananjeva, 2008).[29] Zarudny provided Dresser with many rare eggs that were a prominent feature of his papers and exhibitions on Palaearctic bird eggs.

Sergei Buturlin (see figure 12.3) worked as a justice of the peace in Wesenberg (Ravkere, Estonia). He made ornithological expeditions to the Volga and the Baltic in the 1880s, and to Kolguev and Novaya Zemlya in the Arctic in 1900–2.[30] Dresser and Buturlin became acquainted in the 1890s (see McGhie and Logunov, 2005, 2006). Dresser wrote, in 1903:

12.2 Nikolai Zarudny, from Henry Dresser's album of correspondents.

12.3 Sergei Buturlin, from Henry Dresser's album of correspondents.

> I have a very large collection of Palaearctic eggs but am weakest in those from Siberia and south east Russia. Do you know how I can get some of these for if you could help me I should be greatly obliged.... If I can in any way reciprocate and be of any use to you in this country I shall be very pleased.[31]

This led to a productive collaboration: Dresser sent English publications to Buturlin and checked specimens in the BM(NH) for him, while Buturlin sent Dresser information on birds in Russia and wrote notices on Russian articles for the *Ibis*.

When Dresser was planning a visit to Russia, in 1903, Buturlin helped with arrangements. Dresser's letters reveal his state of health: 'I am rather lame with rheumatism ... and when I get a touch of it I cannot walk far ... in one leg I am always rather lame.'[32] In spite of this he made the trip in May 1904. At the Zoological Museum of the Imperial Academy of Sciences in St Petersburg, Valentin Bianchi showed him the eggs of Arctic-breeding waders – including the Knot, Sanderling and Curlew Sandpiper – that had been collected during the Russian Polar Expedition (1900–3). Dresser also met Alexander Birulya, one of the Expedition's naturalists, who had found the rare waders in the New Siberian Islands and survived a 640-kilometre journey across the frozen sea (Vaughan, 1992: 76–9). Dresser travelled on to Wesenberg (Estonia) to meet Buturlin in early June. They collected birds and eggs on some islands in the Gulf of Finland before visiting Baron Harald Loudon, a veteran traveller

12.4 Photograph of a boy climbing to a Buzzard's nest, taken by Baron Harald Loudon in June 1904; note the boy's flat cap.

in Transcaspia and Central Asia, at his estate in Livonia. Loudon was a keen photographer and some of the photographs he took during Dresser and Buturlin's visit are still in existence (see figures 12.4 and 12.5). Dresser returned from Russia with over 500 eggs from his own collecting and various other collectors, including eggs of the rare Arctic-breeding waders received from Bianchi, for which he must have given some extremely good exchanges.[33] Dresser described the trip as 'An oological journey to Russia' in the *Ibis* (Dresser, 1905b).

Dresser kept abreast of ornithological discoveries in Russia and he was the first to announce (in Britain) the discovery of the Knot's breeding grounds by the Russian Polar Expedition (Dresser, 1904a). He went on to write two lengthy papers in the *Ibis* describing the Expedition and its human cost: the Expedition leader, Baron Toll, and the surgeon–naturalist, Dr Hermann Walter, both died on the Expedition (the latter died of a heart attack after taking an accidental overdose of digitalis) (Dresser, 1904b, 1908a; see also Dresser, 1908b).

The assistance Dresser provided his Russian collaborators was clearly appreciated by them, as he helped make their discoveries known to a wide audience. Buturlin named a subspecies of the Common Starling *Sturnus purpurescens dresseri*; he and Zarudny named an Eastern Rock Nuthatch from South-west Asia *Sitta dresseri*, in recognition of Dresser's assistance (Buturlin, 1904b; Sarudny [sic] and Buturlin, 1906).

12.5 Photograph of Henry Dresser with a Redwing's nest, taken by Baron Harald Loudon in June 1904.

Collecting birds for Manchester Museum

Dresser continued to take a great interest in his old collection. He borrowed birds from it on a number of occasions to settle disputed points.[34] He also borrowed birds for Archibald Thorburn to illustrate, including a rare Siberian Eagle-Owl in 1901 and an Ibisbill the following year.[35] The Eagle-Owl was singled out as it was unlike any in the BM(NH). The Ibisbill lives along fast-flowing shingly rivers in Central Asia and was – and still is – a particularly mysterious bird, with a long down-turned red bill and unusual patterning.

Whenever he was travelling, Dresser sought specimens to enhance his collection of Palaearctic bird skins at Manchester Museum to maintain its – and his – standing. Additions included eight 'rather good larks' in 1901,[36] an almost unknown kind of Central Asian Sparrow (*Passer yatii*),[37] a Sacred Ibis skin from Walter Rothschild,[38] and some 'new' birds from Cyprus obtained from Budapest Museum.[39] Dresser used duplicate Japanese birds that had gone to Manchester to arrange exchanges with museums in Budapest (1902) and St Petersburg (1904).[40]

Notes

1 He served in the Imperial Yeomanry with Paget's Horse, a volunteer force. He was commissioned into the Royal Warwickshire Regiment in 1902.

2 Natural History Museum Library and Archives, London (NHM hereafter), TM1/62/7/55, Tring Museum Correspondence (TMC hereafter): Correspondence D–F, 1900–02, letter from Henry Dresser (HED hereafter) to Ernst Hartert (EH hereafter), 20 January 1902.

3 NHM, TM1/55/14/227, TMC: Correspondence C–F, 1900–02, letter from HED to EH, 17 July 1901.

4 NHM, WP1/1/89, Alfred Russel Wallace Family Papers (ARWFP hereafter): Correspondence, letter from Alfred R. Wallace (ARW hereafter) to Will Wallace, 1 March 1904, sent with letter from HED to ARW, 1 March 1904.

5 NHM, WP1/1/89, ARWFP: Correspondence, letter from HED to ARW, 1 March 1904.

6 Wilson wrote to HED and subsequently sent him a carte-de-visite photograph, which he included in his album of ornithological correspondents, now in John Rylands Library, MS1404.

7 Sclater chaired a session where he apologised for the poor show of British zoologists, which he put down to 'the present aspect of political affairs' (Sclater, 1902: 120).

8 The Rules of Zoological Nomenclature were published as an appendix to Stiles' report in German (pp. 933–46), French (pp. 947–60) and English (pp. 961–72), with an introductory essay by Paul Matschie (pp. 929–32) (Anon., 1902; Matschie, 1902b).

9 'Der Stein ist ins Rollen gekommen, und er rollt weiter. Der Begriff der Subspecies wird jetzt schon nicht mehr auf die geringfügigen geographischen Abweichungen einer Art be- schränkt.… Wir müssen umkehren! Die Species muss bleiben, was sie bisher war, die kleinste unteilbare Einheit im System, der Inbegriff der Einzelwesen.'

10 Manchester Museum (MM hereafter), ZAC/1/23, correspondence between William R. Ogilvie-Grant (WROG hereafter) and William E. Hoyle (WEH hereafter), 12 March–3 June 1901.

11 MM, ZAC/1/23/11, letter from WROG to WEH, 20 March 1902.

12 NHM, DF230/20/314a, Department of Zoology: Bird Section Correspondence, Copybook of out-letters of Richard B. Sharpe (RBS hereafter) 1897–1902, duplicate letter from WROG to HED, 1(?) April 1902.

13 MM, ZAC/1/23/15, letter from WROG to WEH, 7 April 1902.

14 MM, ZAC/1/23/16, duplicate letter from WEH to WROG, 14 April 1902.

15 NHM, DF231/2/457 Department of Zoology: Vertebrate Section, Reports to Trustees and other Official Documents, 1896–1909, duplicate letter from RBS to E. Ray Lankester, 18 April 1902; NHM, DF231/2/469–70, Department of Zoology: Official documents, copies of reports to Trustees by RBS, with

supporting letters, lists and notes, 1899–1903, letter and accompanying report from RBS to E. Ray Lankester, 18 April 1902.

16 The name *Saxicola seebohmi* was given by Charles Dixon, who did some collecting for Seebohm, to two wheatears he and Henry Elwes collected in Algeria in 1882 (Dixon, 1882). Neither was singled out as a type specimen in the article naming the new species. The second specimen that Sharpe referred to was in the collection of Henry Elwes and passed into the collection of Walter Rothschild; it is now in the American Museum of Natural History in New York (specimen AMNH 583232). The label for this bird is also marked 'type' in handwriting that Ernst Hartert considered to have been Dixon's own, while he thought the label on the specimen in the BM(NH) had been written by Seebohm. Hartert considered the specimen in the AMNH was the 'main' type specimen, the basis of the illustration in the original article, while the specimen that was in Dresser's collection (the specimen passed to the BM(NH)) was indeed a 'cotype' (Hartert, 1920: 471; LeCroy, 2005: 25–6).

17 MM, ZAC/1/23/17, letter from RBS to WEH, 30 April 1902; NHM, DF230/20/365–7, Department of Zoology: Bird Section Correspondence, Copybook of out-letters of RBS 1897–1902, duplicate of part of the aforementioned letter.

18 MM, ZAC/1/23/18, letter from WEH to RBS, 2 May 1902 (presumably a duplicate letter).

19 NHM, DF230/20/370–1, Department of Zoology: Bird Section Correspondence, Copybook of out-letters of RBS 1897–1902, duplicate letter from RBS to WEH, 4 May 1902.

20 MM, ZAC/1/23/20 (duplicate) and NHM, DF230/8 (original), letter from WEH to RBS, 21 August 1902.

21 NHM, TM1/12/14/382–3, TMC, Correspondence with HED, 1895, letter from HED to EH, 12 December 1895.

22 NHM, TM1/62/7/61, TMC: Correspondence with HED, 1902, letter from HED to EH, 20 August 1902.

23 The nest of Ross's Gull had been reportedly found in Greenland in 1885 but most ornithologists, including Dresser, did not accept the record and in any case it did not establish where the bird bred regularly (Prŷs-Jones, 2006). The nest of the Knot had been found by Russian explorers in the New Siberian Islands in 1886 but the eggs were destroyed during the homeward journey. A supposed Knot egg collected in Greenland in 1875 and given to Henry Seebohm by the chief tenor of the Copenhagen opera(!) was generally discounted (except by Seebohm) as having been misidentified, as was an egg supposedly laid in Lord Lilford's aviary by a captive Knot and given by Lilford to Alfred Newton (Vaughan, 1992: 162).

24 Witherby described a number of new subspecies himself, using three-part scientific names (Witherby, 1905b,c).

25 WROG described new subspecies under two-part (binomial) scientific names in the *Bulletin of the British Ornithologists' Club* and the *Ibis*, while EH, Rothschild and others used trinomials in the same publications. American and Russian authors were more ready to use trinomials for subspecies and some new subspecies were described with trinomials in the *Ibis* (e.g. Bianchi, 1904; Buturlin, 1904a) and the *Bulletin* (Sushkin, 1904).

26 The image of fruit may well have been a plate by William Kurtz, the first widely reproduced image produced with the three-colour process, in 1893.

27 National Museums Scotland Library, GB 587, JHB 15/240, letter from HED to John A. Harvie-Brown, 5 December 1901.

28 Cambridge University Library, MS.Add.9839/1D/502, L852, letter from HED to Alfred Newton, 3 April 1893.

29 Zarudny's collection of 13,281 bird specimens is in the National University of Uzbekistan in Tashkent.

30 Buturlin's archive of personal papers and his bird collection are in the Museum of Local Lore, History and Economy, Ulyanovsk (MLLU hereafter).

31 MLLU, YKM 27159/1, letter from HED to Sergei A. Buturlin (SAB hereafter), from London, 26 February 1903.

32 MLLU, YKM 27159/3, letter from HED to SAB, 8 April 1903.

33 MM, ZDH/8/6/1, letter from HED to WEH, 27 June 1904.

34 MM, ZDH/8/3/5, letter from HED to WEH, 30 July 1901; MM, ZDH/8/7/3, letter from HED to WEH, 19 May 1905.

35 MM, ZDH/8/3/9, letter from HED to WEH, 25 October 1901; MM, ZDH/8/4/1, letter from HED to WEH, 28 January 1902.

36 MM, ZDH/8/3/6, letter from HED to WEH, 2 August 1901.

37 MM, ZDH/8/4/1, letter from HED to WEH, 28 January 1902.

38 MM, ZDH/8/3/2 letter from HED to WEH, 15 July 1901; MM, ZDH/8/3/4, letter from HED to WEH, 29 July 1901.

39 MM, ZDH/8/4/4, letter from HED to WEH, 2 December 1902.

40 MM, ZDH/8/4/3, letter from HED to WEH, 23 June 1902; MM, ZDH/8/6/1, letter from HED to WEH, 27 June 1904.

13 The grand finale: producing *Eggs of the Birds of Europe*

In spite of ongoing health problems and advancing years – he was sixty-seven in 1905 – Dresser continued to work in business in the City through the week.[1] He wrote to his friend John Harvie-Brown:

> My business is strictly personal so a partner would be no good. When I cannot attend to it personally it drops, as it is doing just now and I am literally losing money and not paying even office expenses, which worries me not a little.... If I were alone I could retire and work at birds only but a wife and family make it impossible.[2]

Dresser and his wife Eleanor moved from Queensborough Terrace to nearby Hornton Court (in affluent Kensington) in March 1908.[3] They had earlier placed Topclyffe Grange for sale or rent again, in 1906.[4]

Preparing *Eggs of the Birds of Europe*

Dresser issued a prospectus for *Eggs of the Birds of Europe* early in 1904 (subsequently announced in the *Ibis*), projecting that it would be issued in twenty parts at two-monthly intervals, at 10s 6d per part (Anon., 1904b). The first part was issued in August 1905, published by special permission at the Office of the Royal Society for the Protection of Birds. Dresser wrote in the preface:

> many, if most, ornithologists commence like myself, when quite young, by collecting eggs, which naturally leads to the study of birds.... Undoubtedly there are cabinet oologists who collect eggs as they would postage stamps, and obtain their specimens chiefly by purchase from dealers and field oologists, but those collections are certainly of the greatest value where the specimens are carefully identified. In egg collecting the chief object should be to obtain only such specimens as are undoubtedly well authenticated, and not merely to amass indiscriminately a large series to the serious detriment of the rarer species.

Here we see Dresser establishing himself as a scientific collector, rather than a mere accumulator of specimens. Bird collecting had gone into decline, but the popularity of egg collecting was increasing and yet to reach its peak. Scientific

13.1 Henry Dresser's album of photographs of nests.

collecting was still generally considered respectable, except so far as rare British and Continental species were concerned. Dresser stringently followed the conservation laws – unsurprisingly, as he had helped draft them – and did not acquire British eggs of rare species, except in a very few instances.

The first part of the book covered twenty-nine species, with text, plates of eggs and a photograph (*in situ* where possible) of the nests of many species. Dresser requested readers to send in photographs of the nests of other species to include in subsequent parts; this met with great success and he received hundreds of photographs from amateur naturalist photographers (many of these photographs are still in existence in Manchester Museum) (see figure 13.1).[5] Among these were some amazing photographs, featuring nests of rarer Scottish birds from G. G. and W. J. Blackwood, and of birds of prey in Gibraltar from Major Cyril Moore. The first part received an extremely favourable review in the *Ibis* (Anon., 1906h). The second part, covering vultures, kites and more warblers, was issued in December 1905. After this time, double parts were issued twice each year.

The timing of *Eggs of the Birds of Europe* could not have been better, as Russian ornithologists were making many spectacular discoveries, from the breeding grounds of the Knot in the high Arctic to Zarudny's discoveries in Persia. Dresser held the privileged position of announcing many of these remarkable discoveries in the West, adding significantly to his own reputation. For example, Valentin Bianchi travelled to London to show off Knot eggs and chicks collected on the Russian Polar Expedition of 1900–3. Bianchi allowed Dresser to have the Knot's eggs photographed for *Eggs of the Birds of Europe* (see figure 13.2).

Several other books on eggs (i.e. competing books) were issued around the same time as Dresser's *Eggs of the Birds of Europe*. Francis Jourdain began issuing

TRINGA CANUTUS.

13.2 Photographic plate of twelve Knot eggs collected on the Russian Polar Expedition, from *Eggs of the Birds of Europe*.

The Eggs of European Birds in 1906. Reviewers praised the text of this book, but the illustrations were based on paintings of eggs and were no comparison with Dresser's photographic plates (Anon., 1906i). Another major book on eggs was Georg Krause's *Oologia Universalis Palaearctica*, issued in Stuttgart from 1905 (Krause, 1905–13). Krause intended it to be 'a fundamental and monumental ideal work on palearctic oology' (quoted in Anon., 1906j: 725).

While Dresser's book on eggs was underway, *A History of the Birds of Europe* had a new lease of life as J. Lewis Bonhote used some of the plates for his *Birds of Britain* (1907). Dresser selected the plates himself. Bonhote's book cost a fraction of the price of *A History* (20s), so it took Dresser's plates to a much wider audience, although the book received a rather stinging review in *British Birds* as including 'nothing novel' (Anon., 1908c).

The Fourth International Ornithological Congress

The Fourth International Ornithological Congress was held in London in June 1905 (Sharpe *et al.*, 1907). Richard Sharpe was the Congress's President; Dresser was on the general organising committee along with Lord Avebury (John Lubbock), the Duke of Bedford (President of the Zoological Society of London) and many of his long-time associates.[6] The meetings were held in the Imperial Institute, close to the British Museum (Natural History) (BM(NH)), built to commemorate Queen Victoria's Golden Jubilee and to promote research that would benefit the British Empire. The Congress was divided into five sections: Systematic Ornithology; Migration; Biology and Oology; Economic Ornithology and Bird Protection; and Aviculture. Sharpe's presidential address gave a blow-by-blow account of the growth of the BM(NH) collection, swelled with generous donations and purchases from leading collectors (Sharpe, 1906). Sharpe noted: 'it will be seen that nearly every great private collection in England has passed with the willing consent of the owners into the British Museum, while the donations of the great collections ... have contributed to the renown of the British Museum' (p. 143). As we have already seen, Henry Tristram's collection in Liverpool and Dresser's in nearby Manchester were the glaring exceptions to this rule. Presentations from Frank Chapman of the American Museum of Natural History in New York and Ernst Hartert of Rothschild's Tring Museum set out the principal aims of museum bird collections: for making systematic investigations of the distribution and variation of birds, and (according to Chapman) for educating and exhibiting to the public (Chapman, 1907; Hartert, 1907). These contributions demonstrated how museums – at least the largest ones – were organising themselves as research institutions, much as Baird had done with the Smithsonian years earlier, and how ornithological research continued to be centred on museums.

The section on Systematic Ornithology, with Sclater as President, included presentations from Walter Rothschild on recently extinct birds and two presentations on the birds of Antarctica, from Edward Wilson who had sailed on the *Discovery* and William Bruce of the Scottish National Antarctic Expedition. The German Rudolf Blasius presented a paper on the scientific naming of birds but some members (notably Reichenow and Stejneger) declined to discuss the subject, as they deemed it to be the business of the International Zoological Congresses.[7]

Dresser was the President of the section on Economic Ornithology and Bird Protection. Sir Digby Pigott reviewed the progress of bird conservation in Britain, including the work of the Close-Time Committee over thirty years earlier (Pigott, 1907). Other papers discussed bird protection in Australia, and the rationale of bird protection in Britain. Walter Rothschild and Baron Hans von Berlepsch advocated the provision of nestboxes for birds and the creation of nature reserves with dedicated wardens. Papers on economic ornithology focused on the food of birds, particularly those accused of being injurious to agriculture such as Rooks and House Sparrows.

Twenty-eight delegates were elected as permanent members of the International Ornithological Congress, including Dresser and three other noted conservationists.[8] Delegates enjoyed a number of social events, including: a 'social gathering' at Earls Court; an evening 'conversazione' at the BM(NH); a day trip to Tring to examine Walter Rothschild's private collections (both dead and alive); and a reception with the Lord Mayor and Lady Mayoress of London at the Mansion House. The week after the Congress, further trips were made, to the Duke of Bedford's private zoo, to Cambridge to visit Alfred Newton's museum and book collection, and to Flamborough Head in Yorkshire to see the seabirds that had benefited from the early bird protection laws.

Birds from Tibet

There was a further dispute between Dresser and the BM(NH) over the ownership of birds. This involved some specimens collected during the British Expedition to Tibet (1903–4), led by Colonel Francis Younghusband, veteran explorer and fervent imperialist. The Expedition's original purpose was to fix part of the Sikkim–Tibet boundary to restrict the spread of Russian influence in the region, but it quickly became an all-out assault on Lhasa. After two bloody episodes in which hundreds of Tibetans were killed, the British triumphantly entered Lhasa in August 1904. The Dalai Lama had already fled the city but the British forced a trade treaty upon the Tibetans, drafted single-handedly by Younghusband, before returning to India via Sikkim (Hopkirk, 1982, 1992; French, 1994).

The Expedition's naturalist, Captain James Walton (1869–1938) of the Indian Medical Service, collected 416 birds, which went to the BM(NH) (Walton, 1906; Mearns and Mearns, 1998: 171). Lieutenant Colonel Laurence Austine Waddell (1854–1938) of the Indian Medical Service served as cultural adviser to Younghusband and as the official collector for the British Museum (for antiquities and ethnographic objects) (Carrington, 2003: 92). Waddell was an expert on Buddhism and an accomplished ornithologist, having written about the birds of Sikkim based on his experiences there (Waddell, 1894, 1899). Following the Expedition he returned to England, taking up a lecturing post at University College London. Waddell published an account of the Expedition, as did several other Expedition members (Waddell, 1905).[9] He had collected some birds during the Expedition but most of these were lost in transit. Dresser examined those specimens that did make it back to Britain and named three of them – two babblers and a shrike – as new species at a meeting of the Zoological Society of London (Dresser, 1905a), describing them as 'the ornithological first-fruits of the recent expedition to Tibet' (quoted in Anon., 1905c). In the end, only one of the three, the Giant Babax that Dresser named *Babax waddelli* after its discoverer, turned out to be new. Dresser persuaded Waddell to let Manchester Museum have the type specimens of these three 'new' species, which Waddell regarded as his personal property (see figure 13.3). Dresser wrote to William Hoyle: 'The Brit. Museum people will be very angry, but I want to make your collection as good a Palaearctic one as possible, and work for you and not for them.'[10] He wrote again soon afterwards:

> I had no end of trouble to secure the types of the Tibetan birds I described. Firstly the Brit. Mus. people considered that all the birds collected on the expedition

13.3 Birds skins Henry Dresser received from Laurence Waddell and named as new species: *Babax waddelli* (front), *Garrulax tibetanus* (middle) and *Lanius lama* (back).

should have gone to them and wanted to claim them – but Col. Waddell said they (viz. those he had himself collected) were his private property and he could dispose of them as he thought fit; and then Rothschild wanted to buy them and was willing to buy them at a good price.[11]

In addition to birds, Waddell had gathered a personal collection of Tibetan artefacts, despite being the official collector for the British Museum, which resulted in parallel disputes over ownership (Carrington, 2003: 109). Dresser, Rothschild and Waddell agreed that Rothschild could buy all of Waddell's birds except the type specimens, 'so the Museum [BM(NH)] got none'.[12] Ogilvie-Grant spoke critically of Dresser's naming of the supposedly new birds from the expedition at a BOC meeting, saying that Dresser had failed to compare them with well-known species (Ogilvie-Grant, 1905).

Waddell put Dresser in touch with Captain Robert Steen, the medical officer at the newly established Gyantse Trade Agency, in 1904 (see McKay, 2005). Steen supplied Dresser with skins and eggs of twenty-eight bird species. Dresser exhibited the eggs at a BOC meeting in 1905 and wrote about them in the *Ibis* in 1906 (Dresser, 1905c, 1906b). He passed the skins to Rothschild, as he was solely (or almost solely) interested in acquiring eggs to help with his book. Dresser negotiated for rare bird skins on Rothschild's and Hartert's behalf on several other occasions. For example, he acquired many rare eggs that Nikolai Zarudny had collected in Persia and Transcaspia, and negotiated with the Zoological Museum of the Imperial Academy of Sciences in St Petersburg for Zarudny's bird skins to go to Rothschild.[13]

Sergei Buturlin and Ross's Gull

In 1905, while most ornithologists were in London for the International Ornithological Congress, Sergei Buturlin was in the Kolyma region in northeast Siberia, investigating a state supply system to the remote outpost. He also had official support from the Imperial Academy of Sciences in St Petersburg for his collecting of birds, eggs and other natural history specimens. Buturlin and two assistants left St Petersburg in January and travelled 10,000 kilometres via Irkutsk and Yakutsk to Srednekolymsk on the Kolyma River (see plate 51). He described the journey in three short articles for the *Ibis*, the first of which was written while he was still in the Kolyma region. In May they set up camp in the village of Pokhodskoe (Pokhodsk); Buturlin spent a week at the remote village of Sukharnoe (Sukharnoye), studying seal hunters' techniques while his companions collected on the nearby Ola and Alazeya Rivers.

Birds arrived in great numbers to breed and the expedition collected many specimens, including 2,000 bird skins, 500 clutches of eggs, over 4,000 insects,

13.4 Photograph showing the inside of Sergei Buturlin's hut in Pokhodskoe, Kolyma, 1905, with dead birds and eight Ross's Gull skins on the windowsill. Note the trap hanging from the wall and stuffing hanging from the ceiling.

2,000 herbarium sheets and some ethnographic collections. In Pokhodskoe, they prepared specimens in a small wooden hut (see figure 13.4). As soon as the ice began to break up on the river, Buturlin had observed Ross's Gulls, the rare and mysterious bird so beloved of Arctic explorers and ornithologists: 'I went to the river – where the fathom-thick ice was still quite safe – and came across several dozens [of Ross's Gulls]. The sun was shining brightly, and in the distance each pair appeared like so many roseate points on the bluish ice of the great stream' (Buturlin, 1906: 133).

Buturlin found three breeding colonies of the rare gull in the swampy tundra (see plate 52). He studied their behaviour closely, and collected eggs and birds in various plumages. Buturlin told Dresser of his fabulous discovery and sent him detailed notes from the Kolyma. Dresser announced the discovery on Buturlin's behalf at a BOC meeting in December 1905 (Dresser, 1905d) and arranged for the publication of Buturlin's notes in the *Ibis*. The first part of Buturlin's article had already been published in the *Ibis* by the time he reached Moscow, following a three-month journey during which he lost two toes to frostbite. Buturlin received his first Ross's Gull eggs from the Kolyma in April 1906; he sent some of these to Dresser, which were exhibited at the May BOC meeting and formed the basis of an article in the *Ibis*, illustrated by a plate prepared using the three-colour process (Dresser, 1906a,c; McGhie and Logunov, 2005, 2006) (see plate 53).

Most of Buturlin's Kolyma collections went to the Zoological Museum of Moscow University but some specimens were traded away. Buturlin collected at least thirty-eight adult Ross's Gulls, several young birds and over fifty eggs. Most of the eggs (thirty-six) were sent to Dresser to trade on Buturlin's behalf, as were some Ross's Gull skins. Some of the birds and eggs went to the collector Otto Ottosson in Stockholm.[14] Dresser fought hard to keep specimens out of dealers' hands, and indeed out of Ottosson's hands after Ottosson gave some to a dealer. Dresser was keenly associated with Buturlin's discovery: as well as making the initial announcement, he exhibited eggs and published illustrations of them and of the plumage of a chick in the *Ibis* (see plate 54). Dresser provided specimens of the rare bird to many European and American museums and private collectors, with eggs and birds going for £5 each, equivalent to almost £500 today.[15] The BM(NH) was unhappy with this price and would not pay more than £4, so it did not receive any of the specimens at the time (McGhie and Logunov, 2005).

Buturlin provided Dresser with specimens of other fabulous birds as well as detailed information on the distribution of birds in Russia.[16] He also put Dresser in touch with a number of his correspondents, who were making their own discoveries. Buturlin was at the very forefront of ornithological discovery and Dresser was his 'publication partner', bringing his discoveries to a wide audience. Dresser arranged for several of Buturlin's papers to be published in Britain, even if he disagreed with his collaborator's views on scientific naming: Buturlin was keen on separating subspecies and gave them the three-part scientific names Dresser could not abide (Dresser, 1907a).

Exhibitions at the BOC and articles in *Ibis*

In April 1907, Dresser announced the discovery of the breeding grounds (and eggs) of the extraordinary Ibisbill at a BOC meeting. He had encouraged Samuel Whymper (1857–1941), the manager of a brewery in the United Provinces (India) and a keen naturalist (and big game hunter), to search for the birds. In the spring of 1906, Whymper found the birds breeding among boulders and shingle on the Bhaghirthi River (a headwater of the Ganges) at 2,500 metres altitude. He gave a brief announcement of his discovery in the *Journal of the Bombay Natural History Society* in 1906 (see Lowe, 1942). Dresser produced a fuller article, accompanied by a plate showing two of Whymper's Ibisbill eggs, together with eggs of other new and rare species he had received from Russia. The final article in the series of articles on rare eggs appeared in 1908; it covered a variety of Asian birds, mostly received from Dresser's Russian collaborators.

Dresser scarcely missed a meeting of the BOC during 1905–10. Meetings were held at Frascati's Restaurant until September 1907 and then at Pagani's Restaurant in Great Portland Street, 'a venue for Freemasons, Oddfellows Clubs and quasi-scientific societies' (Cole and Trobe, 2000: 132). Most of Dresser's discoveries were announced at the BOC meetings, with the more notable being followed up by articles in the *Ibis*. Some of these concerned particularly rare species, such as an egg of the rare Siberian Crane borrowed from Bianchi, while others extended the known breeding range of more familiar species eastwards in Siberia. Among the most interesting were eggs of the Spectacled Eider (a rare high-Arctic duck; see plate 55), received by Buturlin from Indigirka in north-east Siberia; they represented the first confirmed breeding record of the species in the Palaearctic (Dresser, 1908c). In February 1909, Dresser exhibited two skins of an extremely rare wader, the Asian Dowitcher, shot by hunters at Tara (Siberia) in May 1908. The birds had been given to a correspondent of Buturlin, Valentin Ushakov, who found a fully formed egg inside the female bird. The skins and egg were sent to Dresser on loan, the skins being sent to confirm the identification of the egg. The egg was damaged in transit, but was accompanied by a photograph taken when it was intact. Dresser exhibited the specimens at the British Ornithologists' Club and wrote an article on the discovery in the *Ibis* in 1909 (Dresser, 1909a,b), illustrated by a particularly beautiful plate by Keulemans (see plate 56).

Arguably the most remarkable discovery that Dresser reported from far-distant Siberia – possibly even eclipsing that of Ross's Gull – was the finding of the breeding grounds of the Slender-billed Curlew. Although this rare bird was known to winter on the Mediterranean, its breeding grounds were a complete mystery. Two of Buturlin's correspondents discovered the birds breeding in Siberia in 1909. P. A. Schastovski discovered the birds breeding at Lake Chany, 240 kilometres east of Omsk, on 25 May. Valentin Ushakov found the birds at Tara, 300 kilometres north of Omsk at the confluence of the Irtysh and Tara Rivers, breeding in bogs. Dresser exhibited one of two eggs collected by Schastovski, on loan from the Tomsk museum.[17] He received a single egg from Ushakov, which he exhibited at the British Ornithologists' Club (Dresser, 1910a). The bird is now possibly already extinct and the egg in Manchester Museum, from Ushakov, may be the only one in existence (see Gretton *et al.*, 2002; McGhie, 2002).

In addition to his exhibitions and articles, Dresser helped with the running of the BOU and the BOC. He was elected onto the Committee of the BOU in 1905, a select group made up of the President (Frederick Godman), Editors (Philip Sclater and Arthur Evans), Secretary (Howard Saunders) and three others (Anon., 1907a). Dresser dropped out of the Committee after three years, but audited the accounts of the BOC in 1908 and the BOU in 1910, showing his ongoing involvement in the running of these societies (Anon., 1908d; Anon., 1910c: 535).

Completion of *Eggs of the Birds of Europe*

Eggs of the Birds of Europe incorporated large amounts of novel information on the breeding distribution of birds in Asia, largely down to the help of Sergei Buturlin. Alfred Newton and Howard Saunders assisted Dresser by reading through the proofs for the book; Arthur Evans took on the task after both Newton and Saunders passed away in 1907. *Eggs* had been envisaged to take twenty parts; it had to be expanded to include new discoveries. Dresser issued a double part on waders, terns and gulls in February 1910 and an extra double part on gulls, skuas, auks and divers in November that year to complete the work. These birds were particularly desirable to collectors (see chapter 5) as they have particularly beautiful eggs, so Dresser was certainly keeping the best till last. Among the species covered in these last parts were the fabulous extinct Great Auk and Ross's Gull. An appendix covered species whose eggs had been discovered since Dresser commenced the work, including rare thrushes received from Buturlin and the enigmatic Slender-billed Curlew.

In the preface, Dresser explained that he had originally intended to cover bird eggs in *A History of the Birds of Europe* thirty years earlier; the three-colour photographic process, which he had first encountered in 1900 (see chapter 12), had inspired him to prepare *Eggs of the Birds of Europe*. He wrote that the book had been 'a labour of love' and that this was likely to be his last major work on birds. He dedicated the book to the memory of his greatest ornithological friend and mentor, Alfred Newton, 'in recognition of the great assistance received from him during the publication of this work, and in memory of a life-long friendship'. The book received excellent reviews in the *Ibis*, notably on the 'great mass of new information on the range of many species' provided by Buturlin, and the high quality of the plates (Anon., 1910b). The review of the final part noted:

Mr. Dresser concludes his admirable work on Palaearctic Oology, and he is to be congratulated on the successful termination of what he tells us has been a labour of love.... The plates are, as usual, excellent ... and we feel quite sorry to bid farewell to such a fine series of illustrations, accompanied as they are by Mr. Dresser's careful and accurate information. (Anon., 1911c: 383)

On the other side of the Atlantic, William Schufeldt wrote a glowing review in the *Auk*, noting that the plates of eggs 'were the most beautiful things of the kind I had ever seen, and, in fact, I had one or two of them framed for my study' (Schufeldt, 1912: 274).

The two rival books on eggs mentioned earlier, by Jourdain and Krause, were both ill-fated and were never completed. Dresser's *Eggs of the Birds of Europe* was by far the finest work of its kind for many years.

Dresser's egg collection goes to Manchester Museum

After *Eggs of the Birds of Europe* was completed, Dresser wasted no time in securing the future of the egg collection that formed the basis of the book. He gave the bulk of his egg collection – approaching 60,000 eggs – to his nephew (by marriage), George Malcolm (Cole, 2006: 120). As Dresser's main collection included 'select' and well-authenticated specimens, it stands to reason that the 'selecting' had to be done from among a vast number of specimens. Dresser approached Owens College in 1911, offering his main collection for £1,000.[18] Walter Tattersall, the Keeper of Manchester Museum, felt that the Museum could not afford the offer but he also wanted to acquire the collection.[19] The Museum Committee approached Mrs Katharine Thomasson, widow of the benefactor who had paid for Dresser's bird skins:

Dear Madam,

You will no doubt be aware that some years ago the late Mr. Thomasson purchased and presented to the Manchester Museum the fine collection of Palaearctic Bird Skins formed by Mr. Dresser of London. This was one of the most important and valuable gifts the Museum ever received; conferring great distinction on Manchester, and rendering signal service to the University in promoting and facilitating Ornithological study and research.

Mr. Dresser, who is now well advanced in years, has recently offered to the Museum Committee his further remarkable collection of the Eggs of the same group of Birds, together with his extensive Ornithological Library.... It is right I should point out that Mr. Dresser's offer is largely in the form of a gift. His main desire is that the collection should not be dispersed; still less that it should be allowed to leave the country and go to America. His hope – which we all heartily share – is that, if possible, it should find a permanent home in Manchester. In continuation with your late Husband's munificent gift it would form the finest and most complete collection of the kind in existence.[20]

Mrs Thomasson offered to contribute £250 towards the purchase and Tattersall sent a letter to the local newspapers appealing for public donations for the remaining £750.[21] The chairman of the Museum Committee wrote to Sir William Lever (of Lever Bros fame) to try to find the final funds required:

> It [the Museum] is visited every year by a considerable number of classes of children from schools in the district, and thus takes its share in the elementary education of the City. Finally, it is used for purposes of reference and study by the Naturalists and Collectors from all parts of South Lancashire, a district which for many years past has produced enthusiastic and successful students of nature. I venture to emphasise these points in order to show you that the Manchester Museum is performing a public service.[22]

By this time, Dresser's health was declining and he intended to go to the south coast of England or Europe for the winter. He wrote to Tattersall:

> I am still sorely troubled with severe headache and the continual noise of London seems to make it worse, and my doctor tells me that I must have a long rest and quiet or I shall break down. I must where I move either take my collection and library with me, or else store them, and before I decide what to do I would like to decide what has finally to become of them, or more especially of the egg collection, as it is as well not to move an egg collection more than it is quite necessary to do so.[23]

Dresser and Eleanor moved to a flat in Henley soon afterwards.[24] Tattersall drew up an agreement that the Museum would house the egg collection while they tried to raise funds money to buy it and would return it in the event the money could not be raised.[25] Dresser agreed to this arrangement.[26] An assistant was dispatched from the Museum to pack the collection in October 1911. It took him fifteen yards of cotton wool to pack just a portion of the eggs and he had to scour the local shops for more supplies.[27]

Tattersall's plan for raising funds by public appeal did not meet with any success: after the first contribution of £250 from Mrs Katharine Thomasson, he obtained £200 from the Museum's subscription fund (money set aside for new acquisitions).[28] Desperate for money to cover the cost of the collection, the Museum sold its copy of Audubon's *The Birds of America: From Original Drawings* (Audubon, 1827–38), now among the most valuable books in the world. They had asked for £500 but received only £310 (Alberti, 2009: 103, 117).[29] Katharine Thomasson came to the rescue once again in February 1913 and provided the remainder (£240) required for the purchase (see Anon., 1912).[30]

Conclusion

Eggs of the Birds of Europe was the crowning glory of Dresser's career, complementing *A History of the Birds of Europe* and his popular *Manual of Palaearctic Birds*. The *Eggs* book also allowed him to update information in his previous books. Through the project, Dresser had been part of the machinery that brought amazing discoveries to a scientific readership. His old friend Newton, who published the final part of his catalogue of the Cambridge egg collection in 1907, thanked Dresser in particular for his many donations of eggs of recently discovered birds (Newton, 1905–7: iii; see also Anon., 1931). Ornithological society was changing though, as key figures of earlier times passed on: Henry Tristram, a truly veteran ornithologist and one of the founders of the BOU, died in March 1906 at the age of eighty-three; Dresser wrote obituaries of both Tristram and Newton for the *Zoologist* (Dresser, 1906e, 1907c). The BM(NH) continued to consolidate its position, shifting the balance of ornithological power out of the hands of wealthy independent individuals to institutions and the professionals they employed. These changes are explored in the following chapter.

Notes

1   National Museums Scotland Library (NMSL hereafter), GB 587, JHB 15/240, letter from Henry Dresser (HED hereafter) to John A. Harvie-Brown (JAHB hereafter), 20 October 1905.

2   NMSL, GB 587, JHB 15/240, letter from HED to JAHB, 31 October 1905.

3   Manchester Museum (MM hereafter), ZDH/8/8/2, letter from HED to William Hoyle (WEH hereafter), 25 March 1908.

4   HED offered the house for rent to Laurence Waddell. University of Glasgow Special Collections, S Gen 1691/3/28, letter from HED to Laurence A. Waddell, 12 January 1906.

5   MM, ZDH/9.

6   Including William Blanford, Henry Feilden, Allan Hume, Howard Saunders, Ernest Shelley and Henry Tristram.

7   At almost the same time, someone (probably Sclater, as Editor) published a review in the *Ibis* of the 'International Rules of Zoological Nomenclature' (Anon., 1905b), with the disclaimer that zoologists should decide how they wanted to adopt these for themselves – a contrast to the aims of the Commission, of which Sclater had once been a part.

8   Frank Lemon, his wife Elsie Lemon and Edmund Meade-Waldo.

9   Ogilvie-Grant named a new babbler that Waddell had collected in Sikkim *Garrulax waddelli* at the BOC in 1894. Waddell donated approximately 1,600 bird specimens, some mammals and insects from the Sikkim area to the Hunterian Museum in Glasgow, where they still reside (Sclater, 1908).

10   MM, ZDH/8/7/1, letter from HED to WEH, 12 February 1905.

11   MM, ZDH/8/7/2, letter from HED to WEH, 18 May 1905.

12   MM, ZDH/8/7/2, letter from HED to WEH, 18 May 1905. See Vaurie (1972: 66–7) for discussion of Waddell's birds in relation to HED and Rothschild.

13   Natural History Museum Library and Archives, London, TM1/62/7/55, Tring Museum Correspondence: Correspondence with HED, 1902, letter from HED to Ernst Hartert, 20 January 1902.

14   Correspondence in the Museum of Local Lore, History and Economy, Ulyanovsk (MLLU hereafter) from Otto Ottosson (OO hereafter) reveals this. OO received at least two Ross's Gull chicks, three skins of adult gulls and eighteen eggs from

Sergei A. Buturlin (SAB hereafter); he gave five eggs to a dealer in England. Four skins and thirteen eggs are in the Swedish Museum of Natural History (Stockholm). MLLU, YKM 21937.13, letter from OO to SAB, undated (skins); MLLU, YKM 21937.14, letter from OO to SAB, 23 March 1907 (eggs); MLLU, YKM21937.15, letter from OO to SAB, 16 April 1907 (chicks).

15   Calculated as the purchasing power, using www.measuringworth.com, accessed 11 March 2017.

16   SAB sent HED handwritten notes, MM, ZDH/2/2.

17   The egg from Schastovski was evidently returned, after it had been photographed for Dresser's *Eggs of the Birds of Europe*.

18   HED had contemplated his collection going to Manchester Museum since at least 1902: see MM, ZDH/8/4/1, letter from HED to WEH, 28 January 1902. WEH was interested in the collection, but hoped that HED meant to bequeath it: see MM, ZDH/8/4/2, duplicate letter from WEH to HED, 30 January 1902.

19   MM, ZDH/8/10/1, letter from HED to Professor William Boyd Dawkins, 5 May 1911.

20   MM, ZDH/8/10/12, letter from Walter M. Tattersall (WMT hereafter) to Mrs Katharine L. Thomasson (KLT hereafter), 31 May 1911.

21   MM, ZDH/8/11/4, draft letter by WMT, 28 February 1912.

22   MM, ZDH/8/10/28–31, letter from Henry Plummer to Sir William Lever, 3 August 1911.

23   MM, ZDH/8/10/20, letter from HED (from 'Luxore', Shepperton) to WMT, 23 June 1911.

24   MM, ZDH/8/10/29, letter from Eleanor Walmisley Dresser (EWD hereafter) to WMT, 12 August 1911.

25   MM, ZDH/8/10/30, letter from WMT to EWD, 24 August 1911.

26   MM, ZDH/8/10/32, letter from HED to WMT, 26 August 1911.

27   MM, ZDH/8/10/43, letter from Wilfrid Jackson to WMT, 10 October 1911.

28   MM, ZDH/8/11/6, WMT to KLT, 29 February 1912.

29   MM, ZDH/8/11/16, WMT to KLT, 20 May 1912.

30   MM, ZDH/8/11/33, letter from WMT to KLT, 11 February 1913.

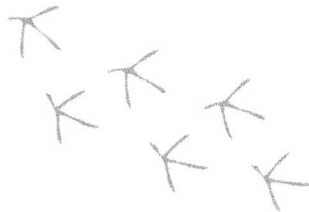

14 Time for a change

The fiftieth anniversary of the BOU

The fiftieth anniversary meeting of the BOU was planned to be held in Cambridge, birthplace of the Union, in November 1908 (Anon., 1907b: 478; Sclater, 1909: 62). Following Alfred Newton's death in June 1907, the event was moved to London, a clear sign of the shift in focus from a Cambridge-dominated scene.[1] Henry Dresser was on the organising committee, along with Walter Rothschild, Richard Sharpe and several others (Sclater, 1909: 63). The anniversary was held on 9 December at the Zoological Society of London's offices. Congratulatory telegrams were read and speeches were given on the progress of ornithology and the phenomenal growth of the BM(NH) (Anon., 1909a). Sclater gave an account of the contents of the *Ibis* over the years (Sclater, 1909) and gold medals were presented to the four remaining original members – 'our ancestors who are still with us' (as described by Henry Upcher, who handed out the medals; see Anon., 1909a: 8). A dinner at the Trocadero Restaurant in Piccadilly Circus was attended by eighty-one members of the Union, together with their guests (see figure 14.1). The original members occupied the top table along with a handful of others. Dresser sat at an adjacent table, between Henry Feilden and Hanbury Barclay, men he had known for forty-odd years; Richard Sharpe sat close by.

The Jubilee Supplement to the *Ibis* (the published proceedings of the fiftieth anniversary meeting) included forty biographical notices of the original members, contributors to the first series of six volumes and the Union's officials. Dresser's biography appeared among those of the officials as he had served as Secretary of the Union (Evans, 1909e). Only eleven of those featured in the biographical entries were still living. Richard Sharpe's biography was included, although he did not fit into any of the categories (another good indication of the changing times); the organisers were reluctant to introduce a new category especially for him. The biographies reveal a great deal about what mattered to the organisers of the anniversary celebrations and the Jubilee Supplement, and to those they sought to impress. We find Edward Newton 'in whose character were blended all the qualities that go to make the careful, truthful naturalist, and the refined Christian gentleman' (Evans, 1909c: 120). Colonel Henry

14.1 The BOU fiftieth anniversary dinner, held at the Trocadero Restaurant in Piccadilly Circus on 9 December 1908, from Henry Dresser's album of correspondents.

Drummond-Hay, first President of the Union, was 'a noble specimen of the true field-naturalist, as well as of the soldier and country gentleman, a keen observer of nature in every department' (Evans, 1909a: 77). The biographical notices said a lot about their subjects' family background (and social class), their education, prowess on the sports field and with a gun; they said comparatively little about ornithology. Dresser's biographical notice – probably written by Dresser himself – mostly covered his well-to-do family background, his many adventurous trips abroad and his lifelong passion for collecting birds and eggs (Evans, 1909e). Dresser was one of the longest-serving members of the BOU, and also one of the oldest (although William Tegetmeier, born in 1816, far exceeded him in years; see figure 14.2).

Much of the BOU anniversary focused on the time when the Union was most influential, earlier in the previous century. There was no mention of the growing rift from American ornithology or the loss of influence over scientific naming practices. Walter Rothschild and Ernst Hartert scarcely got a mention, nor did Henry Seebohm. The section on 'contributors to the first series of *The Ibis*' in the Jubilee Supplement was highly biased, omitting biographies on some foreign ornithologists who contributed far more to the *Ibis* than

14.2 William Tegetmeier 'aged 92 years can still write firmly owing to the absence of the poison known as alcohol', from Henry Dresser's album of correspondents. Photographed around the time of the BOU fiftieth anniversary.

some of the Brits honoured by biographical entries.[2] Bensaude-Vincent has emphasised that such events are 'ceremonies that help scientific communities enhance their social prestige' (1996: 493). Similarly, Johnson (2004: 518–19) emphasises that commemorative events say more about the prevailing concerns and narrative of their own time than of the period they are commemorating.

Growing divisions

By 1906 or so, British ornithology was dominated by at least five powerful factions. Firstly, there were Sclater and other 'old boys', deeply conservative and aligned with the BM(NH). Then there were Richard Sharpe and William Ogilvie-Grant of the BM(NH) itself. Thirdly, there were Rothschild and Hartert with their radical views (and Rothschild's independent means) at Tring. There were also, fourthly, a developing 'new breed', including Harry Witherby, the two Ticehurst brothers and Francis Jourdain; these were not new figures on the scene, but their influence was on the increase.[3] Finally, there was Dresser, who fell into a different category; he allied himself most closely with his foreign correspondents and the old Ibises, but maintained a steady independence.

The factions came into conflict over membership of elitist societies, publishing on birds, scientific naming practices and competition for specimens. The word of prominent figures such as Dresser could open doors; it could also slam them shut. Dresser nominated Sergei Buturlin as a Foreign Member of

14.3 Baron Harald Loudon, from Henry Dresser's album of correspondents.

the BOU in 1906 and he was successfully elected.[4] The following year Dresser arranged for Baron Harald Loudon (see figure 14.3) to be nominated as a Foreign Member.[5] There was a disagreement over this as Sharpe and Ogilvie-Grant had a candidate of their own in mind. Dresser wrote to Buturlin:

> Sharpe, Grant and the museum clique collected all their friends who are members of the BOU, and brought forward the plea that it was not right for the Committee to recommend anyone in particular as foreign member.... Of course this led to a row as it cast a slur on the Committee, and a long and hot discourse followed.... I am very disappointed for had I suspected that there would be an attack at the General Meeting I would have collected my friends, and could, I believe, have outvoted the Museum clique.[6]

Different factions each controlled particular publications: Sclater was still editing the *Ibis* along very traditional lines, William Ogilvie-Grant was Editor of the *Bulletin of the British Ornithologists' Club*, while Rothschild and Hartert had their own journal, *Novitates Zoologicae*. Harry Witherby issued the journal *British Birds* from June 1907 to document sightings of unusual birds and discuss topics such as migration and conservation; the journal quickly became very successful (Anon., 1957; Ogilvie *et al.*, 2007). Dresser continued to use a particularly old-fashioned publishing model – self-publishing books based (largely at least) on his own collection – but with great success.

In 1907, Buturlin sent Dresser an article on new species and subspecies of Siberian birds for the *Bulletin of the British Ornithologists' Club* (for a brief notice of the article's presentation at the Club see Dresser, 1907a). Ogilvie-Grant, the editor, returned the manuscript to Dresser as being too long.

Dresser wrote to Buturlin: 'He [Ogilvie-Grant] is I fear doing as the museum people so often do, giving full scope for his own papers and cutting short all others.'[7] After failing to get the article in the *Ibis*, the Zoological Society of London agreed to publish it (Buturlin, 1907a), but Buturlin published it in the German journal *Ornithologische Monatsberichte* instead (Buturlin, 1907b). Dresser arranged to present any subsequent papers by Buturlin at meetings of the Zoological Society of London, although he did continue to exhibit specimens received from Buturlin at the British Ornithologists' Club (e.g. Dresser, 1907b).[8] Dresser also checked over specimens of Pheasants in the BM(NH) on Buturlin's behalf: 'I send a feather from the side of the lower breast which was loose, and wish a scapular feather had been loose as I could have sent it. You had better not say anything about this as the British Museum authorities (or rather Mr. Ogilvie Grant in the bird room) are so very jealous.'[9] No doubt Ogilvie-Grant would have taken a dim view of even a single feather being removed from the Museum.

Involvement in bird conservation

Dresser maintained a steady involvement in bird conservation, complaining in the *Ibis* about excessive collecting of eggs of the Great Skua in Iceland, for example (Dresser, 1906d; see also Anon., 1906b). He continued to serve on the Council of the Royal Society for the Protection of Birds (the Society received the royal approval in 1904). The Society promoted nature study and education, advocated for legislation to protect birds and campaigned tirelessly against the plume trade (see Doughty, 1975). Horrendous numbers of birds were still being killed for their feathers, notably egrets (the long feathers from their backs were confusingly called 'osprey plumes'). For example, the Commercial Sale Rooms, in Mincing Lane in London (where Dresser had done business many years earlier – see chapter 1), sold feathers from 150,000 egrets and 41,000 birds-of-paradise in 1906 alone (Anon., 1906f). In the face of growing criticism, the feather trade argued that the feathers were artificial (as they were too inexpensive to come from real birds), that the trade supported the lives of poor people and that it enabled Britain to remain competitive in the world market (see Anon., 1906c,e).

In 1906 the RSPB Council gained the support of Queen Alexandra against the wearing of egret feathers, which appeared in an RSPB leaflet, *The Queen and Osprey Plumes*. In 1908, the Importation of Plumage Bill was developed by James Buckland and Lord Avebury and introduced into the House of Lords; a satirical cartoon showing Avebury as 'A Modern St. Francis' appeared in *Punch* (see figure 14.4). A Select Committee was formed to investigate the need for such a Bill, and consulted with experts including Richard Sharpe, scientific societies and the RSPB, as well as representatives of the plume trade. Various claims and counter-claims were made in the press. Dresser entered the fray on the side of the RSPB as 'an ornithologist of world-wide repute', giving

14.4 Lord Avebury as 'A modern St. Francis'.

A MODERN ST. FRANCIS.

evidence based on his experiences in egret breeding colonies in the United States, Bosnia and Croatia (Anon., 1908b: 46).

After the Bill failed to become law, the RSPB produced *Feathers and Facts* in April 1911 to set out its case, including Dresser's comments on egret colonies (Royal Society for the Protection of Birds, 1911). The feather trade industry also raised its game and issued *The Feather Trade: The Case for the Defence* (Downham, 1911). After many failed attempts, the Importation of Plumage (Prohibition) Act was passed in 1921.

Dresser was one of sixty-four(!) naturalists who signed their names to a letter to *The Times* in 1911, requesting that the military operations that took place in the New Forest be held outside the breeding season or in some other place, to preserve the rare birds of the Forest, while accepting the necessity of such manoeuvres 'on patriotic grounds'.[10]

Collecting and controversy

In 1906, the RSPB identified five main 'classes' of people who threatened birds: vandals, bird-catchers, feather traders and feather-wearing women, gamekeepers, and private collectors. It set out different strategies for dealing with each class. In relation to collectors, it said: 'When it is regarded as a shameful and despicable thing to kill and possess the stuffed remains, or the egg, of a vanishing species, the private collector will probably cease to exist' (Anon., 1906a: 2). After the formulation of this strategy, statements deriding collectors and collecting appeared with increasing frequency in the press, notably *The Times*, and in magazines (Anon., 1906d; Fortune, 1906). In February 1908, for example, the Watchers' Committee of the RSPB complained to *The Times* of collectors' 'ceaseless endeavour to obtain rare "British-taken" eggs and birds', and that the high prices paid for eggs induced others to break the law to steal eggs.[11] Shortly afterwards, the Editor of *The Times* commended the RSPB and drew attention to the increasing interest in birds and bird protection 'in many classes of society'.[12]

A couple of months later, at the BOU annual meeting, Harry Witherby proposed a regulation that any member of the Union who had knowingly collected any egg or bird of thirty-one particularly rare British species would be expelled from the Union (Anon., 1908d: 519–21). At the same meeting, Percy Bunyard – a particularly active egg collector who amassed huge series of eggs of common and rare species – was elected as an ordinary member of the BOU. [13] It seems likely that the Union was attempting to curtail the activities of Bunyard and other 'serial collectors', although the proposal was not passed at the meeting. Witherby's proposal was praised in *Bird Notes and News*, which stated that 'men who are doing their best to exterminate rare birds, and to evade the laws of the country, are unfit to be ranked as British naturalists' (Anon., 1908a). The BOU implemented the rule the following year, enabling it to expel any member 'who acted in a manner injurious to the good name of the Union' or who wilfully took rare birds or their eggs (Anon., 1909b: 532–3). It was subsequently known as 'the notorious Rule 7' (Mountfort, 1959: 16).

In 1910, Jamrach, the well-known London dealer in live animals, advertised wild Hooded Cranes (from Africa) for sale at £80 a pair, alive, adding 'these birds are now on the point of extinction'. William Henry Hudson, a devoted conservationist and RSPB committee member, wrote to *The Times* to protest at Jamrach's actions and also those of Walter Rothschild, who, Hudson wrote, had proposed that

> when a species is so rare as to justify us in thinking it near extinction, the only thing to do is to extinguish it altogether, as quickly as possible, so that collections shall not lack specimens. This is, indeed, the system followed by collectors of rare birds generally.... To others it appears a damnable proceeding.[14]

Jamrach responded that he had procured the birds 'at the request of the greatest ornithologists of the day, numbering more than 40 persons'.[15] If he thought this would appease Hudson, he was mistaken. Rothschild also wrote to *The Times* to correct Hudson's assertion; Rothschild explained that what he had said was that, when a bird species was on the verge of extinction, 'it would be better for our successors if these last specimens were preserved in museums', which was more or less what Hudson had complained of in the first place.[16] Hudson attacked Jamrach again:

He is only shooting them, not for gain, you will understand, but solely for the honour of the thing, to supply the specimens ordered by the 40 greatest ornithologists in this country!

Probably the '40 greatest' will not be delighted at these revelations; they will think them injudicious; but the recent final capture of the British Ornithological Union by the private collectors and exterminators of rare birds has made it pretty plain to every one that ornithologists and collectors are not distinguishable in this matter, that they are engaged in the same business…. Those same cranes … will quickly increase in value, and may eventually become as costly in the market as Great Auks and their eggs. As soon, that is, as some important member of the B.O.U. or the B.O.C. can get up to announce that the hooded crane (*Grus monachus*) may now be considered an extinct species.[17]

Hudson was clearly referring to Bunyard's election to the BOU as the 'recent final capture'; he resigned from the BOU and BOC in the same year (1908) as Bunyard was elected. A lengthy editorial on the morality of nature conservation, highly sympathetic to Hudson's point of view, appeared in the same edition of *The Times*.[18] Hudson had brought the hypocrisy of collecting and the morality of nature conservation into a very public arena. The ornithological establishment had to respond to such public criticism. Meade-Waldo, Ogilvie-Grant, Bonhote and some others wrote to *The Times* to point out that Hudson was mistaken and that the BOU was firmly committed to the protection of rare birds.[19]

While bird protection and collecting were being widely discussed in the press, the BOC was having to deal with conflicts between collecting and conservation among its own members. In 1909, Percy Bunyard exhibited a 'large series' of eggs of the Red-backed Shrike, a rare British bird with variably coloured eggs, and series of eggs of the Nightingale and the Garden Warbler, which hardly vary at all, at a BOC meeting (Bunyard, 1909). In October of the following year, he exhibited a clutch of eggs of the Ruff, collected in Britain, where it was an extremely rare breeding bird. The chairman invited a discussion on the matter – which had evidently been prepared in advance – calling firstly on Bonhote. He drew attention to 'Rule 7', passed two years earlier by the BOU: 'yet we now find ourselves looking on, if not with enthusiasm, at least with that silence which gives consent, at a display of clutches of eggs of several distinctly local birds and the exhibition of an extremely rare clutch of eggs.' Bonhote drew attention to Hudson's letter of 15 June in *The Times* 'decrying the present attitude of the Union and stating that it had become a society of exterminating collectors. Such remarks as this, which tend to injure our whole status, must be refuted in no uncertain manner, and if we continue to witness exhibitions such as the present without a protest we are certainly adding an appearance of truth to such remarks.' Bonhote put forward a resolution that the meeting 'strongly disapproves of the collecting and exhibiting of large series of clutches of eggs of British breeding birds' or of the eggs of rare breeding birds, except in order to demonstrate a new scientific fact. The *Bulletin* reported that 'after a somewhat animated discussion in which a number of Members took part, [the motion] was carried almost unanimously' (Bonhote, 1910). Dresser was among the audience; it is almost certain that he supported the motion, as it mirrored the views he had set forth in the preface to *Eggs of the Birds of Europe*.

The tide against collecting had turned so far that even museums were meeting criticism. When the RSPB held a reception at the BM(NH) in November 1911, Edmund Selous (a keen student of living birds) complained that museums were nothing but 'so many centres of destruction, all vying with each other to procure whatever is rare' and that anyone visiting the rows of stuffed carcasses could only conclude 'that slaughter is the one end and aim of natural history, or rather that slaughter <u>in excelsis</u> is natural history'. The RSPB demurred to agree with Selous, but the comment is significant (Anon., 1911a). Growing opposition to collecting and collectors is discussed in the following chapter.

Alfred Wallace and the Darwin anniversary of 1909

The year 1909 marked the centenary of Darwin's birth and the fiftieth anniversary of the publication of *On the Origin of Species by Means of Natural Selection*, one of the most influential books ever published. Dresser had a small involvement in the celebration of this anniversary through his friendship with Alfred Wallace, one of the leading naturalists of the nineteenth century (see figure 14.5). Wallace was idolised by the old-school ornithologists of the BOU as his

14.5 Alfred Wallace, from Henry Dresser's album of correspondents.

work on biogeography was clearly linked to the main traditions of empirical ornithology. Wallace and Dresser were friends for fifty-odd years and wrote to one another fairly regularly.[20] Dresser provided Wallace with information on the numbers of bird species in various parts of the world for his books (Wallace, 1876, vol. 1: xiv, 193, 233; Wallace, 1900, vol. 1: 237), and spoke with some of his business associates to try to find a job for Wallace's son Will.[21] Wallace wrote to his son of 'my old friend Mr. H. E. Dresser (Naturalist) connected with some large Iron Companies in the City, and who knows lots of people'.[22]

Wallace had agreed to give a lecture at the Royal Institution to commemorate the Darwin anniversaries and then, in quick succession, was awarded the prestigious Copley Medal of the Royal Society and the Order of Merit. He wrote to Dresser to ask him to find out who in the government had suggested him for the award, him being 'a red hot <u>Socialist</u>, <u>Land Nationaliser</u>, and <u>Anti-Vaccinationer!</u>'[23] Wallace's letters to Dresser reveal a highly spirited personality, with important phrases emphasised through lots of underlining and exclamation marks. The letter quoted above also give a sense of how well-connected Dresser was with politicians and society figures.[24] Wallace arranged to stay at Dresser's home on the evening of the Royal Institution lecture, arriving on the afternoon of 22 January 1909. Wallace, his daughter, Dresser and his wife Eleanor travelled together to the Institution in Mayfair. Wallace wore the Order of Merit round his neck and gave a lecture entitled 'The world of life: as visualised and interpreted by Darwinism'. The audience included many famous scientists and social figures, and was chaired by the Duke of Northumberland. Dresser and Eleanor had front-row seats, arranged by Wallace (Raby, 2001: 276–8; Slotten, 2004: 477–8; McGhie, 2009).

Dresser and Wallace's visit to the BM(NH)

At Wallace's request, Dresser arranged for the entomologist Edward Poulton, an Oxford-based expert on natural selection, to join them for breakfast on 23 January following his lecture at the Royal Institution (Sir William Preece, whom Dresser had previously approached about employment for Wallace's son, was to join them later that day for lunch). Wallace had also invited Dresser to join him in a visit to the BM(NH) 'to see the wonderful Carnegie beast "Diplodocus!" and a few Birds and Butterflies'.[25] Dresser and Wallace visited the Bird Room of the BM(NH); what happened there is a matter of some interest, as there are varying accounts of the incident, and some of these have been repeated in print (Mearns and Mearns, 1998: 137). Collingwood Ingram (1880–1981), a keen ornithologist and correspondent of Dresser, was present when Dresser and Wallace visited the Bird Room. He told the story of what happened in 1966, after all the main characters had passed away. Dresser's former partner, Richard Sharpe, died in 1909,[26] and was succeeded by William Ogilvie-Grant as head of the Bird Room. Ingram recalled how Ogilvie-Grant had, like Sharpe, an 'intense dislike of H. E. Dresser – a dislike he never made the slightest attempt to disguise'. Concerning the events in the Bird Room on 23 January, he wrote:

> On one occasion when I happened to be working at the same table as Dresser, Ogilvie Grant [sic] unexpectedly appeared from another part of the Bird Room. As he drew near Dresser rose from his chair and with a slight bow and an ingratiating smile politely wished him good afternoon. Ogilvie Grant stopped dead, inserted a monocle in his eye and, staring straight at Dresser and without making the slightest attempt to acknowledge his greeting, called out in a loud voice 'Jimmy'. The said Jimmy was a man named Wells whose duty it was to supply and replace the bird skins required for study by visiting ornithologists. On emerging from behind a cabinet, Wells respectfully enquired if Ogilvie Grant required anything.
>
> 'No, I only wanted to know you were about', and with that Ogilvie Grant let his monocle drop, turned on his heels, and marched off.
>
> I felt sorry for Dresser; to be cut dead and coldly glared at through a monocle was humiliating enough, but how much more galling it must have been to have it imputed, as was done by the calling of Wells, that you were not to be trusted with the specimens you were examining? (Ingram, 1966: 163)

Evidently a complaint about this was made to the Keeper of Zoology of the BM(NH), Sidney Harmer. Unfortunately not all of the details are known, but Ogilvie-Grant's response to the allegations still exists in the archives of the Natural History Museum:

> On Saturday morning, 23rd Jan, about 11 o.c. [o'clock] I received a message to say that Mr. Dresser was in the Bird Room with a Mr. Wallis [sic] and would like to see me. I sent back a perfectly polite message to say that I was engaged, but that my attendant, Wells, would look after him and show him the Birds-of-Paradise, which he had asked to see. Subsequently I heard Mr. Walter Rothschild's voice in the Bird Room, and when Wells returned to ask me some question I learnt that Mr. Dresser's companion was Dr. A. R. Wallace not a Mr. Wallis who belongs to the Brit. Orn. Union. I would have gone out to Dr. Wallace, but as Mr. Rothschild was engaged in showing him the different species of Birds-of-Paradise it seemed useless for me to do so too, especially as I was very busy with other matters. That is all that happened. I neither saw nor spoke to Mr. Dresser. I am of course unaware who the person is who accused me of having been rude to Mr. Dresser. I can only assure you that there is no truth whatever in the statement.[27]

Whatever the truth of all this, Ingram was prepared to go into print on the matter – and in Dresser's defence – many years after the event.

The Fifth International Ornithological Congress

The Fifth International Ornithological Congress was held in Berlin from 30 May to 4 June 1910 at Berlin Zoo. Dresser attended as the representative of both the RSPB and the BOU, accompanied by his wife Eleanor. The report in *Bird Notes and News* tells how Dresser was among the early friends of the Society.... He joined the Council in 1898, and is a constant attendant at its meetings and at those of the Committees. No other British ornithologist, it is safe to say, is held in higher honour and esteem on the Continent, and few have probably so intimate a knowledge of foreign countries and languages. (Anon., 1910a)

Anton Reichenow was President and Dresser was elected as one of the eight Vice-Presidents. Dresser's *Birds of Europe* came in for praise from the President and his name was mentioned in connection with some of the discoveries being made in Siberia by his Russian collaborators. In comparison with previous years, articles on bird migration and bird protection were on the increase (Schalow, 1911). Dresser subsequently wrote the report on the Congress that appeared in the *Ibis* (Dresser, 1910b).

Competing lists of names of British birds

Despite his earlier protestations, *Eggs of the Birds of Europe* was not to be Dresser's last project on birds. In May 1911 the BOU Committee agreed to produce a new edition of the Union's *List of British Birds*. A committee for its development was set up that included the BOU President (Frederick Godman), Editors (Philip Sclater and Arthur Evans), Secretary (Lewis Bonhote), William Ogilvie-Grant, Dresser, William Eagle Clarke and Norman Ticehurst (Anon., 1911d: 555; British Ornithologists' Union, 1915: vii).[28] At its first meeting, the committee agreed to fall in line with the International Rules of Zoological Nomenclature; trinomials were to be used for subspecies (but not for the 'main' form of a species, so there would be no *Pica pica pica*).

At the same time as the BOU list was being prepared, ornithologists associated with the journal *British Birds* began their own project on the same subject. Jourdain had followed the approach to scientific naming taken by Hartert in *Die Vögel der Paläarktischen Fauna* (1903–22) – adhering rigorously to the oldest scientific name used for each species (the law of priority) and using trinomial scientific names to denote subspecies – in his book on eggs (Jourdain, 1906–12: iii), and used the same approach in *British Birds*. Witherby was less of a puritan and favoured a flexible approach to the law of priority to avoid making great changes in naming (Witherby, 1910). This approach elicited strong criticism from Hartert, who called for 'strictest priority in nomenclature' (Hartert, 1910). The list was published in 1912 as *A Hand-list of British Birds*, with revolutionary alterations to the names of many birds (Hartert *et al.*, 1912).[29] The BOU list was published in 1915 as *A List of British Birds* (British Ornithologists' Union, 1915).

In a review of the *Hand-list*, Sclater again bemoaned the alterations of well-known names (Sclater, 1913: 117). He pointed out that the names of over 200 species on the 'British list' were different from those used in the first BOU list of 1883. The BOU list committee took a more conservative standpoint:

while it generally stuck to the rule of absolute priority, it refused to change some well-established names for older ones, in a purely subjective fashion, or to transfer well-known names between species (as being too confusing), on the grounds of common usage (British Ornithologists' Union, 1915: ix–x). This idiosyncratic approach drew criticism from the authors of the *Hand-list*, who wrote that only the International Zoological Congress could introduce new rules for scientific naming (Hartert *et al.*, 1915). This was another reminder that British ornithology was no longer in a position to govern itself.

Closing years

The first decade of the new century had been a particularly tempestuous one for ornithology and ornithologists – and the old guard of British ornithology in particular. The BOU, notably its core members, no longer held sway over ornithology in the way it had in the nineteenth century, and the key figures of the mid–late nineteenth century were rapidly diminishing in number. Daniel Elliot wrote elegiacally of these figures in his obituary of Philip Sclater in 1914:

> The number of eminent naturalists present on those evenings [at the Zoological Society of London] was marvellous, and no such body of celebrated men, all members of one Zoological Society had ever before been assembled together, and we may believe it will be a long time before one equal to it will be again seen, for it was the height of zoological activity in the world, when indeed there were giants in the land. That was a glorious company of eminent men, broad-minded, and far-seeing…. And where now are those brilliant souls! … but two remain with us today, Dresser and Godman long passed the number of years allotted to men upon the earth. The survivors of those meetings stand like lone columns, erect, lifting their heads aloft on a wide deserted plain, surrounded on every side by ruins and when in thought I sweep aside the intervening years and stand again in that once crowded room, and look on row upon row of vacant chairs, which now no man can fill, the heart yearns with a fervent longing for
>
> A touch of a vanished hand
And the sound of a voice that is still.
>
> (Elliott, 1914: 5–7)

Dresser's final years

Dresser withdrew from business in 1911 and gave up his business premises in Cannon Street the following year (see figure 14.6).[30] He and Eleanor moved around a great deal, spending time in Naples, Bournemouth and Maidenhead. In 1914 they lived at the Villa de la Gaité, Boulevard Bru, in Algiers (as recorded in the BOC's list of members) then moved to Cannes, to the Villa Marie-Louise, a *pension* in the Rue des Vosges. During a visit to nearby Monte Carlo in 1915, Dresser died of a heart attack at the Hotel du Littoral, Boulevard

14.6 Henry Dresser, aged around seventy.

des Moulins, at 10 p.m. on 28 November, when he was aged seventy-seven.[31] His funeral was held on 16 December at Christ Church, the oldest Anglican Church in Cannes.[32] He was buried in the Cimetière du Grand Jas (Grand Jas Cemetery) in the 'English Cemetery', among other ex-pats. Eleanor arranged for Dresser's grave to be marked by a simple white marble monument, a slanting slab carved with a Maltese cross and laurel leaves, with the inscription:

> Sacred to the
> Beloved Memory of Henry Eeles Dresser
> ornithologist and author
> who entered the Presence of his Lord
> Nov. 28 1915
> Most dearly loved,
> Deeply mourned
> 'Je ne mettrai pas
> dehors celui qui viendra á Moi'
> 'Praise the Lord, oh my soul, and all that is within me, bless His Holy Name'

Eleanor wrote to Tattersall at Manchester Museum that she kept a laurel wreath with the grave 'to honor his scientific work'. Both she and presumably her late husband wanted his ornithological work to be commemorated; no mention was made of his work in the iron trade, although that was another major aspect of his life.

Dresser's will, dated 10 May 1915, was read in the Court of Probate on 10 June 1916. His estate was valued at £16,849 gross and £12,709 net, equating him with a millionaire by modern standards.[33] Eleanor moved to Bournemouth shortly afterwards but she suffered a further tragedy when her only son was killed in the Battle of the Somme, aged only thirty-seven.

Obituaries

Obituaries to Dresser appeared in numerous bird journals in the UK and abroad, including the United States, Russia, Hungary and Australia. Most of these drew attention to his great publications, to his travelling life, his collecting and involvement in bird conservation. His old-fashioned views about scientific naming practices were also singled out in a number of his obituaries. The following are extracts from some of them (see the Bibliography of Henry Dresser for details of additional obituaries):

Anonymous obituary in the *Ibis*

> Dresser belonged to the old order of systematic ornithologists who did not believe in subspecies or trinomials, and his views on the limits of specific variation and nomenclature would not perhaps commend themselves to present-day workers. All he wrote, however, was marked by a thorough and a rigid accuracy of description and attention to detail, and he took special pains to get his illustrations executed

and reproduced in the most perfect manner possible, so that his monographs and the 'Birds of Europe' were as monuments of ornithological literature. His death is a great loss to us all, and removes one more link in the chain connecting us with the giants of the middle of the nineteenth century. (Anon., 1916b)

Obituary in *Nasha Okhota* [*Our Hunting*]
by Sergei Buturlin (translated from Russian)

I first met him in 1904 when he paid me a visit to Wezenberg on his way back home from Finland to London.... He spent some time with me and we undertook an excursion to islands of the Finnish Bay.... Even at that time, when he was only 67 years old, Dresser had problems with one of his legs, which did not work. Despite this he was enthusiastic and jovial, never dropping behind us, people who were almost half his age. This personal acquaintance left me with the best and the warmest memory, and now, with a great grief, I have found out about the death of this beautiful person, the experienced observer and energetic explorer of nature. However, all his works, especially the 'History' and 'Eggs', will make him remembered for a long time if not forever by all bird-lovers. (Buturlin, 1916a)

Obituary in *Messager Ornithologique*
by Sergei Buturlin (translated from Russian)

Henry Dresser, who was one of the best-known European ornithologists, passed away in his 78th year. According to English standards he was not too old, but he had been unable to work during his last years due to the effects of the yellow fever he had some time ago and of the wounds he got in the 60s during the war against the northern US states. Incidentally, Dresser was completely scalped by one of the frontier Mexican Indian tribes. (Buturlin, 1916b)

A number of inaccuracies creep into Dresser's obituaries: Walter Rothschild (who wrote an obituary for the journal *British Birds*) asserted that Dresser could speak Russian, which is incorrect (Rothschild, 1916). Buturlin's obituary in *Messager Ornithologique* also includes several inaccuracies: Dresser was not scalped in Texas but lost his hair to alopecia; nor did he contract yellow fever, although he was certainly worried about catching it when he was in Texas. Perhaps Dresser told his friend a more colourful story.

Buturlin and Dresser had spent time together in 1904. In the Museum of Local Lore, History and Economy in Ulyanovsk there are some roughly written notes by Buturlin, consisting of many humorous anecdotes that Dresser had told him.[34] These appear to have been hurriedly written down either on the spot or at the end of the day, in Russian and English. They tell how, on building a new house at the age of eighty-two, Alfred Wallace told Dresser 'well, I must now make provisions for my old days', and how Severtzov 'drank a bottle of most awful brandy and was not drunk ... he ate an entire leg of lamb in [Howard] Saunders', who and his daughter were thus left without a dinner'. Buturlin's notes conclude 'today he [Dresser] fell down twice.

Speaking too much.' Of all biographical notices and obituaries, here we may find a clearer insight into Dresser's character, like Philip Manson-Bahr's recollections of Dresser with his 'demoniacal energy, boundless enthusiasm and immense application' referred to in the Introduction, truly 'quite a character in an age of individualism'.

Notes

1   Henry Dresser (HED hereafter) wrote an obituary of Alfred Newton in the *Zoologist*, referring to Newton as 'the most constant and truest friend it has been my good fortune to possess' (Dresser, 1907c).
2   These included Theodor von Heuglin, Daniel Elliot, Johann Blasius, Gustav Hartlaub, Julius von Haast, Jules Verreaux and Charles Abbott.
3   Witherby became Secretary and Treasurer of the BOC in 1905, for example.
4   Museum of Local Lore, History and Economy, in Ulyanovsk (MLLU hereafter), ZKM27159/14, 15, letters from HED to Sergei A. Buturlin (SAB hereafter), 3 May and 13 June 1906.
5   MLLU, ZKM27159/21, letter from HED to SAB, 25 May 1907.
6   MLLU, ZKM27159/22, letter from HED to SAB, 12 June 1907.
7   MLLU, ZKM27159/62, letter from HED to SAB, 31 January 1907.
8   MLLU, ZKM27159/18, letter from HED to SAB, 17 April 1907; MLLU, ZKM27159/19, letter from HED to SAB, 23 April 1907; MLLU, ZKM27159/20, letter from HED to SAB, 26 April 1907.
9   MLLU, ZKM27159/31, letter from HED to SAB, 27 April 1907.
10 S. F. Harmer *et al.*, 'The New Forest: an appeal from naturalists to the Editor of the Times', *The Times*, 2 June 1911, p. 6, col. d.
11 M. Sharpe *et al.*, 'The preservation of British birds', *The Times*, 15 February 1908, p. 15, col. f.
12 Editorial, 'The protection of wild birds', *The Times*, 2 March 1908, p. 9, col. e.
13 Bunyard had been introduced to the BOC by HED, at the 21 November 1906 meeting. Bunyard exhibited the nest, eggs and parent birds of the Grey-headed Wagtail, supposedly collected in Sussex. These were part of the Hastings Rarities fraud and had been collected by George Bristow. See Bunyard (1906).
14 W. H. Hudson, 'Proposed extermination of a crane', *The Times*, 7 June 1910, p. 13, col. c.
15 W. Jamrach, 'Extermination of cranes', *The Times*, 9 June 1910, p. 13, col. d.
16 W. Rothschild, 'Extermination of cranes', *The Times*, 14 June 1910, p. 13, col. c.
17 W. H. Hudson, 'Extermination of cranes', *The Times*, 15 June 1910, p.8, col. f.
18 Editorial, 'The extinction of animals', *The Times*, 15 June 1910, p. 11, col. f and p. 12, col. a.
19 E. G. B. Meade-Waldo *et al.*, 'Extermination of cranes – Mr. Hudson's mistake', *The Times*, 27 June 1910, p. 12, col. f.
20 Letters from HED to Alfred R. Wallace (ARW hereafter) are still in existence in the Natural History Museum Library and Archives, London (NHM hereafter), WP1/1, ARW Family Papers (ARWFP hereafter): Correspondence, and in John Rylands Library (Manchester), JRL-EM 1404, HED's album of portraits and autograph letters from ornithologists.
21 These included Sir William Preece, formerly the engineer-in-chief of the Post

Office – See NHM, WP1/1/89, ARWFP: Correspondence, letter from HED to ARW, 1 March 1904 – and Sir Oliver Lodge, a prominent physicist and telegraphist – see NHM, WP1/1/126, ARWFP, letter from ARW to William Wallace (WW hereafter), 13 June 1907.

22 NHM, WP1/1/73, ARWFP: Correspondence, letter from ARW to WW, 30 March 1903.

23 John Rylands Library JRL-EM 1404, ARW to HED, 18 November 1908.

24 These letters, nine in total, were acquired by John Rylands Library and Manchester Museum (MM hereafter) in 2008.

25 John Rylands Library (Manchester), JRL-EM 1404, ARW to Eleanor Walmisley Dresser, 12 January 1909.

26 HED sent a condolences card to Sharpe's family: McGill University Library, Rare Books and Special Collections, Blacker-Wood Autograph Letter Collection, postcard from HED to Miss Sharpe, 27 December 1909.

27 NHM, DF230/23/84, Department of Zoology: Bird Section Correspondence, Letterbook A–Z 1898, memorandum from William R. Ogilvie-Grant to Sidney Harmer, 5 February 1909.

28 Membership of the committee changed, as Frederick Godman was replaced by Robert Wardlaw-Ramsay as President from 1912; Philip Sclater was replaced by William Sclater (his son) as one of the Editors in 1912; Lewis Bonhote was replaced by Edward Stuart Baker as Secretary in 1912.

29 It was prefaced with the words 'Nomenclature is only "a means, not an end," but without uniformity it is a confusion.' Note that Norman Ticehurst was involved in the development of both this and the BOU list.

30 MM, ZDH/8/11/17, letter from EWD to Walter M. Tattersall, 19 June 1912.

31 Information on the place of HED's death comes from his death certificate, a copy of which was kindly provided by the Mairie de Monaco Etat Civil.

32 London Metropolitan Archives, CLC/340 MS 23,611, 'Christ Church Cannes Register of Baptisms and Burials 1860–1933', p. 34, no. 267.

33 Calculated as the purchasing power, using www.measuringworth.com, accessed 8 April 2017.

34 Undated note in the MLLU, Buturlin archive, YKM 17589.

15 Legacies

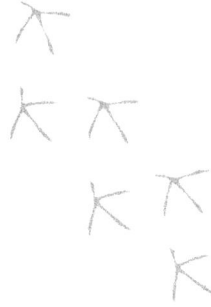

This chapter discusses Dresser's legacies and explores some of the reasons why he and his contemporaries fell from sight, as ornithology shifted from individuals to institutions, and from the study of specimens in collections to studies of living birds.

A History of the Birds of Europe

A History of the Birds of Europe was always Dresser's most famous and influential work on birds. In his biography of Joseph Wolf (who illustrated some of the plates for the book), Alfred Palmer wrote of the relief to be gained from sinking into a chair with a volume of the *Birds of Europe*, describing it as 'one of the most fascinating, comfortable, and useful books a lover of birds can covet' (Palmer, 1895: 114–15). In *Seventy Years of Birdwatching* Horace ('HG') Alexander, a prominent figure in British ornithological society in the mid-twentieth century, described how Dresser's book inspired him and his brother Wilfred when they were boys:

> I took down one volume after another (my hands had to be examined by an adult to see that they were clean each time I took out a volume). I read out the list [of bird species] from Dresser and Wilfred wrote it down.... I laboriously went through each big volume, admiring every picture in turn.

Harry Witherby had also been influenced by the book. He had been in the habit of using the 'Dresser numbers' (a number given by Dresser to identify each species) as a kind of shorthand (see Alexander, 1974: 16).

The book had a lasting importance and was referred to frequently in *British Birds*, especially in articles on identification, habitat, habits and distribution of European birds. The plates were used to identify rare vagrants to Britain until a surprisingly late date. These included the first occurrences of the Siberian Pechora Pipit and Asiatic Isabelline Shrike in Britain (Witherby, 1926, see plate 57; Flower and Kinnear, 1951) and a pair of Mediterranean Rüppell's Warbler supposedly shot in Sussex (Witherby, 1914; Nicholson and Ferguson-Lees, 1962).[1] When a pair of (supposed) Moustached Warblers were found breeding in Cambridgeshire in 1946, the appropriate volume of *A History of the Birds of Europe* was taken into the field(!) along with study skins, to check the identification of the birds (Hinde and Thom, 1947). Other ornithologists examined the skins on which the book was based: a Leach's Petrel (a seabird) blown inland in Lancashire in 1927 and shot was identified using specimens from Dresser's collection, for example (the bird was deposited in Manchester Museum) (Boyd, 1917).

Dresser's *Manual of Palaearctic Birds* also had a long shelf-life, and was used to identify and age the first River Warbler found in Britain, in 1962 (Davis, 1962).

Although *A History of the Birds of Europe* continued to be referred to in *British Birds*, the same could not be said for the more 'scientific' journals – the *Ibis* and *Proceedings of the Zoological Society of London* – where Dresser scarcely got a mention: the scientific study of birds was changing.

Rise and fall of egg collecting

One of the biggest differences between nineteenth-century and twenty-first-century 'ornithology' is the place of collecting and collections. The number of taxidermists in Britain declined progressively from about 1880 (Morris, 2010). The numbers of natural history auctions peaked in the early twentieth century with, rather surprisingly, more collections sold yearly at auction during 1915–30 than 1850–80.[2] Egg collecting, which is often thought of as a nineteenth-century phenomenon, in fact peaked in popularity and activity during the first half of the twentieth century, as increasing numbers of people maintained collections of bird eggs, often made up of series of eggs of each species rather than single or small numbers of eggs. During this period, most egg collectors in Britain obtained at least some of their specimens from any of a small number of commercial egg dealers, notably Herbert Marsden, Watkins and Doncaster, and Charles Jefferys. Charles Gowland came to dominate the British egg-dealing scene through the first half of the century and was practically a household name (Cole, 2006). Dealers bought up older collections to resell and paid others to collect eggs for them. A small magazine, the *Oologists' Exchange and Mart*, was launched in 1919 as many collectors preferred to trade directly with one another. It consisted of subscribers' names and addresses alongside lists of what specimens they had to offer and what they were in search of (their desiderata). The *Oologists' Record*, first issued in 1921, included longer articles on eggs and collecting (Cole, 2006: viii–x). The cover featured an illustration of a Red Kite at its nest (see figure 15.1), a particularly scandalous choice of subject: the Red Kite had been almost exterminated by gamekeepers and the last handful of pairs were harried relentlessly by egg collectors (Lovegrove, 1990).

At the same time that egg collecting rose in popularity, the relationship between collecting and conservation was becoming increasingly untenable.

THE OOLOGISTS' RECORD

A Quarterly Magazine devoted to the advancement of Oology in all parts of the World.

EDITED BY KENNETH L. SKINNER.

VOLUME I - - - - - - - - - 1921.

HARRISON & SONS, Limited, Printers in Ordinary to His Majesty,
45, St. Martin's Lane, London, W.C. 2.

15.1 The cover of *The Oologists' Record* (1921), showing an illustration of the rare Red Kite and its nest.

In 1914 Hector Hugh Munro, better known as Saki, published *The Forbidden Buzzards*, a story concerning a well-heeled egg collector who tried to take the eggs of the only pair of Rough-legged Buzzards in Britain; Dresser's *Birds of Europe* even gets a mention.[3] The tale is typically sardonic: after the hero of the story foils the collector's attempts, 'The buzzards successfully reared two young ones, which were shot by a local hairdresser' (country hairdressers often doubled up as taxidermists). In 1917, Herbert Robinson wrote in the *Ibis* that he had found Green Sandpipers breeding in Westmorland (now part of Cumbria) for the first time (Robinson, 1917). The editors added a footnote that 'until eggs and parents are taken and identified, we feel that we must regard the breeding of the Green Sandpiper in Great Britain as unproven', in old-fashioned 'what's hit is history, what's missed is mystery' style. Robinson wrote to the *Ibis* again, to point out that the editors were practically encouraging him (or anyone else) to break the BOU's own 'Rule 7' established to protect rare birds, and that he could (rightly in his view) be expelled (Robinson, 1918). The letter went unanswered.

The more 'oological' members of the BOU formed an 'Oological Section' of the Union in 1911; this held 'Oological Dinners' from 1915 onwards that featured exhibitions of large numbers of eggs (Anon., 1916a). The earlier meetings and exhibitions were reported in the *Ibis* as little more than enormous lists of eggs (they were reported in the *Bulletin of the British Ornithologists' Club* from 1922). Exhibitions included eggs of many species that were rare (or at least scarce) in Britain – Choughs (a cliff-nesting species of crow), Crossbills, Green-shanks, Peregrines – although it was not always clear (perhaps deliberately) when these had been collected (see plates 22, 38 and 58). In 1922 Earl Buxton, Honorary Treasurer of the RSPB, spoke at the Guildhall on the 'craze for egg-collecting', which was on the increase and which flouted the law openly.[4] He gave the details of the previous year's Ninth Oological Dinner as an example, at which Edgar Chance had exhibited forty clutches of the rare Red-backed Shrike, taken in one year (Anon., 1921). Buxton took the BOU to task on the matter, asking for assurances that it supported bird protection. The Honorary Secretary, Edward Stuart Baker, replied to Buxton's request, to the extent that the Union encouraged bird protection and that it was 'their ambition to limit the collecting of eggs to the taking of such as are required in the interests of Science'. Unconvinced at this, Buxton sought further assurances that the Union deprecated the action of any member who broke the law. Stuart Baker replied that the Oological Dinners were not those of the British Ornithologists' Club (a subset of the BOU), being organised by a separate committee (which was also a subset of the BOU); Clifford Borrer (Secretary of the Oological Dinner Committee) wrote to Earl Buxton to say that the large numbers of eggs exhibited had been collected over many years. When Buxton pointed out that the forty shrikes' nests were categorically stated to have been collected in one year, Borrer replied that this was to settle a 'definite scientific problem'. The oft-voiced appeal to science as a justification for collecting was questioned by the prominent ornithologist William Pycraft (1922) in the *Illustrated London News*, in an article on 'science falsely so called'. Pycraft asked whether a single useful fact had been gathered as a result of 'this orgy of collecting'.

Concerned that the BOU's reputation was being damaged, Henry Elwes, the Union's President, and Edward Stuart Baker wrote to *The Times*. They asserted that the Oological Dinners were not the Union's responsibility (but in fact the dinners were organised by a group of members of the Union), that the Union supported bird protection, and that it supported collecting only for 'the interests of science' and within the law.[5] At a meeting of the British Ornithologists' Club, William Sclater (the chairman) announced that the committee of the BOC had decided not to publish details of the Oological Dinners in its *Bulletin*, as to do so would run counter to the assertions they had made to Buxton (Anon., 1922c).

In spite of the efforts of the BOU and the BOC committee to manage the situation, their rather ambiguous position on collecting was censured in a leader in *The Times*:

> Science is a big word with which to defend the robbing of nests in Great Britain.
> Many ornithologists incline to the view that collections of eggs minister more to

the ambitions of dilettanti than to the advancement of zoology.... We prefer the songs of the birds, and trust that Lord Buxton will not be too readily satisfied with the reply [from the BOU] to his protest.[6]

Some members of the Oological Section broke away to form the British Oological Association, which operated much as the Section had done; Jourdain and Rothschild were leading members. Only Percy Bunyard continued to exhibit large numbers of eggs at BOC meetings (Cole and Trobe, 2000: ix).

Egg collecting was in the spotlight again in 1925, when Edgar Chance (mentioned above in relation to Red-backed Shrike eggs) was found guilty of collecting eggs of the protected Crossbill in Norfolk. The judge castigated Chance, who, as a well-known ornithologist (he was famous for a film and book about the parasitic egg-laying behaviour of Cuckoos), should acquaint himself with the law. Disgraced, Chance offered his resignation to the BOU, which was accepted (Anon., 1926a,b; see also Cole and Trobe, 2000: 35–8). Chance continued to advocate for the value of egg collecting, publishing two booklets on the subject: *Some Observations on Egg Collecting and Birds Protection* (1937) and *An Egg Collector Replies to His Critics* (1938).

Max Nicholson (1904–2003) – a keen ornithologist and able writer – called for more effective bird protection in his book *Birds in England*, published in 1926 (when he was only twenty-two years of age and still a student at the University of Oxford). Collecting came in for a very thorough bashing and Nicholson wrote of the 'leprosy of collecting', a particularly hard-hitting phrase (Nicholson, 1926: 172). Nicholson spared Dresser and Henry Tristram from criticism

> because these men worked on ground where much remained to be done and their published results made a scientific justification for their somewhat wholesale methods. These were true scientists, but for every one of them there have been 200 or more unprofitable squanderers of the wealth of bird-life. (Nicholson, 1926: 250)

The next spat came in 1934, when there was a lengthy correspondence in *The Field* about egg collecting. Those in favour of collecting argued that game-keepers were a worse threat than egg collectors. Through some particularly convoluted reasoning they claimed collectors would help rare birds to persist, by creating a market for their eggs(!) and, in any case, birds would keep relaying eggs after they were taken (which is untrue). Edgar Chance and Francis Jourdain both entered the debate on the side of collecting.[7] One point that was made abundantly clear was that many collectors considered egg collecting to be as much a field sport as a field of scientific enquiry. They were motivated by the challenging search for well-hidden nests (e.g. Meiklejohn, 1922). Rev. James Rashleigh Hale, a Vice-President of the British Oological Association and former committee member of the BOU, wrote: 'egg-collecting is one of the finest sports I know, and I have tried to do most things.... Sir, for sport, give me egg-collecting' (see Parker, 1935: 97).

The pro-egg collecting lobby lost out and the Bird Protection Act was passed in 1954, making it illegal to take, destroy, sell or exchange the eggs of birds that had bred in Britain, or to kill wild birds with the exception of game and pest species. Some of the larger dealers continued to operate, and some (including the leading firm of Charles Gowland) were successfully prosecuted. Egg collecting was no longer a socially acceptable pastime, fulfilling the RSPB's aim of fifty years earlier. Egg collecting dwindled after the 1954 Act but still continued as a popular hobby, even if it was driven underground. Small numbers of fanatical egg collectors continue to harass rare breeding birds in Britain.[8] Egg collecting declined so far in acceptability that egg collectors were reported to have a particularly difficult time in jail: an RSPB investigation officer was quoted as saying 'It's almost seen as paedophilia; it's a crime committed by weirdos in the eyes of the "normal" criminal set.'[9]

From the study to the field, from specimens to hypotheses

The study of eggs in collections – both scientific and pseudoscientific – declined in Britain from the 1920s, having been sidelined by the BOU and the BOC. Studies based on bird skins continued to form the basis of many articles in the *Ibis*, mostly on geographical variation and distribution of species. The study of living birds, on the other hand, grew to greater and greater prominence through the twentieth century. Early examples included the writings of Edmund Selous (e.g. Selous, 1899), Julian Huxley's studies of the courtship behaviour of Redshanks and Great Crested Grebes (Huxley, 1912, 1914), and Elliot Howard's studies of bird territoriality and song (Howard, 1920). Although behavioural and ecological studies multiplied, they were not readily adopted by the leading scientific societies or into their journals. Bernard Tucker (1901–50) established the Oxford Ornithological Society in 1921 while he was still a zoology student (Jourdain was the first President) and, with the aforementioned Max Nicholson, the Oxford Bird Census in 1927. The Census used networks of observers across the country to answer questions about bird distribution; a national census of heronries in 1928 was an early success. The journal *British Birds* was important in communicating the work of the Census, fulfilling the aims Harry Witherby held when he developed the journal. Also at Oxford, Charles Elton (1900–91) established the Bureau of Animal Population in 1932 to study the causes of fluctuations in animal numbers; the British Trust for Ornithology was set up the following year, and the Edward Grey Institute of Field Ornithology (the EGI) in 1938. Around the same time, James Fisher (1912–70) did much to popularise ornithology among the general public, in the sense of watching and trying to understand the lives of birds, rather than shooting them or collecting their eggs (see Fisher, 1940). The net result was a greater pool of volunteers interested in investigations of the lives of birds and willing to take part in surveys and censuses (Allen, 1976: 266). The British government established the Nature Conservancy in 1949, primarily to oversee the protection of habitats. Peter Scott (1909–89), son of Scott of the Antarctic, founded the Severn Wildfowl Trust (1948) (to become the Wildfowl and Wetlands Trust); the International Waterfowl Research Bureau (1954) had similar aims, but shifted focus towards preserving wetland habitats and became Wetlands International.

A new breed emerged with the coming of these organisations and institutions: the professional ornithologist. In the nineteenth and early twentieth centuries, naturalist posts with meagre wages (curators and a handful of academics) were more often positions for wealthy dilettantes (recalling Richard Sharpe's financial struggles). Professionals soon carved out territories distinct from those of amateurs. The most striking example was David Lack (1910–73), appointed Director of the EGI in 1945. He transformed the EGI into a research centre investigating scientific questions concerning natural selection: how animal populations maintained their numbers, how different species lived together and the reasons why different species of birds laid different numbers of eggs (Lack, 1954, 1966, 1968, 1971; Anderson, 2013; Birkhead *et al.*, 2014).

The relationship between behavioural and ecological studies on the one hand, and 'traditional' studies based on description and empiricism on the other, was a rather uncomfortable one. The *Ibis* continued to be run on traditional lines, sticking with factual lists of species of different regions of the world and avoiding hypothesis-driven research. Ernst Mayr wrote to David Lack in 1939: 'it is common knowledge throughout the world that the *Ibis* is full of second-rate papers of colonial officers and faunistic lists of little interest' (see Johnson, 2004: 539). Mayr had been reforming American ornithology and encouraged Lack to do the same. Richard Meinertzhagen, a soldier–naturalist in the 'old' model, complained to Mayr: 'British ornithology is in a bad way … most of the modern generation do field work on somewhat eccentric lines. How many times a Robin wags its tail in an hour or how many times a Blackbird evacuates in 24 h[ours]. That takes us nowhere' (quoted in Johnson, 2004: 542). 'Scientific ornithology' had grown away from classification, taxonomy and museums, although the 'new school' would make use of collections to support their studies of living birds (Johnson, 2004: 543).

By the time of the BOU's centenary in 1959, the balance had shifted further still. Contributors to the anniversary proceedings emphasised the importance of the study of living birds and skimmed over the first fifty years of activity of the Ibises (see Johnson, 2004). Reg Moreau (1897–1970), an English civil servant with a keen interest in the life histories of African birds, noted how the *Ibis* had been 'hide-bound' and had fallen behind the leading American and German ornithological journals by sticking to lists of what species live where (faunistics) and fact-gathering (Moreau, 1959: 33). Hardly any of the contributors had particularly strong links with the British Museum (Natural History). Guy Mountfort reflected on the place of collecting in the modern Union:

> [formerly] the election of potential members was vetoed on the grounds that the candidates were known to collect either skins or eggs without scientific

justification. The cult of collecting for personal gratification or financial gain rapidly declined; today [i.e. 1959] special provision in the Rules [to prevent members threatening rare birds by their collecting] is no longer necessary to enforce a policy which is universally endorsed by members. (Mountfort, 1959: 16)

Anti-collecting, anti-individualism: anti-Victorianism

Writing in 1959, at the time of the BOU's centenary, Reg Moreau attributed the British preoccupation with geographical ornithology to 'the immensity of the areas over which British rule or influence stretched during the 19th century and some of the 20th'. Scientific imperialism had provided a mandate to investigate the birds of these regions (and the imperial power balance enabled the British to build collections from across the Empire), while nationalistic pride promoted the development of collections and museums befitting a global superpower. This mandate produced the collections of the what is now the Natural History Museum, which still has one of the largest collections of bird skins and eggs in the world, mostly dating from the time of the British Empire (see Knox and Walters, 1992; Mearns and Mearns, 1998: 420). Johnson has noted how the products of Empire (such as bird skins) influenced the development of the 'main thrust' of scientific ornithology, namely description and cataloguing, whether relating to the products of the Empire itself or from other regions (Johnson, 2004: 527). With the decline of the Empire, the links between specimens from former colonies and their places of origin is less self-evident and very few British museums continue to collect birds from beyond British shores (or indeed within their own country). While collections mostly date from bygone times, some people have found them difficult to reconcile with more modern sentiments. In an introductory essay to a book on natural history auctions, Simson (in Chalmers-Hunt, 1976: 23–4) described bird study skins:

> They lie, curiously long, thin and unbirdlike, in row upon row in the drawers of museums. In order to sort out juvenile and seasonal changes in plumage as well as sexual differences and sub-specific differences, many skins of the same species were required. Looking at these long arrays from all over the world one wonders if a holocaust on quite such a scale was necessary.

This ambivalence and suspicion towards collections can be problematic when they are left in the care of people who do not understand their potential, as both scientific and historical objects (a subject returned to below).

A notable feature of Dresser's work was his independence from institutions and organisations. He wrote to Samuel Boardman (when Samuel was preparing a biography of his father, Dresser's correspondent George Boardman): 'I have no connection with our universities.... I am only an amateur ornithologist and, like the smith of [Walter] Scott's *Fair Maid of Perth* "I always fight for my own hand only and love my independence'" (Boardman, 1903: 250). Dresser valued his own achievements greatly: he once had some Christmas cards produced showing an owl surrounded by small birds, similar to the owl motif he used in his books.[10] The illustration was accompanied by a quotation in Swedish, 'Hvad ädelt är du vilje, hvad rätt du göre', from Esaias Tegnér's *Fridthjof's Saga* (1825). The quotation is the last line from the verse:

> All men will surely perish with all they prize,
> But one thing know I, Fridthjof, which never dies,
> And that is reputation! therefore, ever
> The noble action strive for, the good endeavour.
> (Holcomb and Holcomb, 1876: 15)

This was to hold with Thomas Carlyle's view of history, as the achievements of great men. By the time of the BOU centenary, such individualism was practically discouraged. Manson-Bahr wrote of Dresser (rather unkindly) as

> an amateur without any scientific training, as ordinarily understood…. He had a striking appearance which was arresting in any company. His countenance was pale, clean-shaven with sharp nose and striking intense features. The oddity of his appearance was heightened by a rather ill-fitting wig, because he was completely bald. He had a rapid method of elocution and the words poured from his mouth and almost stunned his audience. He was in fact quite a character in an age of individualism…. Dresser was as different as can possibly be imagined from other museum ornithologists of the day. (Manson-Bahr, 1959: 59–60)

By the twentieth century, scientific education meant a spell at university and then on-the-job training. Neither of these was readily available in the nineteenth century: even the great Richard Sharpe never went to university. Salvin, Godman and their like received their scientific education in birds at Alfred Newton's Sunday evening meetings rather than in the lecture halls at Cambridge. Progress in serious scientific organisations, too, became increasingly identified with paid positions in institutions. Henry Elwes, who was one of the last of the independent travelling naturalists to attain a Fellowship of the Royal Society, bemoaned the transformations in science, writing at the time of the death of Frederick Godman (his brother-in-law) in 1919:

> I cannot but feel that it is a very great misfortune that so few men of the Godman type are now able to attain the honour of Fellowship in this great Society. If one looks through the list of candidates in recent years it will be seen that country gentlemen, who have trained themselves in scientific work and who could be encouraged to do a great deal for science, are conspicuous by their absence. There seems to be a tendency to limit the elections into the Society to men who, making their living by the teaching or practice of some branch of science, are able to specialise sufficiently, in order to publish what is now considered to be the necessary amount of original work. I feel sure that the Society would gain if more men of the Godman type could be led to believe that the Fellowship of the Society is not closed to them. (Elwes, 1920: v–vi)

Elwes was not entirely correct in his view of what was required to gain entry to the Society: some of those who worked independently produced a great volume of published work, and even the great Richard Sharpe, who did hold a paid position and who produced a huge volume of work, never attained the Fellowship of the Royal Society (although he was a candidate in 1903). Science was changing so that cataloguing was no longer in favour.

Reg Moreau wrote of 'the astonishing rudeness' that was formerly allowed to be included in the *Ibis*, referring to the disagreements between Dresser and Seebohm (see chapter 10). Again and again, the argumentativeness of nineteenth-century ornithologists drew comment. Collingwood Ingram recalled the BOC meetings: 'the obvious boredom with which most of the members of the BOC used to listen to any announcement made by a fellow member was somewhat surprising and not a little disconcerting to the announcer' (Ingram, 1966: 166–7). Such grandiose criticisms were often accompanied by surprisingly personal attacks. Ingram described Dresser as 'that little man who with his sleek auburn toupee and dapper city clothes looked so much more like a successful haberdasher than a world-famous ornithologist' (Ingram, 1966: 25). Ingram recalled Sharpe and Dresser's relationship, describing Dresser as Sharpe's 'arch-enemy', for whom Sharpe held a 'deep and ineradicable detestation'. He recounted a story told to him by Sharpe: Dresser had visited the Bird Room and was engrossed in work at the end of the day. Sharpe tip-toed to the main entrance and was locking the door, so Dresser would have been locked in overnight, but Dresser appeared at the door just as Sharpe was turning the key (Ingram, 1966: 161).

Manson-Bahr, too, recalled those who had gone before him in a short history of the BOC, written in 1952 (the time of the Club's sixtieth anniversary), including:

> those odd productions of the Bird Room at South Kensington, the bustling [Richard] Bowdler Sharpe and the soldierly, monocled Ogilvie[-]Grant…. H. E. Dresser, the timber merchant from the Baltic, was also there with his early Victorian manner and an ear-trumpet, which converted the meetings into a shouting match. Soon too, there appeared that oddly assorted couple from Tring – Walter Rothschild and Ernst Hartert – a masterly association. (Manson-Bahr, 1952: 88)[11]

Manson-Bahr also provides a rare account of Eleanor Walmisley Dresser in scientific society, again probably relating to 1907–10:

> he [Dresser] and his wife had developed different interests. She moved in spiritual circles, while his interests were much more mundane in things ornithological, so that their mutual conversation was a matter of comment and at times amusement to others. (Manson-Bahr, 1959: 59)

The reference to Dresser's 'early Victorian manner' deserves comment; the term 'Victorian' had been used reproachfully for many years (see Young, 1953: 100; also quoted in Gardiner, 2002: 23).[12] If Dresser was a Victorian, this was to mean he was old-fashioned, conservative and prudish: one of those 'eminent Victorians' – to use Lytton Strachey's sarcastic phrase – who were such objects of ridicule. Dresser sat on the far side of the great division of the First World War, an upholder of views and values that many considered to have played a part in precipitating the Great War and all the loss it brought. H. G. Wells wrote in 1918 of those 'base-Imperialists',

> their minds sat crippled with dead traditions … that England of the Victorian old men, and its empire and its honours and its courts and precedences, it is all a dead body now, it has died as the war has gone on, and it has to be buried out of our way lest it corrupt you and all the world again. (Wells, 1918: 569–70; see also Gardiner, 2002: 20–1)

In the same vein, those who clung to old ornithological traditions came in for ridicule in the *Ibis*, a symptom of the changing nature of science and of society. As early as 1918, Edmund Harting complained about changes in the scientific names of British birds and resigned from the BOU (Harting, 1918a,b). Jourdain attacked Harting in the *Ibis*:

> Mr. Harting's claim is based on seniority.… The suggestion is really too puerile for serious consideration, but one would think that even Mr. Harting would have realized by this time that ornithology is not merely the hobby of a clique of English writers…. Recent events [presumably referring to the First World War] might have taught us that there is a world outside the limits of the British Isles. (Jourdain, 1918: 526–8)

Meinertzhagen's reminiscences

One account of Dresser from the centenary *Ibis* of 1959 deserves careful treatment, namely the 'reminiscences' of Richard Meinertzhagen. He wrote of Dresser as:

> a curious character, without much humour, inclined to be secretive and jealous…. Sharpe mistrusted him, accusing him of removing specimens from the national collection, I believe, quite unjustly. Dresser had correspondents all over the world and was probably better known to foreign ornithologists than any other British ornithologist. He collected both skins and eggs and at his death these went to the Manchester Museum. Sharpe inspected the collection there and admitted later that he had been quite wrong about Dresser. On one occasion Dresser wrote to a correspondent in Central Asia asking for some eggs of Pallas' Sandgrouse. A box eventually arrived with over 1000 eggs of this bird. Dresser was aghast and hid the box under his bed, where it was discovered after his death still unpacked. (Meinertzhagen, 1959: 49–50)

This needs careful attention, as Meinertzhagen, for reasons known only to himself, fabricated a number of encounters with famous people, from the Russian royal family to Adolf Hitler. He had also been barred from the Bird Room at the British Museum (Natural History) by 1919, for removing birds without permission, forty years before he wrote his 'recollections'. This was to be the first of a number of such instances, which almost resulted in legal prosecution. However, none of this became public (beyond a handful of leading ornithologists) until after Meinertzhagen's death. He is now known to have swapped labels on specimens to support his ideas on bird distribution and to have fabricated records of species from beyond their usual range. This is not to say that everything he did was fraudulent, but the taint of fraud spreads from the bad to the good, and it is difficult to disentangle fact from fabrication (see especially Cocker, 1989; Knox, 1993; Rasmussen and Prŷs-Jones, 2003). It appears quite in keeping with his unusual character that Meinertzhagen – guilty of removing specimens from the Bird Room – should write about theft from the museum in relation to another ornithologist, who had been accused of the same crime 'quite unjustly'. Possibly Meinertzhagen was thumbing his nose at his contemporaries. Incidentally, there is no evidence that Sharpe ever examined the bird collection at Manchester Museum, nor that Dresser ever had 1,000 Pallas's Sandgrouse eggs. These two 'events' appear to be more of Meinertzhagen's fabrications.

Birds and business

This book has sought to explore Dresser's ornithology in relation to his mercantile work, to examine how much one influenced the other. The intimate relationship that existed between business and ornithology for at least some of the leading collector-naturalists is of the utmost importance, and should not be underestimated. Dresser's ornithological career was launched and maintained by links and collections that grew, at least partly, from his business travels in Scandinavia, New Brunswick and Texas, and latterly through Europe. The language skills that enabled him to communicate with European naturalists had been acquired as a result of his training for business. His success at negotiating and bargaining for specimens was also the result of training for a life of business. Similarly, we find Howard Saunders, who worked for a merchant linked with guano fortunes, taking up seabirds as his own specialist subject. We have the firm Seebohm and Dieckstahl, owned by Dresser's rival Henry Seebohm, in offices right next door to Dresser at 110 Cannon Street (see chapter 9). Dresser may even have been in business with Seebohm and Dieckstahl with the agency of the Anchor Tube Co. (or in competition with Seebohm's firm, as he sold products for other firms). These shared interests in both business and ornithology clearly pitted Dresser and Seebohm against one another.

Looking back

This chapter has attempted to explore some of the main shifts that took place in ornithology during the twentieth century. Dresser and those like him were effectively left behind as 'amateurs', and as ornithology shifted away from museums and collections. Dresser and those like him became less and less recognisable to those who constructed themselves as scientific ornithologists. Hilary Fry wrote in 1984 of Dresser's book on bee-eaters:

> Exactly one hundred years ago there was published in London a sumptuous treatise which has remained until now the only work in any language devoted exclusively to bee-eaters. It was *A Monograph of the Meropidae, or Family of the Bee-Eaters* by H. Dresser, and is nowadays a collector's item. The text is ample, and yet to the modern ornithologist it has hardly anything of interest to say. I mean no disrespect to the memory of Dresser, a great ornithologist in his day – if his monograph expounds little it is for the simple reason that when it was written practically nothing was known about the birds except what could be gleaned from museum specimens and their collectors' scant field notes. (Fry, 1984: 9)

Dresser and those like him fought hard to produce a solid foundation for further ornithological work, even if their names are often forgotten today (see plate 59).

The elitism that characterised the great encyclopaedic projects worked against them and their authors. Enormous, sumptuously illustrated volumes may have ensured fame at the time they were produced, but the high financial value of the morocco-bound volumes with their hand-coloured illustrations meant they were produced in small runs as collectors' items. Even in libraries, Dresser's books are not available on public shelves but are held within library 'special collections'. This inaccessibility contrasts with the smaller (and less expensive) volumes that began to be produced in the earlier part of the twentieth century. In reviewing Thomas Coward's *The Birds of the British Isles and Their Eggs*, issued in 1919, Francis Jourdain wrote: 'The great works of Gould, Dresser and Lilford were issued at prices which were and are still prohibitive, at any rate for beginners, and none of the smaller handbooks contained really good coloured plates' (Jourdain, 1920). Coward's book reproduced plates from Lilford's *Coloured Figures* (just as Lewis Bonhote had included Dresser's plates in his small book).

The works – and lives – of nineteenth- and early twentieth-century ornithologists have seen a great renaissance in recent years. This is at least partly because of the enormous amount of published material, hitherto inaccessible (or as good as inaccessible), that has become freely available on the internet. *A History of the Birds of Europe* can now be accessed by anyone with a computer through the Biodiversity Heritage Library or through Internet Archive.[13]

New uses for old birds

Collections such as Dresser's are products of historical contingency, telling us something about people, their interests in birds and the ways these interests manifested themselves (see plate 60). They also tell us something about birds themselves: of their variation, development, distribution, ecology and other aspects of their lives. Dresser's collections were – and still are – among the finest of their kind. The collections that formed the basis of Dresser's colossal books are in Manchester Museum; significant numbers of birds he collected from America and Japan are to be found in Edinburgh in National Museums Scotland, and smaller numbers of specimens in the University Museum of Zoology in Cambridge. Great numbers of birds that passed through his hands at one time or another are to be found in the Natural History Museum (formerly called the British Museum (Natural History)), and probably in many of the other major museums in Europe and the United States.

Dresser's collection in Manchester is still remarkably intact, considering that it has been through two World Wars and a number of moves. Small numbers of bird skins have disappeared over time as a result of being damaged by insects and thrown out. A small number of eggs were exchanged with other museums.[14] Some also appear to have been stolen from the Museum: there are no Ross's Gull eggs in the collection, for example, when there should be about eight specimens. The collection continues to be used by researchers around the world (see appendix 4). The most frequent topics for studies are biogeographical: checking the occurrence of particular species (or subspecies) in localities and studying variation within particular species. Old specimens are used as sources of tissue to study pollutant levels, which are locked in feathers and skins. Researchers request small slivers of flesh (usually taken from the underside of birds' toes) in order to extract DNA to study genetic diversity and the relationships between different forms of birds. Chemical signatures in feathers can be used to establish where particular specimens came from, enabling researchers to work out the migration routes and diets of the birds that they came from; for example, a study of the stable isotopes in museum specimens of the Slender-billed Curlew suggested that the bird's largely unknown breeding range was further to the south than previously thought (Buchanan *et al.*, 2017). Taking old specimens and comparing them with recent samples enables researchers to reconstruct changes over time. Museum collections of scientific specimens are thus an archive, just as ice cores tell us about the changing climate over time, or tree rings tell us about growing seasons. These are questions that Dresser could not have possibly imagined at the time the specimens were collected, when he and others like him first drew out the rough outline of what birds there were and where they existed.

Closing remarks

Dresser's ornithological output – his writings and the collections that under-pinned them – represents a tremendous legacy. His publications serve as a basis for the understanding of the world's birds, their lives and the threats they face. His collection is important as a historical document in its own right – when one person forged the discipline of European ornithology – and as a resource for modern investigation. While great encyclopaedias such as *A History of the Birds of Europe* may come from a bygone period of ornithology, they can still serve to inspire ornithologists and others who appreciate birds today, both by their factual content and by their beautiful illustrations.

The vast quantities of information Dresser's books contain help us understand and contextualise both the continuity and change that has occurred over the last century and a half. They are a reference point – both literally and metaphorically – connecting us in the long trajectory of environmental change. Dresser produced his books and collections in order to make birds known, as part of the great imperial project of understanding and classifying the world. Since his time, many species have undergone dramatic changes in status. Some species are on the verge of extinction; others, such as the Slender-billed Curlew (see plate 61) and Eskimo Curlew, have probably crossed that never-to-return threshold already. Many others face uncertain – or possibly all too certain – futures in the face of habitat degradation, climate change, pollution and other human impacts (see plate 62).

Dresser's books and collections help us to formulate questions about people's long-standing fascination with birds, and how that fascination has been manifested in different ways by different groups of people. Rather than focusing on continuity, we can appreciate the widely divergent ways in which people have attached importance to birds and collected information about their lives.

A personal view

I have written this book with a purpose, which is to try to understand what Victorian ornithologists were seeking to achieve and why, and to make sense of the specimens in museum collections today: where they came from, who put them there and why ornithology changed over time. My ultimate goal is to find ways to promote the appreciation and understanding of birds and nature, and to find ways to make museums work better in support of nature conservation, through exploring and promoting environmental sustainability. I have not sought to provide simple solutions, but to explore influences and relationships, accepting that others do not hold the same values or opinions as myself. In exploring the past, I have tried not to sit in judgement of it or to project modern viewpoints upon it. Yet, in exploring the past, I know my own values better: of the things that I might do, and would not be prepared to do, and the things that matter to me. To my mind, this is one of the most important potentials of museums. I am faced with the remains of the past in my day-to-day work as a curator; I think about them in the present, and in so doing my present and future choices are shaped. My understanding of birds and ornithology, and indeed of people, is deepened as a result of my interactions with collections, archives and other people in (and beyond) museums.

48 *Lanius funereus*, from *Supplement to A History of the Birds of Europe.*

49 Seebohm's Wheatear *Oenanthe oenanthe seebohmi* as *Saxicola seebohmi*, from *Supplement to A History of the Birds of Europe.*

50 Photographic plate of the eggs of rare thrushes from Dresser, H. E. (1901), 'On some rare or unfigured Palaearctic birds eggs'.

51 Permafrost tundra, Kolyma, Siberia, 1994.

52 Ross's Gull on nest in marsh, Chukochyi river-mouth tundra, Kolyma, June 1994.

53 Photographic plate of Ross's Gull eggs collected by Sergei Buturlin in 1905, from Dresser, H. E. (1906), 'Note on the eggs of Ross's Rosy Gull'.

Ibis. 1907. Pl. XII.

1. RHODOSTETHIA ROSEA, pull.   2. TRINGA MACULATA, pull.

West, Newman Imp.

54 Plate of the chicks of Ross's Gull (foreground) and Pectoral Sandpiper from Buturlin, S. A. (1907), 'On the breeding-habits of the Rosy Gull and the Pectoral Sandpiper', based on specimens sent to Henry Dresser by Sergei Buturlin.

55 Spectacled Eider male (front right) and female, National Petroleum Reserve, near Point Barrow, Alaska.

PSEUDOSCOLOPAX TACZANOWSKII.

56 Asian Dowitcher by J. G. Keulemans, from Dresser, H. E. (1909), 'On the occurrence of *Pseudoscolopax taczanowskii* in Western Siberia'.

57 Pechora Pipit *Anthus gustavi* as *Anthus seebohmi*, from *A History of the Birds of Europe*.

PEREGRINE.
FALCO PEREGRINUS.

58 Peregrine Falcon, from *A History of the Birds of Europe.*

59 Red-bearded and Blue-bearded Bee-eaters from Henry Dresser's collection.

60 Steller's Eider specimens from Henry Dresser's collection. The bird in the foreground is from Varanger Fjord, Norway, and formed the basis of the illustration shown in plate 2.

61 Slender-billed Curlew, Merja Zerga, Morocco, 1995.

62 Steller's Eider males, with Common Eider and King Eider in the background, Båtsfjord, Varanger Peninsula, Norway, 2016.

Museum collections of birds and eggs can appear anachronistic, representing a time of different values and practices. Egg collections in particular are often regarded as challenging for museums to exhibit to the general public, with fears among curators that they will inadvertently encourage egg collecting. This is a pity, as the story of over-collecting can be usefully told to promote bird conservation. Collections of skins and eggs can become regarded solely as 'scientific specimens' all too easily, in the sense of not being suitable for public 'consumption', and effectively kept locked behind the scenes, beyond the public's experience of museums (and collections). Yet this does both collections and visitors a disservice: we should not hide things away just because they might be uncomfortable or, heaven forbid, provoke a reaction. Reactions are good: they mean that we have brought something to the surface, a feeling, an opinion. We do not have to put ourselves in the business of smoothing things over, of wrapping the public in cotton wool, as that is the surest way to becoming irrelevant and redundant. Nor should we allow museums to simply become incredibly bland, sanitised places which solely aim for people to have 'fun', and which repeat things that people have known since they were children. This is one of the reasons why museums are so identified with childhood: places that people visit as children and again with children and grandchildren. There is terrific potential to do much more interesting things than putting a specimen in a case with a label telling you what it is and where it came from.

Museum curators of bird skins and eggs, and of many other things for that matter, perform a rather delicate balancing act. They advocate the value of the collections in their care on the one hand, while they distance themselves from the collecting practices that produced these collections on the other. Many of the collecting practices are illegal today, and for very good reasons. Collectors of the nineteenth and twentieth centuries who provided museums with their specimens may be denigrated by some or celebrated by others, understood or misunderstood, or completely forgotten. Most of the bird study skins in museums come from birds that had been shot, whether specifically for collections or for some other reason. It can be easy to try to cover up this killing but it is not entirely honest to do so. On the other hand, admitting that specimens were killed runs the risk of alienating visitors. Birds in collections were shot? Yes, they were. Would we do that now? No we wouldn't, but it is worth then considering what threats birds face today: insidious and invisible threats from globalisation, pollution, climate change and habitat destruction. Museum curators and educators should be able to tell the story of the collections in their care without feeling obliged to stand up for or defend (or feel guilty for) the practices that lay behind the development of those collections, and they really must find out what contemporary issues the collections they look after can connect with. They should not present a this-then-that view of history (with inevitable progression to the present), or be content with providing museum visitors with a pleasant but rather shallow experience of collections and subjects. The birds in museums may have been shot a century or more ago, but curators and educators do have control over how collections are *used* today. These collections are a fabulously rich resource for presenting and exploring ideas about how people have understood the natural world, and also how those attitudes have contributed to shaping the world we live in today, for better and for worse. Collections should surely be used to help people explore, understand and appreciate the lives of birds (and people), with an eye to the conservation of the diversity of nature still with us today. Even in Britain, some of our rarest birds of prey continue to be threatened by private game-rearing interests. Perhaps if we were bolder in museums in raising conservation issues we would raise people's awareness of them, but we should also balance information with creativity and options for personal action. We could very usefully connect people with broader historical narratives showing that the past and the present were very different, and the future will be different again. We can explore what future visitors want, using collections and historical narratives as prompts, and signpost visitors towards the positive actions they want to take. Curators really must connect with the wider world if they are to help the public connect with the wider world and avoid our museums becoming dusty documents of 'nature past', and to provide people with truly enriching cultural experiences. The collections we look after in museums are there for a purpose, which is surely to make a positive difference to society and the natural environment. Rather than preserving things uncritically because 'that is what museums do', curators should focus their attention on using the collections they have to address social and environmental issues. This does not mean telling people depressing stories of environmental degradation, but it does mean making the most of the museum experience, where people are confronted with unexpected objects and specimens as part of exhibitions and public events. It means getting collections to work for a living, and to connect collections with the issues of today.

Collections, which came from around the world, can be used to bring people face to face with the wider world: with historical narratives with a wide horizon, with a depth of time to explore. Collections from disparate places can also be used to bring people together, whether they be museum curators, academics or the public, to explore narratives of environmental history from beyond a simply nationalistic perspective. We can use collections to break out of existing historical narratives to explore different, more sustainable futures, where people and nature can flourish together.

Notes

1   The records of the Rüppell's Warblers were also a 'first' for Britain, but subsequently rejected as part of the 'Hastings Rarities' affair. See chapter 11 for discussion of the affair.
2   Information extracted from Chalmers-Hunt (1976) and Cole (2006).
3   The story refers to the collector's ruling passion of egg collecting. *The Ruling Passion* is also the title of a painting by J. E. Millais, showing an ornithologist (based on John Gould) with his beloved bird skins.

4   Details of correspondence between Buxton, the BOU and the Oological Dinner Committee can be found in Parker (1935). Earl Buxton's speech was reported in *Bird Notes and News* Anon. (1922a).

5   H. J. Elwes and E. C. Stuart Baker, 'Wild bird protection', *The Times*, 10 April 1922, p. 10, col. d.

6   Editorial, 'Wild birds' eggs', *The Times*, 13 April 1922, p. 13, col. b–c, republished in *Bird Notes and News* (Anon., 1922b).

7   This worked against Chance, as it turned out that his presentation of 1922 referred to the eggs of forty *pairs* of shrikes throughout a breeding season, and included 146 clutches in all. See Parker (1935: 38–55).

8   'Egg society denies aiding nest thefts: an obscure group named after a Victorian clergyman is accused of acting as a front for illegal collectors who damage rare species', *The Independent*, 1 October 1994, www.independent.co.uk/news/uk/ home-news/egg-society-denies-aiding-nest-thefts-an-obscure-group-named-after-a-victorian-clergyman-is-accused-1440402.html, accessed 8 April 2017. Note how Jourdain was referred to as 'a Victorian clergyman' in this headline, while he was most active in the first three decades of the twentieth century, becoming Vice-President of the BOU in 1934, and serving as President of the British Oological Association from 1932 to 1939 (Witherby, 1940).

9   'The egg snatchers', *The Guardian*, 11 December 2006, www.theguardian.com/ environment/2006/dec/11/g2.ruralaffairs, accessed 8 April 2017.

10 One of these Christmas cards was sent to John Harvie-Brown, National Museums Scotland Library, GB 587, JHB 15/240, postcard from HED to John A. Harvie-Brown (JAHB hereafter), 24 December 1908.

11 He referred his early experiences of the BOC to having been 'nearly forty years' previous to 1952. This is not correct as he first attended meetings from October 1906 (Anon., 1906g: 2).

12 Gardiner (2002: 19–36) discusses changing interpretations of Victorianism and explores anti-Victorianism.

13 At www.biodiversitylibrary.org, http://archive.org/index.php, accessed 8 April 2017.

14 A small number of eggs were exchanged with the University of Cambridge Museum of Zoology in 1931 (Anon., 1931).

Appendix 1. Birds mentioned in the text

Common and scientific names follow the BOU British List (Harrop *et al.*, 2013) for species recorded in Britain and the IOC World Birds List, Master IOC list v. 5.4 for other species. Where these differ, the IOC name is in parentheses. The order follows the Master IOC list v. 5.4. For lists containing more than one species, an asterisk indicates the species pictured.

All images are from Dresser, H. E. (initially Sharpe, R. B. and H. E. Dresser) (1871–82), *A History of the Birds of Europe* (from the Biodiversity Heritage Library, digitised by the Smithsonian Institution Libraries, www.biodiversitylibrary.org). See appendices 2 and 3 for additional species.

WATERFOWL

Snow Goose, *Chen caerulescens*
Lesser Snow Goose [subspecies], *Chen c. caerulescens*
American Wigeon, *Anas americana*
Baikal Teal, *Anas formosa*
Eurasian Teal, *Anas crecca*
Steller's Eider, *Polysticta stelleri*
Spectacled Eider, *Somateria fischeri*
King Eider, *Somateria spectabilis*
Common Eider, *Somateria mollissima*
Harlequin Duck★, *Histrionicus histrionicus*

Labrador Duck, *Camptorhynchus labradorus*
Surf Scoter, *Melanitta perspicillata*
Bufflehead, *Bucephala albeola*
Smew, *Mergellus albellus*
Hooded Merganser, *Lophodytes cucullatus*
Goosander (Common Merganser), *Mergus merganser*

GAMEBIRDS

Plain Chachalaca, *Ortalis vetula*
Montezuma Quail, *Cyrtonyx montezumae*
Wild Turkey, *Meleagris gallopavo*

Ruffed Grouse, *Bonasa umbellus*

Hazel Grouse, *Tetrastes bonasia*

Spruce Grouse, *Falcipennis canadensis*

Black Grouse, *Tetrao tetrix* (*Lyrurus tetrix*)

Caucasian Grouse, *Lyrurus mlokosiewiczi*

Greater Prairie Chicken, *Tympanuchus cupido*

Willow Ptarmigan, *Lagopus lagopus*

Red Grouse [subspecies of Willow Ptarmigan]★, *Lagopus lagopus scotica*

Daurian Partridge, *Perdix dauuricus*

Pheasant (Common Pheasant), *Phasianus colchicus*

DIVERS (LOONS)

Great Northern Diver (Great Northern Loon), *Gavia immer*

PETRELS

Leach's Petrel (Leach's Storm Petrel)★, *Oceanodroma leucorhoa*

Fulmar (Northern Fulmar), *Fulmarus glacialis*

Soft-plumaged Petrel, *Pterodroma mollis*

Desertas Petrel, *Pterodroma deserta* [included within Fea's Petrel *Pterodroma feae* by Harrop *et al.*, 2013]

GREBES

Slavonian Grebe (Horned Grebe)★, *Podiceps auritus*

Great Crested Grebe, *Podiceps cristatus*

FLAMINGOS

Greater Flamingo, *Phoenicopterus roseus*

STORKS

Black Stork, *Ciconia nigra*

IBISES, HERONS, PELICANS, CORMORANTS

Sacred Ibis, *Threskiornis aethiopicus*
Northern Bald Ibis, *Geronticus eremita*
American White Ibis, *Eudocimus albus*
Glossy Ibis, *Plegadis falcinellus*
Roseate Spoonbill, *Platalea ajaja*
Bittern (Eurasian Bittern)★, *Botaurus stellaris*
American Bittern, *Botaurus lentiginosus*
Great White Egret (Great Egret), *Ardea alba*
Tricolored Heron, *Egretta tricolor*
Snowy Egret, *Egretta thula*
Brown Pelican, *Pelecanus occidentalis*
Pygmy Cormorant, *Microcarbo pygmeus*

HAWKS, EAGLES

Osprey (Western Osprey)★, *Pandion haliaetus*
Swallow-tailed Kite, *Elanoides forficatus*
Lesser Spotted Eagle, *Clanga pomarina*
Indian Spotted Eagle, *Clanga hastata*
Spotted Eagle (Greater Spotted Eagle), *Aquila clanga* (*Clanga clanga*)
Booted Eagle, *Hieraaetus pennatus*
Tawny Eagle, *Aquila rapax*
Spanish Imperial Eagle, *Aquila adalberti*
Imperial Eagle (Eastern Imperial Eagle), *Aquila heliaca*
Golden Eagle, *Aquila chrysaetos*
Bonelli's Eagle, *Aquila fasciata*
Red Kite, *Milvus milvus*
Bald Eagle, *Haliaeetus leucocephalus*
Rough-legged Buzzard, *Buteo lagopus*
Buzzard (Common Buzzard), *Buteo buteo*

BUSTARDS

Great Bustard, *Otis tarda*

CRANES

Siberian Crane★, *Grus leucogeranus*
Whooping Crane, *Grus americana*
Hooded Crane, *Grus monacha*

WADERS

Canary Islands Oystercatcher,
   Haematopus meadewaldoi
African Oystercatcher, *Haematopus
   moquini*
Ibisbill, *Ibidorhyncha struthersii*
Black-necked Stilt, *Himantopus
   mexicanus*
American Avocet, *Recurvirostra
   americana*
White-tailed Plover (White-tailed
   Lapwing), *Vanellus leucurus*
Pacific Golden Plover, *Pluvialis fulva*
Grey Plover, *Pluvialis squatarola*
Dotterel (Eurasian Dotterel), *Char-
   adrius morinellus*
Rufous-chested Plover, *Charadrius
   modestus*
Snipe (Common Snipe), *Gallinago
   gallinago*
Short-billed Dowitcher, *Limnodromus
   griseus*
Asian Dowitcher, *Limnodromus
   semipalmatus*
Eskimo Curlew, *Numenius borealis*
Slender-billed Curlew, *Numenius
   tenuirostris*
Upland Sandpiper, *Bartramia
   longicauda*
Redshank, *Tringa totanus*
Marsh Sandpiper, *Tringa stagnatilis*
Greenshank, *Tringa nebularia*
Green Sandpiper, *Tringa ochropus*
Spotted Sandpiper, *Actitis macularius*

Knot (Red Knot), *Calidris canutus*
Sanderling, *Calidris alba*
Little Stint, *Calidris minuta*
Temminck's Stint, *Calidris temminckii*
Pectoral Sandpiper, *Calidris melanotos*
Curlew Sandpiper, *Calidris ferruginea*
Dunlin, *Calidris alpina*
Spoon-billed Sandpiper, *Eurynorhyn-chus pygmaeus*
Buff-breasted Sandpiper, *Tryngites subruficollis*
Ruff, *Philomachus pugnax*
Red-necked Phalarope★, *Phalaropus lobatus*
Grey (Red) Phalarope, *Phalaropus hyperboreus*
Collared Pratincole, *Glareola pratincola*

GULLS, TERNS, AUKS

Black Skimmer, *Rhynchops niger*
Ivory Gull, *Pagophila eburnea*
Ross's Gull, *Rhodostethia rosea*
Audouin's Gull, *Larus audouinii*
Sooty Tern, *Onychoprion fuscatus*
Common Tern, *Sterna hirundo*
Arctic Tern, *Sterna paradisaea*
Black Tern, *Chlidonias niger*
Great Skua, *Stercorarius skua*
Great Auk★, *Pinguinus impennis*
Puffin (Atlantic Puffin), *Fratercula arctica*

SANDGROUSE

Pallas's Sandgrouse, *Syrrhaptes paradoxus*

PIGEONS

Bolle's Pigeon★, *Columba bollii*
Passenger Pigeon, *Ectopistes migratorius*

CUCKOOS

Greater Roadrunner, *Geococcyx californianus*
Cuckoo (Common Cuckoo)★, *Cuculus canorus*

OWLS

Barn Owl (Western Barn Owl), *Tyto alba*
Snowy Owl, *Bubo scandiacus*
Eurasian Eagle-Owl★, *Bubo bubo*
Tengmalm's Owl (Boreal Owl), *Aegolius funereus*

ROLLERS, KINGFISHERS, BEE-EATERS

Cuckoo Roller, *Leptosomus discolor*
Purple Roller, *Coracias naevius*
Racket-tailed Roller, *Coracias spatulatus*
European Roller, *Coracias garrulus*
Short-legged Ground Roller, *Brachypteracias leptosomus*
Kingfisher (Common Kingfisher)★, *Alcedo atthis*
Belted Kingfisher, *Megaceryle alcyon*
Red-bearded Bee-eater, *Nyctyornis amictus*
Blue-bearded Bee-eater, *Nyctyornis athertoni*
Arabian Green Bee-eater [subspecies], *Merops orientalis cyanophrys*
Blue-tailed Bee-eater, *Merops philippinus*

WOODPECKERS

American Three-toed Woodpecker,
    Picoides dorsalis
Black Woodpecker★, *Dryocopus
    martius*
Ivory-billed Woodpecker, *Campephi-
    lus principalis*

FALCONS

Lesser Kestrel, *Falco naumanni*
American Kestrel, *Falco sparverius*
Red-footed Falcon, *Falco vespertinus*
Saker Falcon, *Falco cherrug*
Gyr Falcon (Gyrfalcon)★, *Falco
    rusticolus*
Peregrine (Peregrine Falcon), *Falco
    peregrinus*

SHRIKES

Red-backed Shrike, *Lanius senator*
Isabelline Shrike, *Lanius isabellinus*
Great Grey Shrike★, *Lanius excubitor*
Great Grey Shrike [subspecies of],
    Lanius excubitor funereus
Woodchat Shrike, *Lanius senator*

CROWS

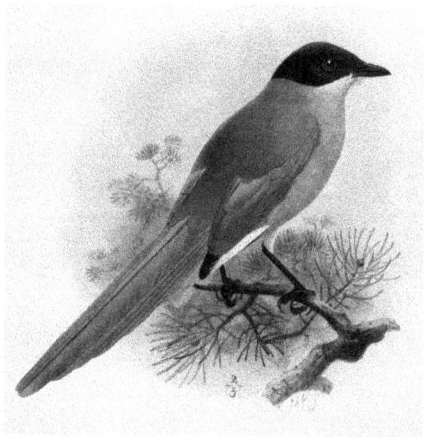

Azure-winged Magpie★, *Cyanopica
    cyanus*
Magpie (Eurasian Magpie), *Pica pica*
Chough (Red-billed Chough),
    Pyrrhocorax pyrrhocorax
Rook, *Corvus frugilegus*
Raven (Northern Raven), *Corvus
    corax*

WAXWINGS

Waxwing (Bohemian Waxwing)★,
   Bombycilla garrulus
Cedar Waxwing, *Bombycilla cedrorum*

TITS, NUTHATCHES

African Blue Tit★, *Cyanistes teneriffae*
Coal Tit, *Periparus ater*
Crested Tit, *Lophophanes cristatus*
Marsh Tit, *Poecile palustris*
Krüper's Nuthatch, *Sitta krueperi*
Eastern Rock Nuthatch, *Sitta
   tephronota*

LARKS

Shore Lark (Horned Lark), *Er-
   emophila alpestris*
'Eastern Shore Lark'/Central Asian
   Horned Lark [subspecies],
   Eremophila alpestris penicillata
White-winged Lark★, *Melanocory-
   pha leucoptera* (*Alauda leucoptera*)

SWALLOWS, MARTINS

Purple Martin, *Progne subis*
Swallow (Barn Swallow)★, *Hirundo
   rustica*

LONG-TAILED TITS

Long-tailed Tit, *Aegithalos caudatus*

WARBLERS

Pallas's Warbler (Pallas's Leaf
  Warbler), *Phylloscopus proregulus*
Yellow-browed Warbler, *Phylloscopus
  inornatus*
Hume's Warbler (Hume's Leaf
  Warbler), *Phylloscopus humei*
Moustached Warbler, *Acrocephalus
  melanopogon*
River Warbler, *Locustella fluviatilis*
Garden Warbler★, *Sylvia borin*
Rüppell's Warbler, *Sylvia ruppelli*
  (*Sylvia ruppeli*)

WRENS

Wren (Common Wren)★, *Troglodytes
  troglodytes*
St. Kilda Wren [subspecies], *Troglodytes troglodytes hirtensis*

STARLINGS

Common Starling, *Sturnus vulgaris*

THRUSHES AND CHATS

Blackbird (Common Blackbird),
   Turdus merula
Redwing★, *Turdus iliacus*
American Robin, *Turdus migratorius*
Robin (European Robin), *Erithacus*
   rubecula
Nightingale (Common Nightingale),
   Luscinia megarhynchos
Canary Islands Stonechat, *Saxicola*
   dacotiae
Seebohm's Wheatear [subspecies],
   Oenanthe oenanthe seebohmi
Black-eared Wheatear, *Oenanthe*
   hispanica

DIPPERS

Dipper (White-throated Dipper),
   Cinclus cinclus

SPARROWS

House Sparrow★, *Passer domesticus*
Dead Sea Sparrow [subspecies
   of, also called Afghan Scrub
   Sparrow], *Passer moabiticus yatii*

PIPITS AND WAGTAILS

Pechora Pipit, *Anthus gustavi*
Yellow Wagtail (left in illustration
   above)★, *Motacilla flava*
Grey-headed Wagtail (right in
   illustration above)★ [subspecies
   of Yellow Wagtail], *Motacilla flava*
   thunbergi

FINCHES

Chaffinch (Common Chaffinch),
   Fringilla coelebs
Pine Grosbeak★, *Pinicola enucleator*
Cinereous Bullfinch [subspecies],
   Pyrrhula pyrrhula cineracea
Azores Bullfinch, *Pyrrhula murina*
Crossbill (Red Crossbill), *Loxia
   curvirostra*
Syrian Serin, *Serinus syriacus*

NEW WORLD WARBLERS

Golden-cheeked Warbler, *Setophaga
   chrysoparia*

BUNTINGS

Pine Bunting, *Emberiza leucocephalos*
Rustic Bunting★, *Emberiza rustica*
Bay-crowned Brush Finch, *Atlapetes
   seebohmi*
Snow Bunting, *Plectrophenax nivalis*

Appendix 2. Birds named by Henry Dresser

Current status in parentheses, following IOC World Birds List v5.4.

Accentor fulvescens pallidus Dresser 1895 (subspecies of the Brown Accentor, renamed *Accentor fulvescens dresseri* by Hartert)

Acredula irbii Sharpe and Dresser 1871 (subspecies of the Long-tailed Tit *Aegithalos caudatus irbii*)

Acredula macedonica Salvadori and Dresser 1892 (subspecies of the Long-tailed Tit *Aegithalos caudatus macedonicus*)

Acrocephalus sogdianensis Dresser 1874 (=Upcher's Warbler *Hippolais languida*)

Anthus seebohmi Dresser 1875 (=Pechora Pipit *Anthus gustavi*)

Aquila leucolena Dresser 1873 (=Spanish Imperial Eagle *Aquila adalberti*)

Babax waddelli Dresser 1905 (Giant Babax *Babax waddelli*)

Calandrella baetica Dresser 1873 (=subspecies of Lesser Short-toed Lark *Alaudala rufescens apetzii*)

Cinclus melanogaster baicalensis Dresser 1892 (subspecies of the White-throated Dipper *Cinclus cinclus baicalensis*)

Cinclus melanogaster pyrenaicus Dresser 1892 (included within another subspecies of the White-throated Dipper *Cinclus cinclus aquaticus*)

Cinclus saturatus Dresser 1895 (dark form of a subspecies of the White-throated Dipper *Cinclus cinclus baicalensis*)

Coracias weigalli Dresser 1890 (included within Racket-tailed Roller *Coracias spatulatus*)

Coracias mosambicus Dresser 1890 (subspecies of the Purple Roller *Coracias naevius mosambicus*)

Garrulax tibetanus Dresser 1905 (=Brown-cheeked Laughingthrush *Trochalopteron henrici*)

Lanius lama Dresser 1905 (=Grey-backed Shrike *Lanius tephronotus*)

Lanius raddei Dresser 1888 (hybrid between Red-backed *Lanius collurio* and Isabelline *Lanius isabellinus* or Red-tailed *Lanius phoenicuroides* Shrikes)

Limicola sibirica Dresser 1876 (=Eastern Broad-billed Sandpiper *Limicola falcinellus sibirica*)

Otocorys brandti Dresser 1874 (subspecies of Horned Lark (Shore Lark) known as Steppe Horned Lark *Eremophila alpestris brandti*)

Parus britannicus Sharpe and Dresser 1871 (subspecies of the Coal Tit *Periparus ater britannicus*)

Parus cypriotes Dresser 1888 (subspecies of the Coal Tit *Periparus ater cypriotes*)

Parus grisescens Sharpe and Dresser 1871 (=Siberian Tit *Poecile cinctus cinctus*)

Phasianus ijimae Dresser 1902 (subspecies of the Copper Pheasant *Syrmaticus soemmeringii ijimae*)

Picus lilfordi Sharpe and Dresser 1871 (subspecies of the White-backed Woodpecker *Dendrocopos leucotos lilfordi*, sometimes treated as a full species)

Saxicola diluta Blanford and Dresser 1874 (=Mountain Wheatear *Myrmecocichla monticola monticola*)

Saxicola griseiceps Blanford and Dresser 1874 (=Mountain Wheatear *Myrmecocichla monticola monticola*)

Tetraogallus tauricus Dresser 1876 (subspecies of the Caspian Snowcock *Tetraogallus caspius tauricus*)

Serinus canonicus Dresser 1876 (=Syrian Serin *Serinus syriacus*)

Appendix 3. Birds named after Henry Dresser

Current status in parentheses, following IOC World Birds List v5.4.

Accentor fulvescens dresseri Hartert 1910 (subspecies of the Brown Accentor *Accentor fulvescens dresseri*)

Carenochrous dresseri Taczanowski 1883 (subspecies of the White-winged Brush Finch *Atlapetes leucopterus dresseri*)

Merops dresseri Shelley 1882 (= Böhm's Bee-eater *Merops boehmi*)

Parus palustris dresseri Stejneger 1886 (subspecies of the Marsh Tit *Poecile palustris dresseri*)

Saxicola dresseri Seebohm (= the dark-throated form of the Western Black-eared Wheatear *Oenanthe hispanica hispanica*)

Sitta dresseri Sarudny (= Zarudny) and Buturlin 1906 (subspecies of the Eastern Rock Nuthatch *Sitta tephronota dresseri*)

Somateria dresseri Sharpe 1871 (subspecies of the Common Eider *Somateria mollissima dresseri*)

Sturnus purpurescens dresseri Buturlin 1904 (included in another subspecies of the Common Starling *Sturnus vulgaris porphyronotus*)

Tetrao dresseri Kleinschmidt 1919 (= Red Grouse (subspecies of Willow Ptarmigan) *Lagopus lagopus scotica*)

Appendix 4. Publications based on Henry Dresser's collections, 1985–2017

Material studied	Investigation and methods	Reference
1. Evolutionary relationships and systematics		
Canary Islands Oystercatcher study skins	Morphological variation, DNA analysis	Valledor de Lozoya, 2013
2. Taxonomy		
Nuthatch study skins	Identifying type specimens as part of a review	Mlíkovský, 2007b
Type specimens of Asian Larks	Establishing the identity and location of type specimens	Dickinson *et al.*, 2001
Type specimen of an Asian Tit (*Parus grisescens*)	Establishing correct scientific names for Asian tits	Dickinson *et al.*, 2006
3. Geographical variation and distributional records		
Slender-billed Curlew egg	Verifying only known breeding record of a highly endangered species (possibly now extinct)	Gretton *et al.*, 2002; McGhie, 2002
Azores Bullfinch study skins	Establishing which specimens are syntypes and reviewing distributional records	Aubrecht, 2000
Study skins and eggs from Turkey	Distributional records of Turkish birds	Kasparek, 1986
Afghan Scrub Sparrow study skins	Study to investigate whether subspecies should be treated as a separate species	Kirwan, 2004
Study skins from Turkey	Subspecific variation and distributional records of Turkish birds	Kirwan *et al.*, 2008
Bullfinches and Shore Larks from Turkey	Reassessment of subspecies, based on examination of additional skins	Kirwan, 2006
Baillon's Crake eggs from the Netherlands	Distributional records	Jansen, 2013
Eggs of waders from Dutch peat bogs	Distributional records to understand former occurrence and timing of disappearance	Jansen, 2014
Northern Long-tailed Tit study skins	Distributional records from the Netherlands	Jansen and Nap, 2008

282

Material studied	Investigation and methods	Reference
Small Buttonquail (Anda-lusian Hemipode) eggs and skins from Spain	Review of ecology and distribu-tion of the Small Buttonquail	Gutiérrez Expósito *et al.*, 2011
Tit, Nuthatch and Treecreeper study skins	Review of geographical variation and identification	Harrap and Quinn, 1996
Study skins and eggs of endangered species	Review of ecology and distribu-tion of endangered species, to inform conservation programmes	Collar and Stuart, 1985
Snowy Owl eggs from Norway	Review of ecology and distribu-tion of Snowy Owl in Norway	Jacobsen, 2005
Pied Raven study skins	Review of the occurrence of the extinct Pied Raven on the Faeroes	Grouw and Bloch, 2015
Spanish Imperial Eagle skins and eggs	Demonstrating range change in an endangered species	Gonzalez *et al.*, 1989

4. Stable isotopes

Common Crossbill study skins	Origins of Common Crossbills from different irruption years to the UK, analysed using stable isotopes in feathers	Marquiss *et al.*, 2012
Slender-billed Curlew study skins	Attempting to identify the species' breeding range from stable isotopes in feathers	Buchanan *et al.*, 2017

5. Population demography

Corncrake study skins	Timing and causes of population declines based on examination of age-classes of Corncrakes, aged according to shape of wing feathers	Green, 2008

6. Egg-shell thinning

Spanish Imperial Eagle eggs	Egg-shell thinning in relation to use of pesticides and other chemical pollutants	Hernández *et al.*, 2008
Eggs of predatory birds, thrushes and birds associated with rivers	Investigation into long-term declines in egg-shell thickness in Britain	Scharleman, 2002

7. Historical studies of collectors' activities

Joseph Baker eggs	Tracing the collecting ac-tivities of a little-known English ornithologist	Jansen and Viek, 2010
William Bridger eggs	Tracing the collecting ac-tivities of a little-known English ornithologist	Jansen, 2012

Material studied	Investigation and methods	Reference
Sergei Buturlin specimens	Tracing collections from various expeditions	McGhie and Logunov, 2005, 2006
Specimens purchased in Leadenhall Market	Leadenhall Market as a source of rare specimens	McGhie, 2012
8. Catalogues of rare and extinct species		
Specimens of extinct and endangered bird species	Catalogue of extinct and endangered bird species to promote greater use by researchers	McGhie, 2005
9. Cultural history and museum studies		
Archives, collections	Postcolonial approaches to exploring bird collections	McGhie, 2010
Ross's Gull specimens	Investigation of development of ideas about birds involving museum specimens, using Ross's Gull as an example	McGhie, 2013
10. Bioarchaeology		
Sacred Ibis	Genetics and pathology of modern and ancient Egyptian Sacred Ibis, using CT scanning and DNA analysis	Atherton, 2008

References

Acheson, T. W. (1985). *Saint John: The Making of a Colonial Urban Community*. University of Toronto Press, London.

Alberti, S. J. M. M. (2009). *Nature and Culture: Objects, Disciplines and the Manchester Museum*. Manchester University Press, Manchester.

Alexander, H. G. (1974). *Seventy Years of Birdwatching*. T. and A. D. Poyser, London.

Allen, D. E. (1976). *The Naturalist in Britain: A Social History*. Allen Lane (Penguin Books), London.

Allen, D. E. (1980). The women members of the Botanical Society of London. *British Journal for the History of Science* 13(3): 240–54.

Allen, J. A. (1896). Sclater on rules of naming animals. *Auk* 13(4): 325–8.

Allen, J. A. (1903). Dresser's *A Manual of Palaearctic Birds*. *Auk* 20(4): 441–2.

Allen, J. A. (1904). Hartert's *Die Vögel der paläarktischen Fauna*. *Auk* 21(1): 94–5.

Allen, J. C. (1861). Beattie against Garbutt. In *Reports of Cases Argued and Determined in the Supreme Court of New Brunswick, Vol. 3*, pp. 1–8. William L. Avery, Saint John.

Allingham, E. A. (1924). *A Romance of the Rostrum*. Witherby, London.

Alston, E. R. (1880). On female deer with antlers. *Proceedings of the Zoological Society of London* 1879(4): 296–9.

Alston, E. R. and J. A. Harvie-Brown (1873). Notes from Archangel. *Ibis* 15(1): 54–73.

American Ornithologists' Union (1886). *The Code of Nomenclature and Check-list of North American Birds Adopted by the American Ornithologists' Union, Being the Report of the Committee of the Union on Classification and Nomenclature*. American Ornithologists' Union, New York.

'An Old Bushman' (H. W. Wheelwright) (1864). *A Spring and Summer in Lapland, with Notes on the Fauna of Luleå, Lapmark*. Groombridge and Sons, London.

'An Old Bushman' (H. W. Wheelwright) (1865). *Ten Years in Sweden, being a description of the landscape, climate, domestic life, forests, mines, agriculture, field sports, and fauna of Scandinavia*. Groombridge and Sons, London

Ananjeva, N. (2008). Zarudniĭ, Nikolai Alexseevich. *Encyclopaedica Iranica*, www.iranicaonline.org/articles/ZARUDNII-NIKOLAI_ALEXSEEVICH, accessed 8 April 2017.

Anderson, A. (1872). Notes on the raptorial birds of India. *Proceedings of the Zoological Society of London* 1871(3): 675–90.

Anderson, R. M. (1928). The work of Bernhard Hantzch in Arctic ornithology. *Auk* 45(4): 450–66.

Anderson, T. R. (2013). *The Life of David Lack: Father of Evolutionary Ecology*. Oxford University Press, Oxford.

Anker, J. (1938). *Bird Books and Bird Art: An Outline of the Literary History and Iconography of Descriptive Ornithology*. Levin and Munksgaard, Copenhagen.

Anon. (T. E. Bowdich) (1820). *Taxidermy or, the Art of Collecting, Preparing, and Mounting Objects of Natural History for the Use of Museums and Travellers*. Longman, Hurst and Rees, London.

Anon. (E. Newman) (1845). *A Catalogue of British Vertebrate Animals, the Names Derived from Bell's British Quadrupeds and Reptiles, and Yarrell's British Birds and Fishes, so Printed as to be Available for Labels*. Van Voorst, London.

Anon. (1860). Remarks on Mr. A. Newton's 'Suggestions for forming collections of birds' eggs'. *Ibis* 2(4): 415–18.

Anon. (1868). Review of J. E. Harting's *Catalogue of ... in the Collection of ... Ibis* 10(3): 340.

Anon. (1878a). Obituary of Robert Swinhoe. *Ibis* 20(1): 126–8.

Anon. (1878b). Proposed B.O.U. list of British birds. *Ibis* 20(3): 386–7.

Anon. (1878c). The Swinhoe collection of birds. *Ibis* 20(4): 492.

Anon. (1879). The Swinhoe collection. *Ibis* 21(1): 117.

Anon. (1882). Review of Dresser's *Birds of Europe*. *Ibis* 24(3): 459–60.

Anon. (1883). Review of Seebohm's *A History of British Birds, with Coloured Illustrations of Their Eggs*. *Ibis* 25(1): 114–15.

Anon. (1884). Zoological nomenclature. *Nature* 30(767): 256–9; 30(768): 277–9.

Anon. (1885a). Dresser's monograph of the bee-eaters – review. *Ibis* 27(1): 103–4.

Anon. (1885b). Obituary – Nikolai Alexsyewich Severtzoff. *Ibis* 27(2): 238–9.

Anon. (1897a). Dresser's *Supplement to the Birds of Europe* – review. *Ibis* 39(2): 273.

Anon. (1897b). The Tristram collection of birds. *Ibis* 39(3): 488–90.

Anon. (1899a). Obituary: William Edwin Brooks. *Ibis* 41(3): 468.

Anon. (1899b). Notice of the acquisition of Mr H. E. Dresser's collection of birds by the Manchester Museum. *Ibis* 41(4): 663–4.

Anon. (1899c) Notice of the acquisition of H. E. Dresser's collection by Manchester Museum. *Science Gossip* (new series) 6(64): 119.

Anon. (1899d). Sale of H. E. Dresser's ornithological collection. *Zoologist* (fourth series) 3(698): 384.

Anon. (1902). Anhang [Rules of Zoological Nomenclature, Adopted by the Fifth International Congress of Zoology, in German, French and English]. In P. Matschie (ed.), *Verhandlungen des V. Internationalen Zoologen-Congresses zu Berlin 12–16 August 1901*, pp. 927–72. Gustav Fischer, Jena.

Anon. (Philip Sclater?) (1903a). Dresser's *Manual of Palaearctic Birds*. *Ibis* 45(4): 610–11.

Anon. (1903b). *A Manual of Palaearctic Birds* [review]. *The Field* 102(2552, 10 October): 626.

Anon. (1903c). *A Manual of Palaearctic Birds*. By H. E. Dresser. *Zoologist* (fourth series) 7(749): 438–9.

Anon. (Philip Sclater?) (1904a). Hartert on the Palaearctic avifauna. *Ibis* 46(2): 291–3.

Anon. (1904b). New books in preparation. *Ibis* 46(2): 298–9.

Anon. (1904c). Obituary: Frederick Dresser. *Proceedings of the Institution of Civil Engineers* 156: 440.

Anon. (1905a). Obituary: William Thomas Blanford, C.I.E., L.L.D., F.R.S. *Geographical Journal* 26(2): 223–5.

Anon. (1905b). Review of *Rules of Zoological Nomenclature*. *Ibis* 47(3): 493–4.

Anon. (1905c). The Zoological Society. *The Field* 105(2717, 21 January): 110.

Anon. (1906a). The Society and Its work. *Bird Notes and News* 2(1): 1–3.

Anon. (1906b). The Great Skua in Iceland. *Bird Notes and News* 2(2): 16

Anon. (1906c). In the Queen's name. *Bird Notes and News* 2(2): 19–20.

Anon. (1906d). Bird protection orders. *Bird Notes and News* 2(3):27–8.

Anon. (1906e). The Royal Society for the Protection of Birds autumn conference – how to approach the wearers of bird millinery. *Bird Notes and News* 2(4): 42–3.

Anon. (1906f). Plume sales. *Bird Notes and News* 2(4): 51–2.

Anon. (1906g). List of attendees. *Bulletin of the British Ornithologists' Club* 119(127): 1–2.

Anon. (1906h). Dresser's 'Eggs of the Birds of Europe'. *Ibis* 48(1): 192–3.

Anon. (1906i). Jourdain on the eggs of European birds. *Ibis* 48(4): 722–3.

Anon. (1906j). Krause's 'Palaearctic Oology'. *Ibis* 48(4): 725–6.

Anon. (1907a). British Ornithologists' Union. *Ibis* 49(3): front matter.

Anon. (1907b). Proceedings at the Annual General Meeting of the British Ornithologists' Union, 1907. *Ibis* 49(3): 476–8.

Anon. (1908a). The B. O. U. and bird protection. *Bird Notes and News* 3(2): 16.

Anon. (1908b). The Importation of Plumage Bill. *Bird Notes and News* 3(4): 45–7.

Anon. (1908c). 'Birds of Great Britain'. *British Birds* 1(9): 300–1.

Anon. (1908d). [Financial accounts of the British Ornithologists' Club]. *Bulletin of the British Ornithologists' Club* 23(146): 28.

Anon. (1908e). Proceedings at the annual general meeting of the British Ornithologists' Union, 1908. *Ibis* 50(3): 517–21.

Anon. (1909a). Proceedings of the special jubilee meeting of the British Ornithologists' Union. *Ibis* 50, Supplement 1 (Jubilee Supplement): 1–18.

Anon. (1909b). Proceedings at the annual general meeting of the British Ornithologists' Union, 1909. *Ibis* 51(3): 532–6.

Anon. (1910a). International Ornithological Congress. *Bird Notes and News* 4 (2): 13.

Anon. (1910b). Dresser on Palaearctic birds' eggs. *Ibis* 52(4): 739–40.

Anon. (1910c). Proceedings of the annual general meeting. *Ibis* 52(3): 535–7.

Anon. (1911a). National collections. *Bird Notes and News* 4(8): 92.

Anon. (1911b). George Ernest Shelley (obituary). *Ibis* 53(2): 369–76.

Anon. (1911c). Dresser on Palaearctic birds' eggs. *Ibis* 53(2): 382–3.

Anon. (1911d). Proceedings at the annual general meeting of the British Ornithologists' Union, 1911. *Ibis* 53(3): 553–6.

Anon. (1912). The Dresser collection of birds' eggs. *Ibis* 54(1): 215–16.

Anon. (1916a). Oological dinner. *Ibis* 58(1): 186–91.

Anon. (1916b). Obituary – H. E. Dresser. *Ibis* 58(2): 340–2.

Anon. (1921). Ninth oological dinner. *Bulletin of the British Ornithologists' Club* 41(261): 143–50.

Anon. (1922a). The B. O. U. and the egg-collector. *Bird Notes and News* 10(1): 3–4.

Anon. (1922b). Wild birds eggs. *Bird Notes and News* 10(2): 25–6.

Anon. (1922c). (Oological Club). *Bulletin of the British Ornithologists' Club* 42(268): 114–15.

Anon. (1922d). Obituary: John Biddulph. *Ibis* 64(2): 348–9.

Anon. (1926a). Mr. E. P. Chance. *Ibis* 64(3): 636.

Anon. (1926b). The resignation of Mr. Chance. *Ibis* 64(4): 841–3.

Anon. (1931). The oological collection at Cambridge. *Ibis* 73(2): 355–7.

Anon. (1957). The first fifty years. *British Birds* 50(6): 213–23.

Aplin, O. (1890). The Pine Grosbeak as a British bird. *Zoologist* (third series) 14(161): 184–5.

Armstrong, P. (2000). *The English Naturalist Parson: A Companionship Between Science and Religion*. Gracewing, Leominister.

Atherton, S. (2008). *An Evolutionary and Forensic Study of the Sacred Ibis in Ancient Egypt Using Specimens from the Manchester Museum Collection*. Unpublished MSc thesis, University of Manchester.

Aubrecht, G. (2000). The Azores Bullfinch – *Pyrrhula murina* Godman, 1866. The history of a bird species: persecuted – missing – rediscovered – protected (?)

(including a list of all known specimens and syntypes). *Annalen des Natur-historischen Museums in Wien* 102B: 23–62.

Audubon, J. J. (1827–38). *The Birds of America; From Original Drawings.* Privately published, London.

Baigent, E. (2004). Lansdell, Henry (1841–1919). *Oxford Dictionary of National Biography*, Oxford University Press (online edition, 2006), www.oxforddnb.com/view/article/45627, accessed 8 April 2017.

Baird, S. F. (1852). *Directions for Collecting, Preserving, and Transporting Specimens of Natural History, Prepared for the Use of the Smithsonian Institution.* Smithsonian Institution, Washington, DC.

Baird, S. F., T. M. Brewer and R. Ridgway (1874). *A History of North American Birds.* Three vols. Little, Brown and Co., Boston.

Balfour, F. R. S. (rev. E. Baigent) (2004). Elwes, Henry John (1846–1922). *Oxford Dictionary of National Biography*, Oxford University Press (online edition, 2006), www.oxforddnb.com/view/article/33019, accessed 8 April 2017.

Bannerman, D. A and W. M. Bannerman (1963–68). *The Birds of the Atlantic Islands.* Three vols. Oliver and Boyd, Edinburgh.

Barber, L. (1980). *The Heyday of Natural History, 1820–70.* Jonathan Cape, London.

Barclay-Smith, P. (1959). The British contribution to bird protection. *Ibis* 101(1): 115–22.

Barnston, G. (1860). Recollections on the swans and geese of Hudson's Bay. *Ibis* 2(3): 253–9.

Barr, W. (1985). *The Expeditions of the First International Polar Year, 1882–83.* AINA Technical Paper 29. Arctic Institute of North America, Calgary.

Barrow, M. V. (1998). *A Passion for Birds: American Ornithology After Audubon.* Princeton University Press, Princeton.

Bassin, M. (1983). The Russian Geographical Society, the 'Amur Epoch', and the Great Siberian Expedition 1855–1863. *Annals of the Association of American Geographers* 73(2): 240–56.

Bavington Jones, T. and W. T. Pike (1904). *Kent at the Opening of the Twentieth Century.* W. T. Pike, Brighton.

Bensaude-Vincent, B. (1996). Between history and memory: centennial and bi-centennial images of Lavoisier. *Isis* (thirteenth series) 1: 255–302.

Bernath, S. L. (1970). *Squall Across the Atlantic: American Civil War Prize Cases and Diplomacy.* University of California Press, Berkeley and Los Angeles.

Berthold, P. (2001). *Bird Migration: a General Survey.* (Second edition.) Oxford University Press, Oxford.

Bewick, T. (1797–1804). *A History of British Birds.* Two vols. Hodgson, Beilby and Bewick, Newcastle.

Bianchi, V. A. (1904). Key to the Palaearctic species of larks of the genus *Otocorys*. *Ibis* 46(3): 370–2.

Biddulph, J. (1880). *Tribes of the Hindoo Koosh.* Superintendent of Government Printing, Calcutta.

Biddulph, J. (1881–82). On the birds of Gilgit. *Ibis* 23(1): 35–102; 24(2): 266–90.

Binnema, T. (2014). *Enlightened Zeal: The Hudson's Bay Company and Scientific Networks, 1670–1870.* University of Toronto Press, London.

Bircham, P. (2007). *A History of Ornithology.* Collins New Naturalist. Harper Collins, London.

Birkhead, T. R. and P. T. Gallivan (2012). Newton's contribution to ornithology: a conservative quest for facts rather than grand theories. *Ibis* 154(4): 887–905.

Birkhead, T. R., J. Wimpenny and B. Montgomerie (2014). *Ten Thousand Birds: Ornithology Since Darwin.* Princeton University Press, Princeton.

Bishop, P. (2003). Tibet. In J. Speake (ed.), *Literature of Travel and Exploration: An Encyclopedia, Vol. 3*, pp. 1178–80. Three vols. Fitzroy Dearborn (Taylor and Francis Group), London.

Black, B. J. (2000). *On Exhibit: Victorians and Their Museums*. University Press of Virginia, Charlottesville.

Blakiston, T. A. (1883). Zoological indications of ancient connection of the Japan islands with the continent. *Transactions of the Asiatic Society of Japan* 11: 126–40.

Blakiston, T. A. (1884). *Amended List of the Birds of Japan, According to Geographical Distribution: With Notes Concerning Additions and Corrections Since January 1882*. Taylor and Francis, London.

Blakiston, T. and H. Pryer (1878). A catalogue of the birds of Japan. *Ibis* 20(3): 209–50.

Blanford, W. T. (1870). *Observations on the Geology and Zoology of Abyssinia*. Macmillan, London.

Blanford, W. T. (1873). Notes on *Stray Feathers*. *Ibis* 15(2): 211–25.

Blanford, W. T. (1874). On the eggs of warblers and the British Association rules of zoological nomenclature. *Ibis* 16(2): 300–2.

Blanford, W. T. (1876). *Eastern Persia, An Account of the Journeys of the Persian Boundary Commission 1870–71–72, Vol. 2: The Zoology and Geology*. Two vols. Macmillan, London.

Blanford, W. T. and H. E. Dresser (1874). Monograph of the genus *Saxicola*, Bechstein. *Proceedings of the Zoological Society of London* 1874(2): 213–41.

Blasius, R. von (1904). Gustav Radde – ein Lebensbild. *Journal für Ornithologie* 52(1): 1–49.

Bloch, D. (2012). Beak tax to control predatory birds in the Faroe Islands. *Archives of Natural History* 39(1): 126–35.

Boardman, S. L. (1903). *The Naturalist of the Saint Croix – Memoir of George A. Boardman*. Privately published, Bangor (Maine).

Boase, G. C. (rev. E. Baigent) (2004). Brady, Sir Antonio (1811–1881). *Oxford Dictionary of National Biography*, Oxford University Press (online edition, 2006), www.oxforddnb.com/view/article/3214, accessed 8 April 2017.

Boelens, B. and M. Watkins (2003). *Whose Bird? Common Bird Names and the People They Commemorate*. Yale University Press, London.

Bonhote, J. L. (1907). *Birds of Britain, with 100 Illustrations in Colour Selected by H. E. Dresser from His 'Birds of Europe'*. A. and C. Black, London.

Bonhote, J. L. (1910). Remarks on P. Bunyard's exhibition of eggs. *Bulletin of the British Ornithologists' Club* 27(163): 18–19.

Bourne, W. R. P. (1988). In memory of 'The Zoologist'. *British Birds* 88(1): 1–4.

Boyd, A. W. (1927). Leach's Fork-tailed Petrel inland in Lancashire and Cheshire. *British Birds* 21(7): 185.

Bree, C. R. (1859–67). *A History of the Birds of Europe, Not Observed in the British Isles*. Four vols. Groombridge and Sons, London.

Brewer, T. M. (1877). A run through the museums of Europe. *Popular Science Monthly* 11(25): 472–81.

British Association for the Advancement of Science (1869). Applications for reports and researchers and involving grants of money. *Report of the British Association* 1868: xlvii–xlviii.

British Ornithologists' Union (Committee of) (1883). *A List of British Birds, Compiled by a Committee of the British Ornithologists' Union*. Van Voorst, London.

British Ornithologists' Union (Committee of) (1915). *A List of British Birds, Compiled by a Committee of the British Ornithologists' Union*. Second and revised edition. BOU, London.

Brooke, V. (1879). On the classification of the Cervidae, with a synopsis of the existing species. *Proceedings of the Zoological Society of London* 1878(4): 883–928.

Brooks, W. E. (1868). Letter on European eagles. *Ibis* 10(3): 349–52.

Brooks, W. E. (1871). Notes on the ornithology of Cashmir. *Proceedings of the Asiatic Society of Bengal* 9: 209–10.

Brooks, W. E. (1872). On the breeding of *Reguloides superciliosus, Reguloides proregulus, Reguloides occipitalis*, and *Phylloscopus tytleri*. *Ibis* 14(1): 24–31.

Brooks, W. E. (1873a). Notes upon some of the Indian and European eagles. *Stray Feathers* 1(2, 3, 4): 290–4.

Brooks, W. E. (1873b). Notes upon some of the Indian and European eagles, No. II. *Stray Feathers* 1(5): 325–31.

Brooks, W. E. (1873c). Notes upon some of the Indian and European eagles, No. III. *Stray Feathers* 1(6): 463–4.

Brooks, W. E. (1875). Notes on 'The Spotted Eagle' (*Aquila naevia*). *Stray Feathers* 3(4): 304–13.

Brooks, W. E. (1876). A few ornithological notes and corrections. *Ibis* 18(4): 499–504.

Brooks, W. E. (1884). A few ornithological notes and corrections. *Ibis* 26(3): 234–40.

Brown, T. (1820). *The Taxidermist's Manual; or, the Art of Collecting, Preparing and Preserving Objects of Natural History*. Archibald Fullarton, Glasgow.

Browne, J. (1983). *The Secular Ark: Studies in the History of Biogeography*. Yale University Press, Newhaven.

Bruce, R. V. (1987). *The Launching of Modern American Science, 1846–1876*. Knopf, New York.

Bruner, K. F., J. K. Fairbank and R. J. Smith (1986). *Entering China's Service: Robert Hart's Journals, 1854–63*. Harvard East Asian Monographs, 125. Harvard University Press, Cambridge.

Buchanan, G. M., A. L. Bond, N. J. Crockford, J. Kamp, J. W. Pearce-Higgins and G. M. Hilton (2017). The potential breeding range of Slender-billed Curlew *Numenius tenuirostris* identified from stable-isotope analysis. *Bird Conservation International* 2 March: 1–10, https://doi.org/10.1017/S0959270916000551, accessed 8 April 2017.

Buckland, A. R. (rev. R. A. Butlin) (2004). Tristram, Henry Baker (1822–1906). *Oxford Dictionary of National Biography*, Oxford University Press (online edition, 2006), www.oxforddnb.com/view/article/36560, accessed 8 April 2017.

Bunyard, P. F. (1906). Exhibition of a pair of Grey-headed Wagtails together with their nest and eggs, taken at Rye, Sussex [introduced by H. E. Dresser]. *Bulletin of the British Ornithologists' Club* 19(128): 23.

Bunyard, P. F. (1909). 'Exhibition of eggs of Red-backed Shrike' and 'Exhibition of eggs of Nightingale and Garden Warbler'. *Bulletin of the British Ornithologists' Club* 23(148): 53.

Burkhalter, L. W. (1965). *Gideon Lincecum, 1793–1874: A Biography*. University of Texas Press, Austin.

Burkhardt, F. and J. Secord (eds) (2015). *The Correspondence of Charles Darwin, Vol. 23: 1875*. Cambridge University Press, Cambridge.

Burkhardt, F. and S. Smith (eds) (1985). *The Correspondence of Charles Darwin, Vol. 1: 1821–36*. Cambridge University Press, Cambridge.

Burton, A. (2010). *The Development of Museums in Victorian Britain and the Contribution of the Society of Arts*. William Shipley Group for RSA History, London.

Buturlin, S. A. (1904a). On the geographical distribution of the true Pheasants (genus *Phasianus* sensu stricto). *Ibis* 46(3): 377–414.

Buturlin, S. A. (1904b). Über neue formen der echten Stare. *Ornithologisches Jahrbuch* 15: 205–13.

Buturlin, S. A. (1906). On the breeding grounds of the Rosy Gull. *Ibis* 48(1): 131–9; 48(2): 333–7, 400; 48(4): 661–6.

Buturlin, S. A. (1907a). On some new or little-known Siberian birds (Communication of a paper by S. A. Buturlin, containing descriptions of three new species and five new subspecies of Siberian birds). *Abstracts of Proceedings of the Zoological Society of London* 43 (9 April): 20.

Buturlin, S. A. (1907b). Neue ost-Asiatische formen. *Ornithologische Monatsberichte* 15(5): 79–80.

Buturlin, S. A. (1916a). [Henry E. Dresser – obituary.] *Nasha Okhota [Our Hunting]* 1916(3): 3–4 (in Russian).

Buturlin, S. A. (1916b). [Henry Eeles Dresser – obituary.] *Messager Ornithologique [Ornitologiskii Vyestnik]* 7: 71–4 (in Russian).

Carrington, M. (2003). Officers, gentlemen and thieves: the looting of monasteries during the 1903/04 Younghusband mission to Tibet. *Modern Asian Studies* 37(1): 81–109.

Casto, S. D. (1995a). Patrick Duffy and his collection of birds from Fort Stockton, Texas. *Bulletin of Texas Ornithological Society* 28(1): 23–6.

Casto, S. D. (1995b). A. L. Heermann and his natural history collections from San Antonio and the Medina River Valley. *La Tierra* 22: 19–24.

Casto, S. D. (1997). The birds collected at San Antonio by A. L. Heermann. *Bulletin of Texas Ornithological Society* 30(1): 2–10.

Caswell, J. E. (1977). The RGS and the British Arctic Expedition, 1875–76. *Geographical Journal* 143(2): 200–10.

Cavanagh, T. (1997). *Public Sculpture of Liverpool.* Liverpool University Press, Liverpool.

Chalmers-Hunt, J. M. (1976). *Natural History Auctions 1700–1972: A Register of Sales in the British Isles.* Sotheby Parke Bernet, London.

Chance, E. P. (1937). *Some Observations on Egg Collecting and Bird Protection.* Privately published.

Chance, E. P. (1938). *An Egg Collector Replies to His Critics.* Privately published.

Chapman, E. (1907). What constitutes a museum collection of birds? In R. B. Sharpe, E. J. O. Hartert and J. L. Bonhote (eds), *Proceedings of the Fourth International Ornithological Congress, London June 1905*, pp. 144–56. Dulau and Co., London.

Chilton, G. (1997). Labrador Duck (*Camptorhynchus labradorius*). In A. Poole and F. Gill (eds), *The Birds of North America, No. 307*, pp. 1–12. Academy of Natural Sciences and American Ornithologists' Union, Philadelphia, and Washington, DC.

Chisholm, A. H. (1976). Wheelwright, Horace William (Horatio) (1815–1865). In B.Nairn (ed.), *Australian Dictionary of Biography, Vol. 6*, pp. 383–4, Melbourne University Press, Melbourne.

Chisholm, A. H. (1979). Foreword to H. W. Wheelwright ('An Old Bushman'), *Bush Wanderings of a Naturalist.* Oxford University Press, Oxford. (*Bush Wanderings of a Naturalist* first published 1861 by Routledge, Warne and Routledge, London.)

Chohan, A. S. (1984). *The Gilgit Agency, 1877–1935.* Atlantic Publishers and Distributors, New Delhi.

Clancey, P. A. (1969). On the status of *Coracias weigalli* Dresser, 1890. *Ostrich: Journal of African Ornithology* 40(4): 156–62.

Clifton, Lord (Edward Henry Stuart) (1879). Letter on some British birds. *Ibis* 21(3): 368–71.

Close-Time Committee (1870). Report on the practicability of establishing a 'close-time' for British animals. *Report of the British Association for the Advancement of Science* 1869: 91–6.

Close-Time Committee (1871). Report on the practicability of establishing 'A Close Time' for the protection of indigenous animals. *Report of the British Association for the Advancement of Science* 1870: 13–14.

Close-Time Committee (1875). Report of the Committee, consisting of the Rev. H. F. Barnes, H. E. Dresser (Secretary), T. Harland, J. E. Harting, Professor Newton, and the Rev. Canon Tristram, appointed for the purpose of inquiring into the possibility of establishing a 'Close Time' for the protection of indigenous animals. *Report of the British Association for the Advancement of Science* 1874: 264–6.

Close-Time Committee (1879). Report of the committee … appointed for the purpose of inquiring into the possibility of establishing a 'close time' for indigenous animals. *Report of the British Association for the Advancement of Science* 1878: 146–9.

Coates, P. D. (1988). *The China Consuls: British Consular Officers, 1843–1943.* Oxford University Press, Oxford.

Cocker, M. (1989). *Richard Meinertzhagen: Soldier, Scientist, Spy.* Martin Secker and Warburg, London.

Cole, A. C. (2006). *The Egg Dealers of Great Britain.* Peregrine Books, Horsforth (Leeds).

Cole, A. C. and W. M. Trobe (2000). *The Egg Collectors of Great Britain and Ireland.* Peregrine Books, Horsforth (Leeds).

Collar, N. J. (2004). Pioneers of Asian ornithology: Robert Swinhoe. *Birding Asia* 1: 49–53.

Collar, N. J. and S. N. Stuart (1985). *Threatened Birds of Africa and Related Islands: The ICBP/IUCN Red Data Book.* Third edition, part 1. ICBP, Cambridge.

Collett, R. (1869). *Norges Fugle, og deres geographiske Udbredelse i Landet.* Videnskabs-Selskabet Forhandlinger (1868), Oslo.

Collinson, J. M. (2012). Leadenhall Market as a historical source of rare bird specimens. *British Birds* 105(6): 293–350.

Cortazzi, H. (1999). Thomas Wright Blakiston (1832–91). In J. E. Hoare (ed.), *Britain and Japan: Biographical Portraits, Vol. 3*, pp. 52–65. Japan Library, Richmond.

Coues, E. (1872). *Key to North American Birds, Including a Concise Account of Every Species of Living and Fossil Bird at Present Known from the Continent North of the Mexican and United States Boundary.* Naturalists' Agency, Salem.

Coues, E. (1874). *Field Ornithology.* Naturalists' Agency, Salem.

Coues, E. (1876). The Labrador Duck (letter). *American Naturalist* 10(5): 303.

Coues, E. (1884). On the application of trinomial nomenclature to zoology. *Zoologist* (third series) 8(91): 241–7.

Cowell, F. R. (1975). *The Athenaeum – Club and Social Life in London 1824–1974.* Heinemann, London.

Cropley, R. A. (1874). *Thirty-Ninth Report of the Diocesan Church Society of New Brunswick.* H. A. Cropley, Fredericton.

Cudworth, W. (1891). *Histories of Bolton and Bowling (Townships of Bradford), Historically and Topographically Treated.* Thomas Brear, Bradford.

Dance, S. P. (1978). *The Art of Natural History: Animal Illustrators and Their Work.* Overlook Press, Woodstock.

Danford, C. G. (1877). Ornithology of Asia Minor. *Ibis* 19(3): 260–74.

Danford, C. G. (1878). Ornithology of Asia Minor (continued). *Ibis* 20(1): 1–35.

Danford, C. G. (1880). A further contribution to the ornithology of Asia Minor. *Ibis* 22(1): 81–99.

Danford, C. G. and E. R. Alston (1877). On the mammals of Asia Minor. *Proceedings of the Zoological Society of London* 1877(2): 270–82.

Danford, C. G. and E. R. Alston (1880). On the mammals of Asia Minor. Part 2. *Proceedings of the Zoological Society of London* 1880(1): 50–64.

Darwin, C. R. (1859). *On the Origin of Species by Means of Natural Selection, or the Preservation of Favoured Races in the Struggle for Life.* John Murray, London.

Darwin, C. R. (1958). *The Autobiography of Charles Darwin 1809–82, with Original Omissions Restored. Edited and with Appendix and Notes by His Grand-Daughter Nora Barlow.* Collins, London.

Davis, P. (1962). River Warbler on Fair Isle: a bird new to Britain. *British Birds* 55(4): 137–8.

Davison, G. W. H. (2013). Dresser, Seebohm, and the scope of Palaearctic ornithology. *Raffles Bulletin of Zoology*, Supplement 29: 259–68.

Deiss, W. A. (1980). Spencer F. Baird and his collectors. *Journal of the Society for the Bibliography of Natural History* 9(4): 635–45.

Demata, M. (2003). 'Murray Handbooks'. In J. Speake (ed.), *Literature of Travel and Exploration: An Encyclopedia, Vol. 2*, pp. 830–1. Three vols. Fitzroy Dearborn (Taylor and Francis Group), London.

Dement'ev, G. P. (1948). [*N. A. Severtsov.*] Moscow (in Russian).

Densley, M. (1999). *In Search of Ross's Gull.* Peregrine Books, Horsforth (Leeds).

Diamond, J. W. (1940). Imports of the Confederate Government from Europe and Mexico. *Journal of Southern History* 6(4): 470–503.

Dickinson, E. C., R. W. R. J. Dekker, S. Eck and S. Somadikarta (2001). Systematic notes on Asian birds. 12. Types of the Alaudidae. *Zoologische Verhandlungen Leiden* 335: 85–126.

Dickinson, E. C., V. M. Loskot, H. Morioka, S. Somadikart and R. van den Elzen (2006). Systematic notes on Asian birds. 50. Types of the Aegithalidae, Remizidae and Paridae. *Zoologische Mededelingen, Leiden* 80–5(2): 65–111.

Dike, C. (1983). *Cane Curiosa, From Gun to Gadget.* ACC Distribution, Woodbridge.

Dixon, C. (1882). Notes on the birds of Constantine, Algeria. *Ibis* 24(4): 550–79.

Dixon, C. (1893). *The Nests and Eggs of British Birds: When and Where to Find Them.* Chapman and Hall, London.

Dodsworth, C. (1969). Further observations on the Bowling Ironworks. *Industrial Archaeology* 6(2): 114–23.

Doughty, R. B. (1975). *Feather Fashions and Birds Preservation: A Study in Nature Protection.* University of California Press, Berkeley.

Downham, C. F. (1911). *The Feather Trade: The Case for the Defence.* F. Howard Doulton and Co., London.

Dresser, H. E. (1863). Occurrence of a white Redwing in Norfolk. *Zoologist* (first series) 21: 8484.

Dresser, H. E. (1865). Notes on the birds of Southern Texas. *Ibis* 7(3): 312–30; 7(4): 466–95.

Dresser, H. E. (1866a). Notes on the birds of Southern Texas (concluded). *Ibis* 8(1): 23–46.

Dresser, H. E. (1866b). Notes on the breeding of the Booted Eagle (*Aquila pennata*). *Proceedings of the Zoological Society of London* 1866(3): 377–80.

Dresser, H. E. (1867a). List of birds noticed in East Finmark, with a few short remarks respecting some of them. By Ch. Sommerfeldt, parish priest of Naesseby. Translated and communicated by H. E. Dresser. *Zoologist* (second series) 2(April): 692–700; 2(June): 761–78.

Dresser, H. E. (1867b). Notes on the breeding of the Booted Eagle (*Aquila pennata*) (communicated by the author and reprinted from the *Proceedings of the Zoological Society of London*). *Zoologist* (second series) 2(July): 803–7.

Dresser, H. E. (1869). Cuckow's eggs [letter]. *Nature* 1(8): 218.

Dresser, H. E. (1870). Exhibition of some eggs of the Little Gull (*Larus minutus*). *Proceedings of the Zoological Society of London* 1869(3): 530–1.

Dresser, H. E. (initially Sharpe, R. B. and H. E. Dresser) (1871–82). *A History of the Birds of Europe, Including All the Species Inhabiting the Western Palaearctic Region.* Eighty-four parts. Privately published, London. (Note that Sharpe assisted only with the first thirteen parts of this work, although his name appeared as lead author on the paper covers of parts 1–17.)

Dresser, H. E. (1872). Exhibition of, and remarks upon, some skins and eggs of various species of *Reguloides* and *Phylloscopus*. *Proceedings of the Zoological Society of London* 1872(1): 25–6.

Dresser, H. E. (1873). Exhibition of, and remarks upon, the skins of various eagles (*Aquila*). *Proceedings of the Zoological Society of London* 1872(3): 863–5.

Dresser, H. E. (1875a). Notes on Severtzoff's 'Fauna of Turkestan' (Turkestanskie Jevotnie). *Ibis* 17(1): 96–112; 17(2): 236–50; 17(3): 332–2.

Dresser, H. E. (1875b). Notes on *Falco labradorus*, Aud., *Falco sacer*, Forster, and *Falco spadicus*, Forster. *Proceedings of the Zoological Society of London* 1875(2): 114–17.

Dresser, H. E. (1875c). Letter: On *Syvia rama* and an occurrence of the Eastern Golden Plover in Britain. *Ibis* 17(4): 513–14.

Dresser, H. E. (1876a). Notes on Severtzoff's *Fauna of Turkestan* (Turkestanskie Jevotnie). *Ibis* 18(1): 77–94; 18(2): 171–91; 18(3): 319–30; 18(4): 410–22.

Dresser, H. E. (1876b). *Falco labradorus* (Labrador Falcon). *Ornithological Misellany* 1: 185–91.

Dresser, H. E. (1876c). Remarks on a hybrid between the Black Grouse and the Hazel Grouse. *Proceedings of the Zoological Society of London* 1876(2): 345–7.

Dresser, H. E. (1876d). On a new species of *Tetraogallus*. *Proceedings of the Zoological Society of London* 1876(3): 675–7.

Dresser, H. E. (1880). *A History of the Birds of Europe, Including All the Species Inhabiting the Western Palaearctic Region*. Parts 77–82. Privately published, London.

Dresser, H. E. ('1881' = 1882). *A History of the Birds of Europe, Including All the Species Inhabiting the Western Palaearctic Region*. Part 83–84 (final part). Privately published, London.

Dresser, H. E. (1883a). Exhibition of, and remarks upon, the identity of *Melittophagus boehmi* and *Merops dresseri*. *Proceedings of the Zoological Society of London* 1882(4): 634.

Dresser, H. E. (1883b). Exhibition of, and remarks upon, a specimen of *Merops philippinus*, stated to have been obtained near the Snook, Seaton Carew. *Proceedings of the Zoological Society of London* 1883(1): 1.

Dresser, H. E. (1884). Letter: On *Otocorys brandti* and *Otocorys longirostris*. *Ibis* 26(1): 116–18.

Dresser, H. E. (1884–6). *A Monograph of the Meropidae, or Family of the Bee-eaters*. Five parts. Privately published, London.

Dresser, H. E. (1885a). Exhibition of, and remarks upon, specimens of *Sylvia nisoria* and *Hypolais icterina*, killed in Norfolk. *Proceedings of the Zoological Society of London* 1884(4): 477–8.

Dresser, H. E. (1885b). The species of British-killed Spotted Eagles determined. *Zoologist* (third series) 9(102): 230–1.

Dresser, H. E. (1886). On the Wren of St Kilda. *Ibis* 28(1): 43–5.

Dresser, H. E. (1890a). Three weeks on the Guadalquivir. *The Naturalist* 174: 17–38.

Dresser, H. E. (1890b). Notes on the Racquet-tailed Rollers. *Annals and Magazine of Natural History* (sixth series) 6: 350–1.

Dresser, H. E. (1890c). Letter: On different forms of African Rollers. *Ibis* 32(3): 384–6.

Dresser, H. E. (1892). Remarks on the Palaearctic White-breasted Dippers. *Ibis* 34(3): 380–7.

Dresser, H. E. (1893a). Exhibition and description of *Cryptolopha xanthopygia* obtained from the Island of Palawan by Mr J. Whitehead. *Bulletin of the British Ornithologists' Club* 1(6): 31.

Dresser, H. E. (1893b). *A Monograph of the Coraciidae, or Family of the Rollers.* Privately published, Farnborough.

Dresser, H. E. (1895–96). *Supplement to A History of the Birds of Europe, Including All the Species Inhabiting the Western Palaearctic Region.* Nine parts. Privately published, London.

Dresser, H. E. (ed.) (1896–97). Society for the Protection of Birds. Educational Series leaflets. Published by the Society for the Protection of Birds at the *Knowledge* office, London.

Dresser, H. E. (1897a). Pallas's Willow Warbler shot at Cley-next-the-sea, Norfolk. *Proceedings of the Zoological Society of London* 1896(4): 856.

Dresser, H. E. (1897b). Osprey in Dorset. *Zoologist* (fourth series) 1(677): 508.

Dresser, H. E. (1898). Rare partridges in Leadenhall Market. *Zoologist* (fourth series) 2(683): 215.

Dresser, H. E. (1901a). Exhibition of specimens of 'three-colour' printing. *Bulletin of the British Ornithologists' Club* 11(79): 59.

Dresser, H. E. (1901b). Notice of reproductions of an illustration of the Labrador Falcon by Joseph Wolf. *Zoologist* (fourth series) 5(724): 400.

Dresser, H. E. (1901c). On some rare or unfigured Palaearctic birds' eggs. *Ibis* 43(3): 445–9.

Dresser, H. E. (1901d). Local variation in Grouse [letter]. *The Field* 98(2552, 23 November): 829.

Dresser, H. E. (1902). On a new pheasant from Japan. *Ibis* 44(4): 656–7.

Dresser, H. E. (1902–3). *A Manual of Palaearctic Birds.* Two parts. Privately published, London.

Dresser, H. E. (1903). Birdsnesting in lower Hungary, Bosnia, &c. *The Field* 101(2615, 7 February): 222.

Dresser, H. E. (1904a). Breeding of the Knot on the Taimyr Peninsula. *Bulletin of the British Ornithologists' Club* 14(102): 32. (Meeting date 16 December 1903, abstract dated 29 December, but probably not widely available until early 1904.)

Dresser, H. E. (1904b). On the late Dr. Walter's Ornithological researches in the Taimyr Peninsular. *Ibis* 46(2): 228–35.

Dresser, H. E. (1905a). Descriptions of three new species of birds obtained during the recent expedition to Lhassa. *Proceedings of the Zoological Society of London* 1905(1): 54–5.

Dresser, H. E. (1905b). An oological journey to Russia. *Ibis* 47(2): 149–58.

Dresser, H. E. (1905c). Exhibition of some Tibetan eggs. *Bulletin of the British Ornithologists' Club* 16(120): 38.

Dresser, H. E. (1905d). On Mr. Buturlin's discovery of the breeding-place of Ross's Rosy Gull in NE Siberia. *Bulletin of the British Ornithologists' Club* 16(120): 41.

Dresser, H. E. (1905–10). *Eggs of the Birds of Europe, Including All the Species Inhabiting the Western Palaearctic Area.* Twenty-four parts. Privately published, London.

Dresser, H. E. (1906a). Exhibition of eggs of Ross's Rosy Gull. *Bulletin of the British Ornithologists' Club* 16(125): 97.

Dresser, H. E. (1906b). On some Palaearctic birds' eggs from Tibet. *Ibis* 48(2): 337–47.

Dresser, H. E. (1906c). Note on the eggs of Ross's Rosy Gull. *Ibis* 48(3): 610–11.

Dresser, H. E. (1906d). Letters on the taking of the eggs of the Great Skua in Iceland. *Ibis* 48(3): 611–12; 48(4): 737–8.

Dresser, H. E. (1906e). Obituary of Canon Henry Baker Tristram, D.D., F.R.S., etc. etc. *Zoologist* (fourth series) 10(778): 155–6.

Dresser, H. E. (1907a). On behalf of Mr. S. A. Buturlin exhibited and made remarks on examples of nine species of Siberian birds. *Bulletin of the British Ornithologists' Club* 19(130): 43.

Dresser, H. E. (1907b). Exhibition of young in down of *Rhodostethia rosea*, *Tringa maculata*, and *Limosa novae zealandiae* from N. E. Siberia. *Bulletin of the British Ornithologists' Club* 19 (135): 109.

Dresser, H. E. (1907c). Obituary of Professor A. Newton. *Zoologist* (fourth series) 11(793): 272–3.

Dresser, H. E. (1908a). On the Russian Arctic Expedition of 1900–1903. *Ibis* 50(3): 510–17; 50(4): 593–9.

Dresser, H. E. (1908b). Exhibition of some rare eggs, viz. *Lampronetta fischeri*, Brandt, and *Phylloscopus viridianus*. *Bulletin of the British Ornithologists' Club* 23(146): 39.

Dresser, H. E. (1908c). Baron Toll's Russian Arctic Expedition. *The Field* 112(2911, 10 October): 630–1.

Dresser, H. E. (1909a). Exhibition of two examples of a rare wader (*Pseudoscolopax taczanowskii*, Verr.) from Western Siberia [together with egg]. *Bulletin of the British Ornithologists' Club* 23(149): 60–1.

Dresser, H. E. (1909b). On the occurrence of *Pseudoscolopax taczanowskii* in Western Siberia. *Ibis* 51(3): 418–21.

Dresser, H. E. (1910a). Exhibition of eggs of the Slender-billed Curlew (*Numenius tenuirostris*), the Siberian form of the Common Curlew (*N. lineatus*), and Swinhoe's Snipe (*Gallinago megala*). *Bulletin of the British Ornithologists' Club* 25(156): 38–9.

Dresser, H. E. (1910b). Proceedings of the Fifth International Congress of Ornithologists. *Ibis* 52(4): 710–13.

Dresser, H. E. and W. T. Blanford (1874). Notes on the specimens in the Berlin Museum collected by Hemprich and Ehrenberg. *Ibis* 16(4): 335–43.

Dresser, H. E. and H. Trueman Wood (1902). The reproduction of colours by photography. *Nature* 67(1728): 127–9.

Drewitt, C. M. P. (1900). *Lord Lilford: A Memoir by His Sister*. Smith, Elder and Co., London.

Driver, F. (1998). Scientific exploration and the production of geographical knowledge: *Hints to Travellers*. Finisterra 33(65): 21–30.

Dubrovin, N. F. (1890). *N. M. Przheval'skii. Biograficheskii ocherk*. Voennaia tipografiia, St Petersburg (in Russian).

Elliot, D. G. (1914). In memoriam: Philip Lutley Sclater. *Auk* 31(1): 1–12.

Elwes, H. J. (1920). Obituary notices of Fellows deceased – Frederick DuCane Godman, 1834–1919. *Proceedings of the Royal Society of London* (series B) 91(641): i–vi.

Elwes, H. J. and T. E. Buckley (1870). A list of the birds of Turkey. *Ibis* 12(1): 59–77; 12(2): 188–201; 12(3): 327–41.

Endersby, J. (2008). *Imperial Nature: Joseph Dalton Hooker and the Practices of Victorian Science*. Chicago University Press, Chicago.

Evans, A. H. (ed.) (1909a). Colonel H. M. Drummond-Hay. In Biographical notices of the original members of the British Ornithologists' Union, of the principal contributors to the first series of 'The Ibis' and of the officials, pp. 75–7. *Ibis* 50, Supplement 1 (Jubilee Supplement): 71–232.

Evans, A. H. (ed.) (1909b). F. D. Godman. In Biographical notices of the original members of the British Ornithologists' Union, of the principal contributors to the first series of 'The Ibis' and of the officials, pp. 81–92. *Ibis* 50, Supplement 1 (Jubilee Supplement): 71–232.

Evans, A. H. (ed.) (1909c). Sir Edward Newton. In Biographical notices of the original members of the British Ornithologists' Union, of the principal contributors to the first series of 'The Ibis' and of the officials, pp. 117–20. *Ibis* 50, Supplement 1 (Jubilee Supplement): 71–232.

Evans, A. H. (ed.) (1909d). Osbert Salvin. In Biographical notices of the original members of the British Ornithologists' Union, of the principal contributors to the first series of 'The Ibis' and of the officials, pp. 127–8. *Ibis* 50, Supplement 1 (Jubilee Supplement): 71–232.

Evans, A. H. (ed.) (1909e). Henry Eeles Dresser. In Biographical notices of the original members of the British Ornithologists' Union, of the principal contributors to the first series of 'The Ibis' and of the officials, pp. 219–20. *Ibis* 50, Supplement 1 (Jubilee Supplement): 71–232.

Evans, A. H. (ed.) (1909f). Howard Saunders. In Biographical notices of the original members of the British Ornithologists' Union, of the principal contributors to the first series of 'The Ibis' and of the officials, pp. 223–6. *Ibis* 50, Supplement 1 (Jubilee Supplement): 71–232.

Evans, D. (1992). *A History of Nature Conservation in Britain*. Routledge, London.

Export Merchant Shippers (1873). *The Export Merchant Shippers of London Afterw. the Export Merchant Shippers and Manufacturers of Great Britain & Ireland*. London.

Fan, F.-T. (2004). *British Naturalists in Qing China: Science, Empire and Cultural Encounter*. Harvard University Press, Cambridge.

Farber, P. L. (1976). The type-concept in zoology during the first half of the nineteenth century. *Journal of the History of Biology* 9(1): 93–119.

Farber, P. L. (1977). The development of taxidermy and the history of ornithology. *Isis* 68(4): 550–66.

Farber, P. L. (1980). The development of ornithological collections in the late eighteenth and early nineteenth centuries and their relationship to the emergency of ornithology as a scientific discipline. *Journal of the Society for the Bibliography of Natural History* 9(4): 391–4.

Farber, P. L. (1982). *Discovering Birds: The Emergence of Ornithology as a Scientific Discipline. 1760–1850*. Studies in the History of Modern Science, 12. D. Reidel, Dordrecht.

Farman, C. (1868–69). On some birds of prey of central Bulgaria (communicated by H. E. Dresser). *Ibis* 10(4): 406–14; 11(2): 199–204.

Feilden, H. W. (1872). The birds of the Faeroe Islands. *Zoologist* (second series) 7(September): 3210–25; 7(October): 3245–57; 7(November): 3277–94.

Feilden, H. W. (1877). List of birds observed in Smith Sound and in the Polar Basin during the Arctic Expedition of 1875–76. *Ibis* 19(4): 401–12.

Feilden, H. W. (1898). Visits to Barents and Kara Seas, with rambles in Novaya Zemlya, in 1895 and 1897. *Geographical Journal* 11(4): 333–65.

Finsch, O. (1870). On a collection of birds from north-eastern Abyssinia and the Bogos country. With notes by the collector, William Jesse, C.M.Z.S., zoologist to the Abyssinian Expedition. *Transactions of the Zoological Society of London* 7(4): 197–331.

Finsch, O. and G. Hartlaub (1870). *Die Vögel Ost-Afrikas. Baron Carl Claus von der Decken's Reisen in Ost-Afrika, Bd. 4*. C. F. Wintersche Verlagshandlung, Leipzig and Heidelberg.

Fischer, D. L. (2001). *Early Southwest Ornithologists, 1528–1900.* University of Arizona Press, Tucson.

Fisher, C. T. (2004a). Salvin, Osbert (1835–1898). *Oxford Dictionary of National Biography*, Oxford University Press (online edition, 2006), www.oxforddnb.com/view/article/24586, accessed 8 April 2017.

Fisher, C. T. (2004b). Swinhoe, Robert (1836–1877). *Oxford Dictionary of National Biography*, Oxford University Press (online edition, 2006), www.oxforddnb.com/view/article/38460, accessed 8 April 2017.

Fisher, J. (1940). *Watching Birds.* Pelican Books/Penguin Books, London.

Flower, W. and M. I. Kinnear (1951). Isabelline Shrike on the Isle of May: a new British bird. *British Birds* 44(7): 217–19.

Flower, W. H. (1898). Preface. In R. B. Sharpe and W. R. Ogilvie-Grant, *Catalogue of the Birds in the British Museum, Vol. 26: Catalogue of the Plataleae, Herodiones, Steganopodes, Pygopodes, Alcae, and Impennes in the Collection of the British Museum*, pp. v–viii. Trustees of the British Museum, London.

Forsyth, T. D. (ed.) (1875). *Report of a Mission to Yarkund in 1873, Under Command of Sir T. D. Forsyth, K.S.C.I., C.B., with Historical and Geographical Information Regarding the Possessions of the Ameer of Yarkund.* Foreign Department Press, Calcutta.

Fortune, R. (1906). Birds requiring protection in Yorkshire (reprinted from the *Zoologist*). *Bird Notes and News* 2(2): 18.

Forwood, W. B. (1910). *Recollections of a Busy Life, Being the Reminiscences of a Liverpool Merchant, 1840–1910.* Henry Young and Sons, Liverpool.

Fredeman, W. E. (2006). *The Correspondence of Dante Gabriel Rossetti 6. The Last Decade, 1873–1882, Vol. 1: 1873–1874.* Cromwell Press, Trowbridge.

French, P. (1994). *Younghusband: The Last Great Imperial Adventurer.* Harper Collins, London.

Fry, C. H. (1984). *The Bee-eaters.* T and A. D. Poyser, Calton.

Fuller, E. (1999). *The Great Auk.* Harry N. Abrams, New York.

Gabriel, R. S. (2011). *American and British 410 Shotguns.* Krause Publications, Iola.

Gadow, H. (1883). *Catalogue of the Birds in the British Museum, Vol. 8: Catalogue of the Passeriformes or Perching Birds in the Collection of the British Museum. Cinchlomorphae Part V and Certhiomorphae.* Trustees of the British Museum, London.

Galton, F. (1855). *The Art of Travel.* John Murray, London.

Gardiner, J. (2002). *The Victorians: An Age in Retrospect.* Hambledon , London.

Garfield, B. (2007). *The Meinertzhagen Mystery: The Life and Legend of a Colossal Fraud.* Potomac Books, Washington, DC.

Gätke, Heinrich (1895). *Heligoland as an Ornithological Observatory; The Result of Fifty Years' Experience.* David Douglas, Edinburgh.

Gittings, J. (1973). *A Chinese View of China.* BBC, London.

Godman, F. duC. (1907–10). *Monograph of the Petrels.* Two vols. Witherby, London.

Godman, F. duC. and O. Salvin (1918). *Biologia Centrali-Americana: Zoology, Botany and Archaeology.* Sixty-three vols. B. Quaritch, London.

Goldsmid, Sir F. J., O. St John, B. Lovett and E. Smith (1876). *Eastern Persia, an Account of the Journeys of the Persian Boundary Commission 1870–71–72, Vol. 1. The Geography with Narratives.* Two vols. Macmillan, London.

Gollop, J. B, T. W. Barry and E. H. Iverson (1986). *Eskimo Curlew: A Vanishing Species?* Nature Saskatchewan (Saskatchewan Natural History Society), Regina.

Gonzales, L. M., F. Hiraldo, M. Delibes and J. Calderon (1989). Reduction in the range of the Spanish Imperial Eagle (*Aquila adalberti* Brehm, 1861) since AD 1850. *Journal of Biogeography* 16(4): 305–15.

Gould, J. (1832–37). *The Birds of Europe.* Five vols. Privately published, London.

Gould, J. (1840–48). *The Birds of Australia.* Seven vols. Privately published, London.

Gould, J. (1849–83). *The Birds of Asia.* Seven vols. Privately published, London.

Gould, J. (1862–73). *The Birds of Great Britain.* Five vols. Privately published, London.

Gould, J. (1875–88). *The Birds of New Guinea and the Adjacent Papuan Islands, Including Many New Species Recently Discovered in Australia.* Five vols. H. Sotheran, London.

Graber, R. R. and J. W. Graber (1954). Yellow-headed Vulture in Tamaulipas, Mexico. *Condor* 56(3): 165–6.

Graham, J. M. (2000). *The Millennium Book of Topcliffe.* Published by the author, Topcliffe.

Gratzl, K. (1971). Preface to the 1971 edition. In J. Biddulph, *Tribes of the Hindoo Koosh*, pp. v–xii. Akademische Druck-u. Verlagsanstalt, Graz. (First published in 1880 by the Office of the Superintendent of Government Printing, Calcutta.)

Green, R. E. (2008). Demographic mechanism of a historical bird population collapse reconstructed using museum specimens. *Proceedings of the Royal Society* (series B) 275 (1649): 2381–7.

Gretton, A., A. K. Yurlov and G. C. Boere (2002). Where does the Slender-billed Curlew nest, and what future does it have? *British Birds* 95(7): 334–44.

Griscom, L. W. (1920). Notes on the winter birds of San Antonio, Texas. *Auk* 37(1): 49–55.

Griscom, L. W. and M. S. Crosby (1925–26). Birds of the Brownsville region, southern Texas. *Auk* 42(3): 432–40; 43(1): 18–36.

Grouw, H. van and D. Bloch (2015). History of the extant museum specimens of the Faroese white-speckled raven. *Archives of Natural History* 42(1): 23–38.

Günther, A. (1888). Preface. In R. B. Sharpe, *Catalogue of the Birds in the British Museum, Vol. 12: Catalogue of the Passeriformes, or Perching Birds, in the Collection of the British Museum. Fringilliformes: Part III. Containing the Family Fringillidae,* pp. v–vi. Trustees of the British Museum, London.

Gunther, A. E. (1975). *A Century of Zoology at the British Museum Through the Lives of Two Keepers 1815–1914.* Dawsons, London.

Gurney, J. H. (junior) (1870). Birds in Leadenhall Market. *Zoologist* (second series) 5(December): 2393–4.

Gurney, J. H. (junior) (1876). *Rambles of a Naturalist in Egypt and Other Countries, with an Analysis of the Claims of Certain Foreign Birds to be Considered British and Other Ornithological Notes.* Jarrold, London.

Gurney, J. H. (junior) (1883). Imported gamebirds in English markets. *Zoologist* (third series) 7(79): 300–1.

Gurney, J. H. (junior) (1890). On the claim of the Pine Grosbeak to be regarded as a British bird. *Zoologist* (third series) 14(160): 125–9.

Gurney, J. H. (senior) (1872). Letter on the nostrils of certain Aquilae. *Ibis* 14(4): 472.

Gurney, J. H. (senior) (1876). Notes on a *Catalogue of the Accipitres in the British Museum. Ibis* 18(2): 230–43.

Gutiérrez Exposito, C., J. L. Copete, P.-A. Crochet, A. Qninba and H. Garrido (2011). History, status and distribution of Andalusian Buttonquail in the WP. *Dutch Birding* 33(2): 75–93.

Haffer, J. (1992). The history of species concepts and species limits in ornithology. *Bulletin of the British Ornithologists' Club (Centenary Volume)* 112A: 107–58.

Hall, P. B. (1987). Robert Swinhoe (1836–77), FRS, FZS, FRGS: a Victorian naturalist in treaty port China. *Geographical Journal* 153(1): 37–47.

Hancock, J. (1874). A catalogue of the birds of Northumberland and Durham with 15 photographic copper-plates from drawings by the author. *Transactions of the Natural History Society of Northumberland and Durham* 6.

Hargitt, E. (1890). *Catalogue of the Birds in the British Museum, Vol. 18: Catalogue of the Picariae in the Collection of the British Museum. Scansores, Containing the family Picidae.* Trustees of the British Museum, London.

Harrap, S. and D. Quinn (1996). *Tits, Nuthatches and Treecreepers.* Christopher Helm, London.

Harrison, J. M. (1968). *Bristow and the Hastings Rarities Affair.* A. H. Butler, St Leonards-on-Sea.

Harrison, J. M. (1971). The Hastings Rarities: further comments. *British Birds* 64(2): 61–7.

Harrop, A. H. J., J. M. Collinson, S. P. Dudley and C. Kehoe (The British Ornithologists' Union Records Committee) (2013). The British List: a checklist of birds of Britain. Eighth edition. *Ibis* 155(3): 635–76.

Harrop, A. H. J., J. M. Collinson and T. Melling (2012). What the eye doesn't see: the prevalence of fraud in ornithology. *British Birds* 105(5): 236–57.

Hart, E. P. (1936). *Merchant Taylors' School Register 1561–1934.* Two vols. Eastern Press and Merchant Taylors' School, London.

Hart-Davis, D. (2004). *Audubon's Elephant: The Story of John James Audubon's Epic Struggle to Publish* The Birds of America. Orion, London.

Hartert, E. (1892). On new forms from the Dutch East Indies. *Bulletin of the British Ornithologists' Club* 1(3): 12–13.

Hartert, E. (1896). Notes on some species of the families Cypselidae, Caprimulgidae, and Podargidae, with remarks on subspecific forms and their nomenclature. *Ibis* 38(3): 362–76.

Hartert, E. (1903–22). *Die Vögel der Paläarktischen Fauna.* Three vols. Friedländer, Berlin.

Hartert, E. (1904). Some anticriticisms. *Ibis* 46(4): 542–51.

Hartert, E. (1907). The principal aims of modern ornithology. In R. B. Sharpe, E. J. O. Hartert and J. L. Bonhote (eds), *Proceedings of the Fourth International Ornithological Congress, London June 1905*, pp. 265–70. Dulau and Co., London.

Hartert, E. (1910). Strictest priority in nomenclature. *British Birds* 3(10): 327–9.

Hartert, E. (1920). Types of birds in the Tring Museum. B. Types in the general collection (contd.). *Novitates Zoologicae* 27: 425–505.

Hartert, E., F. C. R. Jourdain, N. F. Ticehurst and H. F. Witherby (1912). *A Hand-List of British Birds With an Account of Each Species in the British Isles and Abroad.* Witherby, London.

Hartert, E., F. C. R. Jourdain, N. F. Ticehurst and H. F. Witherby (1915). The B.O.U. List of British Birds. *British Birds* 8(12): 278–86.

Harting, J. E. (1865). Is the Great Black Woodpecker a British bird? *Zoologist* (first series) 23: 9730–2.

Harting, J. E. (1868). *Catalogue of … in the Collection of …* R Hardwicke, London.

Harting, J. E. (1871). *Hints on Shore Shooting; With a Chapter on Skinning and Preserving Birds.* Van Voorst, London.

Harting, J. E. (1872). *A Handbook of British Birds, Showing the Distribution of the Resident and Migratory Species in the British Islands, With an Index to the Records of the Rarer Visitants.* Van Voorst, London.

Harting, J. E. (1918a). Modern methods in nomenclature. *Ibis* 60(2): 334–8.

Harting, J. E. (1918b). The BOU and modern nomenclature. *Ibis* 60(3): 524–5.

Harvie-Brown, J. A. (communicated by) (1877). 'The avifauna of the Ural', translated from the Russian of Leonida Sabanaeff, by F.C. Craemers, and communicated by J.A. Harvie-Brown. *Proceedings of the Natural History Society of Glasgow* 3: 282–316.

Harvie-Brown, J. A. (1905). *Travels of a Naturalist in Northern Europe: Norway 1871, Archangel 1872, Petchora 1875.* Two vols. T. Fisher Unwin, London.

Hayes, L. M. (1905). *Reminiscences of Manchester*. Sherratt and Hughes, London.

Heermann, A. L. (1853). Notes on the birds of California, Observed during a residence of three years in that country. *Proceedings of the Academy of Natural Sciences of Philadelphia* 2: 259–72.

Heermann, A. L. (1854). Additions to North American ornithology, with descriptions of new species of the genera *Actidurus*, *Podiceps*, and *Podilymbus*. *Proceedings of the Academy of Natural Sciences of Philadelphia* 7: 177–80.

Heermann, A. L. (1859). General report upon the zoology of the several Pacific Railroad routes. No. 1, Report upon birds collected on the survey. *Pacific Railroad Reports* 10(3): 9–21; 10(4): 29–77.

Hernández M, L. M. Gonzalez, J. Oria, R. Sánchez and B. Arroyo (2008). Influence of contamination by organochlorine pesticides and polychlorinated biphenyls on the breeding of the Spanish Imperial Eagle (*Aquila adalberti*). *Environmental Toxicology and Chemistry* 27(2): 433–41.

Herschel, J. (ed.) (1849). *Admiralty Manual of Scientific Enquiry: Prepared for the Use of Officers in Her Majesty's Navy; and Travellers in General*. HMSO, London.

Hinde, R. A. and A. S. Thom (1947). The breeding of the Moustached Warbler in Cambridgeshire. *British Birds* 40(4): 98–104.

Hobson, P. (ed.) (1992). Unpublished transcript of Henry Seebohm's 'first journal' (1869–71). Sheffield City Museum.

Holcomb, T. A. E. and M. A. Lyon Holcomb (translators) (1905). *Fridthjof's Saga; A Norse Romance, by Esaias Tegnér, Bishop of Wexiö*. Sixth edition. Scott, Foresman and Company, Chicago.

Holloway, S. (1996). *The Historical Atlas of Breeding Birds in Britain and Ireland 1875–1900*. T. and A. D. Poyser, London.

Hooson, D. J. M. (1968). The development of geography in pre-Soviet Russia. *Annals of the Association of American Geographers* 58(2): 250–72.

Hopkins, B. D. (2007). The bounds of identity: the Goldsmid Mission and the delineation of the Perso-Afghan boundary in the nineteenth century. *Journal of Global History* 2007(2): 233–54.

Hopkirk, P. (1982). *Trespassers on the Roof of the World: The Race for Lhasa*. Oxford University Press, Oxford.

Hopkirk, P. (1992). *The Great Game: On Secret Service in High Asia*. John Murray, London.

Howard, E. H. (1920). *Territory in Bird Life*. John Murray, London.

Hruby, J. (2005). Ferdinand Stoliczka. *Birding Asia* 3: 50–6.

Hume, A. O. (1869). Stray notes on ornithology in India, No. 2. Birds'-nesting in Bareilly in the early rains. *Ibis* 11(1): 1–20.

Hume, A. O. (1874). *The Indian Ornithological Collector's Vade Mecum: Containing Brief Practical Instructions for Collecting, Preserving, Packing, and Keeping Specimens of Birds, Eggs, Nests, Feathers, and Skeletons*. Central Press Company, Calcutta, and Bernard Quaritch, London.

Hunt, G. (2001). *Hustlers, Rogues and Bubble Boys: White-Collar Mischief in New Zealand*. Reed, Auckland.

Hunter, A. A. (ed.) (1890). *Cheltenham College Register 1841–1889*. George Bell and Sons, London.

Huse, C. (1904). *Supplies for the Confederate Army*. T. R. Marvin and Son, Boston.

Huxley, J. S. (1912). A first account of the courtship of the Redshank (*Totanus calidris* L.). *Proceedings of the Zoological Society of London* 1912(3–4): 647–55.

Huxley, J. S. (1914). The courtship habits of the Great Crested Grebe (*Podiceps cristatus*); with an addition to the theory of sexual selection. *Proceedings of the Zoological Society of London* 1914(3–4): 491–562.

Ingram, C. (1966). *In Search of Birds*. Witherby, London.

IOC World Bird List, Master IOC list v5.4, www.worldbirdnames.org/ioc-lists/crossref (accessed 8 April 2017).

Irby, L. H. L. (1875). *Ornithology of the Straits of Gibraltar*. R. H. Porter, London.

Jackson, C. E. (1975). *Bird Illustrators: Some Artists in Early Lithography*. Witherby, London.

Jackson, C. E. (1994). Richard Bowdler Sharpe and his ten daughters. *Archives of Natural History* 21(3): 261–9.

Jacobsen, K.-O. (2005). *Snøugle (Bubo scandiacus) I Norge. Hekkeforekomster I perioden 1868–2005*. NINA rapport 84. Norsk Institutt for Naturforskning, Tromsø.

Jansen, J. J. F. J. (2012). William Bridger (1832–1870), collector of birds eggs in Australia and the Netherlands. *Archives of Natural History* 39(1): 174–6.

Jansen, J. J. F. J. (2013). Kleinst Waterhoen in Nederland: voorkomen en herziening van gevallen in 1800–2006 [Baillon's Crake in the Netherlands: occurrence and revision of records in 1800–2006]. *Dutch Birding* 35(5): 311–22.

Jansen, J. J. F. J. (2014). Former breeding of Eurasian Golden Plover, Dunlin and Wood Sandpiper in Limburg and Noord-Brabant. *Dutch Birding* 36: 9–19.

Jansen, J. J. F. J. and W. Nap (2008). Identification of White-headed Long-tailed Bushtit and occurrence in the Netherlands. *Dutch Birding* 30: 293–308.

Jansen, J. J. F. J. and R. Viek (2010). Joseph Baker, een Engelse vogelverzamelaar in Nederland in het midden van de negentiende eeuw [Joseph Baker, an English ornithologist in the Netherlands in the middle of the nineteenth century]. *Limosa* 83: 176–82.

Jardine, N., J. A. Secord and E. C. Spary (1996). *Cultures of Natural History*. Cambridge University Press, Cambridge.

Jerdon, T. A. (1863–64). *The Birds of India: Being a Natural History of All the Birds Known to Inhabit Continental India, with Descriptions of the Species, Genera, Families, Tribes, and Orders, and a Brief Notice of Such Families as Are Not Found in India, Making It a Manual of Ornithology Specially Adapted for India*. Military Orphan Press, Calcutta.

Johansen, H. (1952). Ornithology in Russia. *Ibis* 94(1): 1–48.

Johnson, H. (1907). *The Life and Voyages of Joseph Wiggins, F.R.G.S.: Modern Discoverer of the Kara Sea Route to Siberia Based on His Journals and Letters*. E. P. Dutton and Co., New York.

Johnson, K. (2004). *The Ibis*: transformations in a twentieth century British natural history journal. *Journal of the History of Biology* 37(3): 515–55.

Johnson, K. (2005). Type-specimens of birds as sources for the history of ornithology. *Journal of the History of Collections* 17(2): 173–88.

Jones, T. R. (ed.) (1875). *Manual of the Natural History, Geology and Physics of Greenland and the Neighbouring Regions, Together with Instructions Suggested by the Arctic Committee of the Royal Society for the Use of the Expedition*. HMSO, London.

Jourdain, F. C. R. (1906–12). *The Eggs of European Birds*. Four parts. R. H. Porter, London.

Jourdain, F. C. R. (1918). Reply to Mr. Harting on modern methods in nomenclature. *Ibis* 60(3): 526–8

Jourdain, F. C. R. (1920). *The Birds of the British Isles and Their Eggs*, by T. A. Coward [review]. *British Birds* 14(2): 47–8.

Kasparek, M. (1986). On records of the Pine Bunting, *Emberiza leucocephalos*, in Turkey from the last century. *Zoology in the Middle East* 1: 56–9.

Kearney, M. and A. Knopp (1991). *Boom and Bust: The Historical Cycles of Matamoros and Brownsville*. Eakin Press, Austin.

Keay, J. (1979). *The Gilgit Game: The Explorers of the Western Himalayas 1865–95*. Readers Union, Newton Abbot.

Keegan, N. M. (2005). *Consular Representation in Britain: Its History, Current Status, and Personnel.* Unpublished PhD thesis, University of Durham.

Keene, H. G. (rev. Y. Foote) (2004). Hay, Arthur, ninth marquess of Tweeddale (1824–1878). *Oxford Dictionary of National Biography*, Oxford University Press (online edition, 2006), www.oxforddnb.com/view/article/12710, accessed 8 April 2017.

Keulemans, T. (1982). *Feathers to Brush: The Victorian Bird Artist, John Gerrard Keulemans, 1842–1912.* C. J. Colderrey, Epse.

Killick, J. R. (2004). Forwood, Sir William Bower (1840–1928). *Oxford Dictionary of National Biography*, Oxford University Press (online edition, 2006), www.oxforddnb.com/view/article/47004, accessed 8 April 2017.

Kinnear, N. B. (1931). Some additional notes on James Hepburn. *Condor* 33(4): 169–71.

Kirwan, G. (2004). The taxonomic position of the Afghan Scrub Sparrow *Passer (moabiticus) yatii*. *Sandgrouse* 26: 105–11.

Kirwan, G. (2006). Comments on two subspecies of passerine birds recently described from Turkey, *Eremophila alpestris kumerloevei* and *Pyrrhula pyrrhula paphlagoniae*, with remarks on geographical variation in related forms of Bullfinch from the Balkans and Caucasus. *Sandgrouse* 28(1): 12–23.

Kirwan, G., B. Demirci, H. Welch, K. Boyla, M. Özen, P. Castell and T. Marlow (2008). *The Birds of Turkey.* Christopher Helm, London.

Kisling, V. N. (1994). *The Naturalists' Directory* and the evolution of communication among American naturalists. *Archives of Natural History* 21(3): 393–406.

Knox, A. G. (1993). Richard Meinertzhagen – a case of fraud examined. *Ibis* 135(3): 320–5.

Knox, A. G. (2007). Order or chaos? Taxonomy and the British list over the last hundred years. *British Birds* 100(10): 609–23.

Knox, A. G. and M. P. Walters (1992). Under the skin: the bird collections of the Natural History Museum. In J. F. Monk (ed.), *Avian Systematics and Taxonomy. Bulletin of the British Ornithologists' Club* 112A (Centenary Supplement), pp. 169–90.

Krause, G. A. J. (1905–13). *Oologia Universalis Palaearctica.* Seventy-eight parts. Lehman, Stuttgart.

Krider, J. (1860). *An Ornithological and Oological List of North America.* W. W. Maybury, Philadelphia.

Kropotkin, P. and D. W. Freshfield (1903). Obituary – Dr. Gustav Radde. *Geographical Journal* 21(5): 563–4.

Kumerloeve, H. (1975). The history of ornithology in Turkey. *Ornithological Society of Turkey Bird Report* 3: 289–319.

Kunakhovich, K. (2006). Nikolai Mikhailovich Przhevalsky and the politics of Russian Imperialism. *IDP (International Dunhuang Project) News* 26: 3–9.

Kynaston, D. (1994). *The City of London, Vol. 1: A World of Its Own.* Four vols. Chatto and Windus, London.

Lack, D. (1954). *The Natural Regulation of Animal Numbers.* Oxford University Press, Oxford.

Lack, D. (1966). *Population Studies of Birds.* Oxford University Press, Oxford.

Lack, D. (1968). *Ecological Adaptations for Breeding in Birds.* Methuen, London.

Lack, D. (1971). *Ecological Isolation in Birds.* Oxford University Press, Oxford.

Lambourne, M. (1990). *The Art of Bird Illustration.* Wellfeet Press, Secaucus.

Lansdell, H. (1882). *Through Siberia.* Two vols. Sampson Low, Marston, Searle and Rivington, London.

Lansdell, H. (1885). *Russian Central Asia: Including Kuldja, Bokhara, Khiva and Merv.* Two vols. Sampson Low, London.

Lansdell, H. (1887). *Through Central Asia.* Two vols. Sampson Low, London.

Lansdell, H. (1893). *Chinese Central Asia: A Ride to Little Tibet.* Two vols. Sampson Low, London.

Lascelles, H. (1915). *Thirty-Five Years in the New Forest.* Edward Arnold, London.

Layard, E. L. and R. B. Sharpe (1875–84). *The Birds of South Africa.* Quaritch, London.

LeCroy, M. (2005). Type specimens of birds in the American Museum of Natural History. Part 6, Passeriformes: Prunellidae, Turdidae, Orthonychidae, Timaliidae, Paradoxornithidae, Picathartidae, and Polioptilidae. *Bulletin of the AMNH* 292.

Letch, R. (1927). The London Commercial Sale Rooms. *PLA Monthly* (November): 3–12, 18.

Levere, T. H. (1988). Henry Wemyss Feilden, naturalist on H.M.S. Alert 1875–76. *Polar Record* 24: 307–12.

Levere, T. H. (1993). *Science and the Canadian Arctic: A Century of Exploration 1818–1918.* Cambridge University Press, Cambridge.

Lewis, D. (2012). *The Feathery Tribe: Robert Ridgway and the Modern Study of Birds.* Yale University Press, Yale.

Lilford, Lord (T. L. Powys) (1865). Notes on the ornithology of Spain. *Ibis* 7(2): 166–77.

Lilford, Lord (1866). Notes on the ornithology of Spain. *Ibis* 8(2): 173–87; 8(4): 377–91.

Lilford, Lord (1875). The cruise of the '*Zara*', R.Y.S. in the Mediterranean. *Ibis* 17(1): 1–35.

Lilford, Lord (1885–98). *Coloured Figures of the Birds of the British Islands.* Seven vols. R. H. Porter, London.

Lilford, Lord (1887). Notes on Mediterranean ornithology. *Ibis* 29(3): 261–83.

Lindsay, D. (1993). *Science in the Subarctic: Trappers, Traders and the Smithsonian Institution.* Smithsonian Institution Press, Washington, DC.

Lloyd, C. (1985). *The Travelling Naturalists.* Croom Helm, London.

Lockwood, M. W. and B. Freeman (2014). *The TOS Handbook of Texas Birds.* Second edition. Texas A. and M. University Press, College Station.

Loftus, H. (1997). *The Tetley Affair: Colonial Dreams and Nightmares.* Heritage Press, Waikanae.

Lomonossof, A. (1880). M. Severtsof's journey in Ferghana and the Pamir in 1877–78. *Proceedings of the Royal Geographical Society and Monthly Record of Geography* (new monthly series) 2(8): 499–506.

Long, H. (1968). The Bowling Ironworks. *Industrial Archaeology* 5(2): 171–7.

Lovegrove, R. (1990). *The Kite's Tale: The Story of the Red Kite in Wales.* RSPB, Sandy.

Lovegrove, R. (2008). *Silent Fields: The Long Decline of a Nation's Wildlife.* Oxford University Press, Oxford.

Lowe, P. R. (1942). Obituary – Samuel Leigh Whymper. *Ibis* 84(2): 278–81.

Lucier, P. (2009). The professional and the scientist in nineteenth-century America. *Isis* 100: 699–732.

MacFarlane, R. R. (1891). Notes on and list of birds and eggs collected in Arctic America, 1861–66. *Proceedings of the United States National Museum* 14 (865): 413–46.

MacGillivray, W. (1837–52). *A History of British Birds, Indigenous and Migratory.* Five vols. Scott, Webster and Geary, London.

MacGillivray, W. (1840–42). *A Manual of British Ornithology.* Scott, Webster and Geary, London.

MacKenzie, J. M. (1988). *The Empire of Nature: Hunting, Conservation and British Imperialism.* Manchester University Press, Manchester.

Manson-Bahr, P. (1951). A short history of the Club. *Bulletin of the British Ornithologists' Club* 71(1): 2–4.

Manson-Bahr, P. (1952). Jubilee address. *Bulletin of the British Ornithologists' Club* 72(8): 87–9.

Manson-Bahr, P. (1959). Recollections of some famous British ornithologists. *Ibis* 101(1): 53–64.

Marquiss, M., I. Newton, K. A. Hobson and Y. Kolbeinsson (2012). Origins of irruptive migrations by Common Crossbills *Loxia curvirostra* into northwestern Europe revealed by stable isotope analysis. *Ibis* 154(2): 400–9.

Martorelli, G. (1898). *Commemorazione Scientifica del Conte Ercole Turati.* Pirola di Rubini Enrico, Milan.

Mather, J. R. (1986). *The Birds of Yorkshire.* Croom Helm, London.

Matschie, P. (ed.) (1902a). *Verhandlungen des V. Internationalen Zoologen-Congresses zu Berlin 12–16 August 1901.* Gustav Fischer, Jena.

Matschie, P. (1902b). Vorbemerkung. In Anhang [Rules of Zoological Nomenclature, Adopted by the Fifth International Congress of Zoology, in German, French and English], pp. 929–32. In P. Matschie (ed.), *Verhandlungen des V. Internationalen Zoologen-Congresses zu Berlin 12–16 August 1901,* pp. 927–72. Gustav Fischer, Jena.

McGhie, H. A. (2002). The egg of the Slender-billed Curlew at the Manchester Museum: a unique specimen? *British Birds* 95(7): 359–60.

McGhie, H. A. (2005). Specimens of extinct and endangered birds in the collections of the Manchester Museum, the University of Manchester, UK. *Bulletin of the British Ornithologists' Club* 125(4): 247–52.

McGhie, H. A. (2006). [Henry Dresser and his Russian correspondents.] *Our Birds* 2–3 (32–3): 54–5 (in Russian).

McGhie, H. A. (2009). Letters from Alfred Russel Wallace concerning the Darwin commemorations of 1909. *Archives of Natural History* 36(2): 352–4.

McGhie, H. A. (2010). Contextual research and the postcolonial museum – the example of Henry Dresser. In E. Bauernfield, A. Gamauf, H.-M. Berg and Y. Muraoka (eds), *Collections in Context,* pp. 49–65. Annalen des Naturhistorischen Museums Wien (Proceedings of the Fifth International Conference of European Bird Curators), Vienna.

McGhie, H. A. (2011). Dresser, H. E. (1871–'81' = 1871–82) [Initially Sharpe, R. B. and H. E. Dresser] *A History of the Birds of Europe, Including All the Species Inhabiting the Western Palaearctic Region.* In E. C. Dickinson, L. K. Overstreet, R. J. Dowsett and M. D. Bruce (eds), *Priority! The Dating of Scientific Names in Ornithology,* pp. 89–90. Aves Press, Northampton.

McGhie, H. A. (2012). Nineteenth-century ornithology, Leadenhall Market and fraud. *British Birds* 105(11): 678–82.

McGhie, H. A. (2013). Images, ideas and ideals: thinking with and about Ross's Gull. In L. E. Thorsen, K. A. Rader and A. Dodd (eds), *Animals on Display: The Creaturely in Museums, Zoos and Natural History,* pp. 101–27. Penn State University Press, Pennsylvania.

McGhie, H. A. and D. V. Logunov (2005). Discovering the breeding grounds of Ross's Gull: 100 years on. *British Birds* 98(11): 589–99.

McGhie, H. A. and D. V. Logunov (2006). [Henry Dresser and Sergius Buturlin: friends and colleagues.] In O. E. Borodina *et al.* (eds), *Buturlinski Sbornik. Materialy II Mezhdunarodnykh Buturlinskikh Chtenii, Ulyanovsk, 21.09–24.09.2005,* pp. 40–53. Korporatsiya Tekhnologii Prodvizheniya, Ulyanovsk (in Russian).

McKay, A. (2005). 'The birth of a clinic'? The IMS dispensary in Gyantse (Tibet). *Medical History* 49(2): 135–54.

McVean, C. (1877). Notes on the ornithology of Yedo. *Proceedings of the Royal Physical Society of Edinburgh* 4: 144–54.

Mead, W. R. (1968). *Finland*. Ernest Benn, London.

Mearns, B. and R. Mearns (1988). *Biographies for Birdwatchers: The Lives of Those Commemorated in Western Palaearctic Bird Names*. Academic Press, London.

Mearns, B. and R. Mearns (1992). *Audubon to Xantus: The Lives of Those Commemorated in North American Bird Names*. Academic Press, London.

Mearns, B. and R. Mearns (1998). *The Bird Collectors*. Natural World/Academic Press, London.

Mearns, B. and R. Mearns (2007). *John Kirk Townsend: Collector of Audubon's Western Birds and Mammals*. Privately published, Dumfries.

Meiklejohn, R. F. (1922). A defence of egg-collecting. *Ibis* 64(4): 746–8.

Meinertzhagen, R. (1959). Nineteenth century recollections. *Ibis* 101(1): 46–52.

Melling, T. (2005). The Tadcaster Rarities. *British Birds* 98(5): 230–7.

Melville, R. V. (1995). *Towards Stability in the Names of Animals: A History of the International Commission on Zoological Nomenclature 1895–1995*. International Commission for Zoological Nomenclature/NHM, London.

Meves, W. (1886). *Die Grösse und Farbe der Augen aller Europäischen Vögel, sowie der in der palaearctischen Region vorkommenden Arten in systematischer Ordnung nach Carl J. Sundevall's Versuch einer natürlichen Aufstellung der Vogelkasse von Wilhelm Meves*. Wilhelm Schlüter, Halle.

Meyer, A. B. (1887). *Unser Auer-, Rackel- und Birkwild und seine Abarten*. A. W. Künast, Vienna.

Miers, J. (1881). On a collection of crustacea made by Baron Hermann-Maltzam at Goree Island, Senegambia. *Annals and Magazine of Natural History* (fifth series) 8: 259–81.

Miller, J. (2006). *Fertile Fortune: The Story of Tyntesfield*. National Trust, London.

Mitchell, P. C. (1929). *Centenary History of the Zoological Society of London*. Zoological Society of London, London.

Mlíkovský, J. (2007a). Types of birds in the collections of the Museum and Institute of Zoology, Polish Academy of Sciences, Warszawa, Poland. Part 2: Asian birds. *Journal of the National Museum (Prague), Natural History Series* 176(4): 33–79.

Mlíkovský, J. (2007b). Type specimens and type localities of Rock Nuthatches of the *Sitta neumayer* species complex (Aves: Sittidae). *Journal of the National Museum (Prague), Natural History Series* 176(6): 91–115.

Moore, D. T. (2004). Blanford, William Thomas (1832–1905). *Oxford Dictionary of National Biography*, Oxford University Press (online edition, 2006), www.oxforddnb.com/view/article/31923, accessed 8 April 2017.

Moreau, R. E. (1959). The centenarian *Ibis*. *Ibis* 101(1): 19–38.

Morgan, G. (1973). Myth and reality in the Great Game. *Asian Affairs* 60: 55–65.

Morris, P. A. (1993). An historical review of bird taxidermy in Britain. *Archives of Natural History* 20(2): 241–55.

Morris, P. A. (2004). *Edward Gerrard and Sons: A Taxidermy Memoir*. MPM, Ascot.

Morris, P. A. (2010). *Taxidermy: Science, Art and Bad Taste*. MPM, Ascot.

Moulton, E. C. (2004). Hume, Allan Octavian (1829–1912). *Oxford Dictionary of National Biography*, Oxford University Press (online edition, 2006), www.oxforddnb.com/view/article/34049, accessed 8 April 2017.

Mountfort, G. (1959). One hundred years of the British Ornithologists' Union. *Ibis* 101(1): 8–18.

Moxham, R. (2001). *The Great Hedge of India*. Constable, London.

Mullens, W. H. and H. Kirke Swann (1917). *A Bibliography of British Ornithology from the Earliest Times to the End of 1912*. Macmillan, London. Facsimile reprint (1986), Wheldon and Wesley, Herts.

Müller, H. C. (1862). *Faerøernes Fuglefauna og Bemærkninger om Fuglefangsten. Videnskabelige Meddelelser fra den naturhistoriske Forening i København* 4: 1–78.

Murdoch, J. (1885). Natural history: birds. In P. H. Ray (ed.), *Report on the International Polar Expedition to Point Barrow, Alaska*, pp. 104–128. Government Printing Office, Washington, DC.

Murdoch, J. (1899). An historical notice on Ross's Gull. *Auk* 16(2): 146–55.

Mussell, J. E. P. (2009). '*Knowledge* (1881–1918)'. In L. Brake and M. Demoor (eds), *Dictionary of Nineteenth-Century Journalism in Great Britain and Ireland*, pp. 335–6. Academia Press, Ghent.

Nares, G. S. (1878). *Narrative of a Voyage to the Polar Sea During 1875–76 in H.M. Ships 'Alert' and 'Discovery'*. Two vols. Samson, Low, Marston, Searle and Rivington, London.

Nelder, J. A. (1962). A statistical examination of the Hastings Rarities. *British Birds* 55(8): 283–98.

Nethersole-Thompson, D. and M. Nethersole-Thompson (1986). *Waders: Their Breeding, Haunts and Watchers*. T. and A. D. Poyser, Calton.

Newton, A. (1860a). Suggestions for forming collections of birds' eggs. Appendix to T. M. Brewer and S. F. Baird. *Instructions in Reference to Collecting Nests and Eggs of North American Birds*. Reprinted in *Smithsonian Miscellaneous Collections* 1862 (2): 10–22.

Newton A. (1860b). Suggestions for forming collections of birds' eggs. Reprinted, with additions from the Circular of the Smithsonian Institution. *Zoologist* (first series) 18: 7189–201.

Newton, A. (1860c). Memoir of the late John Wolley. *Ibis* 2(2): 172–85.

Newton, A. (1861). Particulars of Mr. J. Wolley's discovery of the breeding of the Waxwing (*Ampelis garrulus*, Linn.). *Ibis* 3(1): 92–106.

Newton, A. (1864). *Ootheca Wolleyana: An Illustrated Catalogue of the Collection of Birds' Eggs Begun by the Late John Wolley, Jun., M.A., F.Z.S., and Continued with Additions by the Editor Alfred Newton*. Part 1 (rebound as vol. 1, along with part 2). Van Voorst, London.

Newton, A. (1894). Notes on 'A Bill to amend the Wild Birds' Protection Act, 1880'. *Annals of Scottish Natural History* 10: 76–82.

Newton, A. (1902). *Ootheca Wolleyana: An Illustrated Catalogue of the Collection of Birds' Eggs Begun by the Late John Wolley, Jun., M.A., F.Z.S., and Continued with Additions by the Editor Alfred Newton*. Part 2 (conclusion of vol. 1). R. H. Porter, London.

Newton, A. (1905–7). *Ootheca Wolleyana: An Illustrated Catalogue of the Collection of Birds' Eggs Begun by the Late John Wolley, Jun., M.A., F.Z.S., and Continued with Additions by the Editor Alfred Newton*. Parts 3–4 (rebound as vol. 2). R. H. Porter, London.

Newton, A., H. Gadow, R. Lydekker, C. S. Roy and R. W. Schufeldt (1893–96). *A Dictionary of Birds*. A. and C. Black, London.

Newton, A. and H. Saunders (eds) (1871–85). *A History of British Birds by William Yarrell*. Four vols. Van Voorst, London.

Nichols, J. L. (1964). *The Confederate Quartermaster in the Trans-Mississippi*. University of Texas Press, Austin.

Nicholson, E. M. (1926) *Birds in England: An Account of the State of Our Bird Life and a Criticism of Bird Protection*. Chapman and Hall, London.

Nicholson, E. M. (1970). *The Environmental Revolution.* Hodder and Stoughton, London.

Nicholson, E. M. and I. J. Ferguson-Lees (1962). The Hastings Rarities. *British Birds* 55(8): 299–384.

Nicholson, E. M. and I. J. Ferguson-Lees (1971). (Commentary.) *British Birds* 64(2): 67–8.

Nicholson, E. M., I. J. Ferguson-Lees and J. Nelder (1969). The Hastings rarities again. *British Birds* 62(9): 364–81.

Oates, E. W. (1883). *A Handbook to the Birds of British Burmah, Including Those Found in the Adjoining State of Karennee.* Two vols. R. H. Porter, London.

Oberholser, H. C. (1900). Catalogue of a collection of birds from Madagascar. *Proceedings of the United States National Museum* 22(1197): 235–48.

Ogilvie, M., I. J. Ferguson-Lees and R. Chandler (2007). A history of *British Birds. British Birds* 100(1): 3–15.

Ogilvie-Grant, W. R. (1893). *Catalogue of the Birds in the British Museum, Vol. 22: Catalogue of the Gamebirds (Pterocletes, Gallinae, Opisthocomi, Hemipodii) in the Collection of the British Museum.* Trustees of the British Museum, London.

Ogilvie-Grant, W. R. (1905). On birds collected by Colonel Waddell. *Bulletin of the British Ornithologists' Club* 15(117): 94.

Oldham, W. S. and C. E. Jewett (eds) (2006). *Rise and Fall of the Confederacy: The Memoir of Senator Williamson S. Oldham, CSA.* University of Missouri Press, Columbia.

'Oologicus' (1863). Reply to 'Oophilus'. *Ibis* 5(4): 478–9.

'Oophilus' (1863). Indignant letter from Oophilus on egg collecting and dealers. *Ibis* 5(3): 372–6.

Orwell, G. (1936). Shooting an elephant. *New Writing* 2(autumn): 501–6.

Owsley F. L. (1969). *King Cotton Diplomacy.* Third impression. University of Chicago Press, Chicago.

Palmer, A. H. (1895). *The Life of Joseph Wolf.* Longmans, Green and Co., London.

Palmer, T. S. (1922). Obituary – Dr. Theobald Johannes Krüper. *Auk* 39(1): 148–9.

Parker, E. (1935). *Ethics of Egg Collecting.* Field, London.

Parsons, T. H. (1999). *The British Imperial Century, 1815–1914: A World History Perspective.* Rowman and Littlefield, Lanham.

Pearson, H. J. (1898). Notes on the birds observed on Waigats, Novaya Zemlya, and Dolgoi Island, in 1897. *Ibis* 40(2): 185–208.

Pearson, H. J. and H. W. Feilden (1899). *Beyond Petsora Eastward: Two Summer Voyages to Novaya Zemlya and the Islands of the Barents Sea.* R. H. Porter, London.

Pemberton, H. (1963). Two hundred years of banking in Leeds. *Thoresby Miscellany* 13(46): 54–86. Phillips, M. (1894). *A History of Banks, Bankers, and Banking in Northumberland, Durham, and North Yorkshire.* Effingham Wilson and Co., London.

Pickstone, J. V. (2001). *Ways of Knowing: A New History of Science, Technology and Medicine.* Manchester University Press, Manchester.

Pigott, D. (1907). *The Wild Birds Protection Acts as Administered by Orders in Great Britain and Ireland.* In R. B. Sharpe, E. J. O. Hartert and J. L. Bonhote (eds), *Proceedings of the Fourth International Ornithological Congress, London June 1905*, pp. 594–608. Dulau and Co., London.

Pinn, F. (1985). *L. Mandelli – Darjeeling Tea Planter and Ornithologist.* Privately published, London.

Pittie, A. (2011). The dates of seven new taxa described by W. E. Brooks. *Indian Birds* 7(2): 54–5.

Popham, H. (1897). Notes on birds observed on the Yenisei, Siberia. *Ibis* 39(1): 89–108.

Popham, H. (1901). Supplementary notes on the birds of the Yenisei River. *Ibis* 43(3): 449–58.

Postnikov, A. V. (2003a). Central Asia: Russian exploration. In J. Speake (ed.), *Literature of Travel and Exploration: An Encyclopedia, Vol. 1*, pp. 214–17. Three vols. Fitzroy Dearborn (Taylor and Francis Group), London.

Postnikov, A. V. (2003b). Przhevalskii, Nikolai (1839–1888). In J. Speake (ed.), *Literature of Travel and Exploration: An Encyclopedia, Vol. 2*, pp. 980–2. Three vols. Fitzroy Dearborn (Taylor and Francis Group), London.

Postnikov, A. V. (2003c). Semenov, Petr (1827–1914). In J. Speake (ed.), *Literature of Travel and Exploration: An Encyclopedia, Vol. 3*, pp. 1075–6. Three vols. Fitzroy Dearborn (Taylor and Francis Group), London.

Postnikov, A. V. (2003d). Turkestan. In J. Speake (ed.), *Literature of Travel and Exploration: An Encyclopedia, Vol. 3*, pp. 1201–4. Three vols. Fitzroy Dearborn (Taylor and Francis Group), London.

Preble, E. A. (1922). Roderick Ross Macfarlane, 1833–1920. *Auk* 39(2): 203–10.

Prejevalski, N. M. (1876). *Mongolia, the Tangut Country and the Solitudes of Northern Tibet*. Two vols. Samson, Low, Marston, Serle and Rivington, London.

Prejevalski, N. M. (1879). *From Kulja, across the Tian-Shan to Lob Nor*. Samson, Low, Marston, Serle and Rivington, London.

Prejevalski, N. M. (1888). [*From Kyakhta to the Source of the Yellow River: The Exploration of the Northern Boundary of Tibet and the Way Over Lob-Nor Along the Banks of the Tarim River.*] V. S. Balashev, St Petersburg (in Russian).

Prŷs-Jones, R. (2006). The 1885 Greenland breeding record of Ross's Gull. *British Birds* 99(4): 208–10.

Pycraft, W. P. (1922). 'Oology' and science. *Illustrated London News*, 4 March (issue 4324), p. 326.

Raby, P. (1986). *Bright Paradise: Victorian Scientific Travelers*. Princeton University Press, Princeton.

Raby, P. (2001). *Alfred Russel Wallace: A Life*. Chatto and Windus, London.

Radclyffe, H. (1994). Notes on Lord Lilford's *Coloured Figures of the Birds of the British Islands* London, 1885–1897[98] and 1891–1897[98]. *Archives of Natural History* 21(1): 11–16.

Radde, G. (1858). Notes on the River Amur and the adjacent districts – Extract from Mr. Radde's communication on the Hing-gan Range. *Journal of the Royal Geographical Society of London* 28: 418–25.

Radde, G. (1861). *Bericht über Reisn im Süden von Ost-Sibirien ausgeführt in den Jahren 1855 bis incl. 1859*. In the series Beiträge zur Kenntnis des Russischen Reiches und der angrenzenden Ländern Asiens. Bd. 23 plus atlas. Kaiserliche Akademie der Wissenschaft, St Petersburg.

Radde, G. (1862–63). *Reisen im Süden von Ost-Sibirien in den Jahren 1855–1859 incl.* Two vols. Kaiserliche Akademie der Wissenschaft, St Petersburg.

Radde, G. (1884). *Ornis Caucasica*. Theodor Fischer, Kassel.

Rasmussen, P. C. and R. Prŷs-Jones (2003). History vs. mystery: the reliability of museum specimen data. In N. Collar, C. T. Fisher and C. Feare (eds), *Why Museums Matter: Avian Archives in an Age of Extinction*. Bulletin of the British Ornithologists' Club, Supplement 123A, pp. 66–94.

Ratcliffe, D. A. (2005). *Lapland: A Natural History*. T. and A. D. Poyser, London.

Rayfield, D. (1976). *The Dream of Lhasa, the life of Nikolay Przhevalsky (1839–88), Explorer of Central Asia*. Ohio University Press, Athens.

Reichenow, A. (1902). Über begriff und benennun von subspecies. In P. Matschie (ed.), *Verhandlungen des V. Internationalen Zoologen-Congresses zu Berlin 12–16 August 1901*, pp. 910–915. Gustav Fischer, Jena.

Rhodes, R. (2004). *John James Audubon: The Making of an American*. Alfred A. Knopf, New York.

Richardson, C. (1976). *A Geography of Bradford*. University of Bradford, Bradford.

Ridgway, R. (1873). On some new forms of American birds. *American Naturalist* 7(10): 602–19.

Ridgway, R. (1880). A catalogue of the birds of North America. *Proceedings of the United States National Museum* 3: 163–246.

Robinson, H. W. (1917). Some evidence corroborating the supposed breeding of the Green Sandpiper in the British Isles. *Ibis* 59(4): 611–12.

Robinson, H. W. (1918). Letter on the breeding of the Green Sandpiper in the British Isles. *Ibis* 60(1): 178.

Rookmaaker, L. C. (2011). The early attempts of Hugh Strickland to establish a code for zoological nomenclature in 1842–42. *Zoological Bulletin* 68(1): 29–40.

Rookmaaker, L. C., P. A. Morris, I. E. Glenn and P. J. Mundy (2006). The ornithological cabinet of Jean-Baptiste Bécoeur and the secret of the arsenical soap. *Archives of Natural History* 33(1): 146–58.

Rothschild, M. (1983). *Dear Lord Rothschild: Birds, Butterflies, and History*. Hutchinson, London.

Rothschild, M. (2008). *Dear Lord Rothschild: The Man, the Museum and the Menagerie*. NHM, London.

Rothschild, W. (1904). On the Barn-Owls. *Bulletin of the British Ornithologists' Club* 14(108): 87–90.

Rothschild, W. (1916). Henry Eeles Dresser [obituary]. *British Birds* 9(8): 194–6.

Royal Society for the Protection of Birds (1911). *Feathers and Facts: A Reply to the Feather-Trade, and Review of Facts with Reference to the Persecution of Birds for Their Plumage*. RSPB and Witherby and Co., London.

Russell, R. (2011). *The Business of Nature: John Gould and Australia*. National Library of Australia, Canberra.

St John, C. (1849). *A Tour in Sutherlandshire, with Extracts from the Field-Books of a Sportsman and Naturalist*. Two vols. John Murray, London.

St John, O. (1876). Narrative of a journey through Baluchistan and Southern Persia, 1. Gwadar to Pishin. In Sir F. J. Goldsmid, O. St John, B. Lovett and E. Smith, *Eastern Persia, an Account of the Journeys of the Persian Boundary Commission 1870–71–72, Vol. 1: The Geography with Narratives*, pp. 18–33. Two vols. Macmillan, London.

Salvadori, T. and H. E Dresser (1893). Exhibition and description of a new species of *Acredula* from Macedonia. *Bulletin of the British Ornithologists' Club* 1(4): 15. (Abstract dated 31 December 1892, but presumably not widely available until early 1893. Note, the original description was published under Dresser's name alone in error by the publishers: see *Bulletin of the British Ornithologists' Club* 1(5): 23(n).)

Salvin, O. (1859). Five months' birds'-nesting in the eastern Atlas. *Ibis* 1(2): 174–91; 1(3): 302–28; 1(4): 352–65.

Salvin, O. (1860). History of the Derbyan Mountain-Pheasant (*Oreophasis derbianus*). *Ibis* 2(3): 249–53.

Salvin, O. (1861). Quesal-shooting in Vera Paz. *Ibis* 3(2): 138–49.

Salvin, O. (1874). Remarks on W. T. Blanford's letter on scientific nomenclature. *Ibis* 16(3): 302–3.

Salvin, O. (as Editor of the *Ibis*) (1875). (Footnote, on the scientific names of owls.) In G. E. Shelley, Three months on the coast of South Africa. *Ibis* 17(1): 59–87, pp. 66–7.

Salvin, O. (1876). On *Dendroica chrysoparia*. *Ornithological Miscellany* 1: 181–4.

Samstag, T. (1988). *For Love of Birds: The Story of the RSPB*. RSPB, Sandy.

Sarudny, N. A. and S. A. Buturlin (1906). *Sitta dresseri*, spec. nov. *Ornithologische Monatsberichte* 14(7): 132.

Sauer, G. C. (1982). *John Gould the Bird Man: A Chronology and Bibliography*. Henry Sotheran, London.

Saunders, H. (1869). Ornithological rambles in Spain. *Ibis* 11(2): 170–86.

Saunders, H. (1872). On the introduction of *Anser albatus* of Cassin into the British avifauna. *Proceedings of the Zoological Society of London* 1872(2): 519–21.

Schalow, H. (ed.) (1911). *Verhandlungen des V. Internationalen Ornithologen-Kongresses in Berlin 30. Mai bs 4. Juni 1910*. Deutsche Ornithologische Gesellschaft, Berlin.

Scharleman, J. P. W. (2002). *Factors Affecting Long-Term Changes in Eggshell Thickness and Laying Dates of Some European Birds*. Unpublished PhD thesis, University of Cambridge.

Scherren, H. (1905). *The Zoological Society of London: A Sketch of Its Foundation and Development and the Story of Its Farm, Museum, Gardens, Menagerie and Library*. Cassell and Co., London.

Schufeldt, R. W. (1912). The photography of birds' eggs. *Auk* 29(2): 274–6.

Schulze-Hagen, K. (2001). Zoological Society of London, Zoo, zoological research – their importance for Joseph Wolf. In K. Schulze-Hagen and A. Geus (eds), *Joseph Wolf (1820–1899): Animal Painter*, pp. 189–207. Basiliken-Presse, Marburg an der Lahn.

Schulze-Hagen, K., F. D. Steinheimer, R. Kinzelbach and C. Gasser (2003). Avian taxidermy in Europe from the Middle Ages to the Renaissance. *Journal für Ornithologie* 144(4): 459–78.

Sclater, P. L. (1858). On the general geographical distribution of the members of the class Aves. *Journal of the Proceedings of the Linnean Society. Zoology* 2: 130–6.

Sclater, P. L. (1865). Note on two rare species of the American genus *Dendroeca*. *Ibis* 7(1): 87–9.

Sclater, P. L. (1874). New and forthcoming bird-books. *Ibis* 15(2): 172–81.

Sclater, P. L. (1875). On the present state of our knowledge of geographical zoology. *Nature* 12(305): 374–82.

Sclater, P. L. (1876). Presidential address – Biology section. In *Report of the 45th Meeting of the British Association for the Advancement of Science 1875*, pp. 85–133.

Sclater, P. L. (1877). Letter on Bonaparte's *Lophorina respublica*. *Ibis* 19(4): 493–4.

Sclater, P. L. (1879). Remarks on the nomenclature of the British owls and on the arrangement of the order Striges. *Ibis* 21(3): 346–52.

Sclater, P. L. (1885). Report on the additions to the Society's menagerie in December 1884, and description of a new species of *Cervulus*. *Proceedings of the Zoological Society of London* 1885(1): 1–2.

Sclater, P. L. (1896). Remarks on the divergencies between the 'Rules for naming animals' of the German Zoölogical Society and the Stricklandian Code of Nomenclature. *Proceedings of the Zoological Society of London* 1896(2): 306–19.

Sclater, P. L. (1902). Note on address to Fifth International Zoological Congress. P. 120 in P. Matschie (ed.), *Verhandlungen des V. Internationalen Zoologen-Congresses zu Berlin 12–16 August*. Gustav Fischer, Jena.

Sclater, P. L. (1908). Remarks on a collection of birds from the Sikhim Himalayas. *Ibis* 50(1): 116–17.

Sclater, P. L. (1909). A short history of the British Ornithologists' Union. *Ibis* 50, Supplement 1 (Jubilee Supplement): 19–65.

Sclater, P. L. (1913). Commentary on the new 'Hand-list of British birds'. *Ibis* 55(1): 113–27.

Scully, J. (1881–82). A contribution to the ornithology of Gilgit. *Ibis* 23(3): 415–53; 23(4): 567–94.

Sedgwick, A. (ed.) (1899). *Proceedings of the Fourth International Congress of Zoology, Cambridge, 22–27 August, 1898*. C. J. Clay and Sons, London.

Seebohm, H. (1877a). On the Phylloscopi or Willow Warblers. *Ibis* 19(1): 66–108.

Seebohm, H. (1877b). Letter on warblers and *Anthus seebohmi*. *Ibis* 19(1): 128.

Seebohm, H. (1877c). On the Salicariae of Dr. Severtzoff. *Ibis* 19(2): 151–6.

Seebohm, H. (1877d). Supplementary notes on the ornithology of Heligoland. *Ibis* 19(2): 156–65.

Seebohm, H. (1879a). Remarks on Messrs. Blakiston and Pryer's *Catalogue of the Birds of Japan*. *Ibis* 21(1): 18–43.

Seebohm, H. (1879b). Remarks on the genus *Sylvia* and on the synonomy of the species. *Ibis* 21(3): 308–17.

Seebohm, H. (1879c). Remarks on certain points in ornithological literature. *Ibis* 21(4): 428–37.

Seebohm, H. (1880a). Corrections of synonymy in the family Sylviidae. *Ibis* 22(3): 273–9.

Seebohm, H. (1880b). *Siberia in Europe: A Visit to the Valley of the Petchora, in North-East Russia*. John Murray, London.

Seebohm, H. (1881). *Catalogue of the Birds in the British Museum, Vol. 5: Catalogue of the Passerines or Perching Birds in the Collection of the British Museum. Cinchlomorphae Part II*. Trustees of the British Museum, London.

Seebohm, H. (1882a). *Siberia in Asia: A Visit to the Valley of the Yenesay*. John Murray, London.

Seebohm, H. (1882b). Further notes on the ornithology of Siberia. *Ibis* 24(3): 419–28.

Seebohm, H. (1882c). On the interbreeding of birds. *Ibis* 24(4): 546–50.

Seebohm, H. (1883a). Letter on the Stricklandian Code of nomenclature. *Ibis* 25(1): 121–4.

Seebohm, H. (1883b). *A History of British Birds, with Coloured Illustrations of Their Eggs, Vol. 1*. R. H. Porter, and Dulau and Co., London.

Seebohm, H. (1884a). On the East Asiatic Shore-Lark (*Otocorys longirostris*). *Ibis* 26(1): 184.

Seebohm, H. (1884b). On a new species of British wren. *Zoologist* (third series) 8(92): 333–5.

Seebohm, H. (1884c). *A History of British Birds, with Coloured Illustrations of Their Eggs, Vol. 2*. R. H. Porter, and Dulau and Co., London.

Seebohm, H. (1885). *A History of British Birds, with Coloured Illustrations of Their Eggs, Vol. 3*. R. H. Porter, and Dulau and Co., London.

Seebohm, H. (1886). On the Black-throated Wheatear, *Saxicola stapazina*, and its allies. *Zoologist* (third series) 10(113): 193–5.

Seebohm, H. (1893). *Geographical Distribution of British Birds*. R. H. Porter, London.

Seebohm, H. (1901). *Birds of Siberia*. John Murray, London.

Self, A. (2014). *The Birds of London*. Bloomsbury, London.

Selous, E. (1899). An observational diary of the habits of nightjars (*Caprimulgus europaeus*) mostly of a sitting pair. *Zoologist* (fourth series) 3(699): 388–402, 486–505.

Severtsof, N. A. and R. Michell (1870). A journey to the western portion of the Celestial Range (Thian-Shan), or 'Tsun-Lin' of the Ancient Chinese, from the western limits of the Trans-Ili region to Tashkend. By N. Severtsof. Translated from the *Journal of the Russian Imperial Geographical Society*, 1867, by Robert Michell. *Journal of the Royal Geographical Society of London* 40: 343–419.

Severtzov, N. A. (1873). [Vertical and horizontal distribution of Turkestan animals.] *Izdanie obshchestva. Izvestiya imperatorskago osshch. lyibitdei estestvoznaniya, Antropologii i Etnografii, Moskva* 8(2): 1–157 (in Russian).

Severtzova, L. B. (1946). [*Alexei Nikolaevich Severtzov: Biographical Essays.*] Akademii Nauk Sojus SSR, Moscow (in Russian).

Severtzow, N. (1883). On the birds of the Pamir range. *Ibis* 25(1): 48–83.

Sharpe, R. B. (1868–71). *A Monograph of the Alcedinidae: or, Family of Kingfishers.* Fifteen parts. Privately published, London.

Sharpe, R. B. (1871). On the American Eider Duck. *Annals and Magazine of Natural History* (fourth series) 8: 51–3.

Sharpe, R. B. (1874). *Catalogue of the Birds in the British Museum, Vol. 1: Catalogue of the Accipitres in the Collection of the British Museum.* Trustees of the British Museum, London.

Sharpe, R. B. (1881). *Catalogue of the Birds in the British Museum, Vol. 6: Catalogue of the Passeriformes, or Perching Birds, in the Collection of the British Museum. Cinchlomorphae: Part III.* Trustees of the British Museum, London.

Sharpe, R. B. (1885). The Hume collection of Indian birds. *Ibis* 27(4): 456–63.

Sharpe, R. B. (1886). The A.O.U. code and check-list of American birds. *Nature* 34 (869): 168–70.

Sharpe, R. B. (1891). *Aves.* Part 5 of *Scientific Results of the Second Yarkand Expedition.* Government of India, Calcutta.

Sharpe, R. B. (1892). *Catalogue of the Birds in the British Museum, Vol. 17: Catalogue of the Birds in the British Museum [Coraciae, part II].* Trustees of the British Museum, London.

Sharpe, R. B. (1898). *Wonders of the Bird World.* F. A. Stokes, New York.

Sharpe, R. B. (1906). Birds. In A. Günther (ed.), *The History of the Collections Contained in the Natural History Departments of the British Museum, Vol. 2,* pp. 79–515. Trustees of the British Museum, London.

Sharpe, R. B. and H. E. Dresser (1870a). On some new or little-known points in the economy of the Common Swallow (*Hirundo rustica*). *Proceedings of the Zoological Society of London* 1870(2): 244–9.

Sharpe, R. B. and H. E. Dresser (1870b). *Aves. Zoological Record* 7.

Sharpe, R. B. and H. E. Dresser (1871a). Notes on *Lanius excubitor* and its allies. *Proceedings of the Zoological Society of London* 1870(3): 590–600.

Sharpe, R. B. and H. E. Dresser (1871b). On a new species of Long-tailed Titmouse from Southern Europe. *Proceedings of the Zoological Society of London* 1871(2): 312–13.

Sharpe, R. B. and H. E. Dresser (1871c). On two undescribed species of European birds (Picidae and Paridae). *Annals and Magazine of Natural History* (fourth series) 8: 436–7.

Sharpe, R. B., E. J. O. Hartert and J. L. Bonhote (eds) (1907). *Proceedings of the Fourth International Ornithological Congress, London June 1905.* Dulau and Co., London.

Shelley, G. E. (1872). *A Handbook to the Birds of Egypt.* Van Voorst, London.

Shelley, G. E. (1879). On a collection of birds from the Comoro Islands. *Proceedings of the Zoological Society of London* 1879(3): 673–9.

Shelley, G. E. (1882). List of birds sent home by Mr. Joseph Thomson from the River Rovuma, East Africa. *Proceedings of the Zoological Society of London* 1882(2): 302–4.

Simpson, W. H. (1859). Narrative of the discovery of some nests of the Black Woodpecker (*Dryocopus martius*) in Sweden. *Ibis* 1(3): 264–73.

Sitwell, S., H. Buchanan and J. Fisher (1990). *Fine Bird Books, 1700–1900.* Second edition. Witherby, London.

Skinner, W. R. and W. E. Skinner (1891). *Mining Manual Containing Full Particulars of Mining Companies, Vol. 4.* Financial Times, London.

Skipwith, P. (1979). *The Great Bird Illustrators and Their Art, 1730–1930.* A. and W. Publishing, New York.

Slotten, R. A. (2004). *The Heretic in Darwin's Court: The Life of Alfred Russel Wallace.* Columbia University Press, Chichester.

Smiles, S. (1859). *Self-Help: With Illustrations of Character and Conduct.* John Murray, London.

Smith, G. (ed A. Haultain) (1910). *Reminiscences.* Macmillan, New York.

Smith, R. J., J. K. Fairbank and K. F. Bruner (1991). *Robert Hart and China's Early Modernization: His Journal, 1863–66.* Harvard East Asian Monographs No. 15. Harvard University Press, Cambridge.

Snow, D. W. (ed.) (1992). *Birds, Discovery and Conservation: 100 Years of the British Ornithologists' Club.* Helm Information, Mountfield.

Speake, J. (ed.) (2003). *Literature of Travel and Exploration: An Encyclopedia.* Three vols. Fitzroy Dearborn (Taylor and Francis Group), London.

Stager, J. (1967). Fort Anderson: the first post for trade in the western Arctic. *Geographical Bulletin* 9(1): 45–56.

Steinheimer, F. (2005). The whereabouts of pre-nineteenth century bird specimens. *Zoologische Mededelingen, Leiden* 79: 45–67.

Steinitz, R. (2003). Diaries. In J. Speake (ed.), *Literature of Travel and Exploration: An Encyclopedia, Vol. 1*, pp. 331–4. Three vols. Fitzroy Dearborn (Taylor and Francis Group), London.

Stejneger, L. (1887). The British Marsh Tit. *Proceedings of the United States National Museum* 9: 200–1. Reprinted (1887) in the *Zoologist* (third series) 11(130): 379–81.

Stejneger, L. (1891). Seebohm's *Birds of the Japanese Empire* (review). *Auk* 8(1): 99–101.

Stiles, C. W. (1902). Kommission für Nomenklatur [Report of the International Commission on Nomenclature]. In P. Matschie (ed.), *Verhandlungen des V. Internationalen Zoologen-Congresses zu Berlin 12–16 August 1901*, pp. 882–90. Gustav Fischer, Jena (in German).

Stone, I. R. (1994). Joseph Wiggins. *Arctic* 47(4): 405–10.

Stone, W. (1907). Adolphus L. Heermann, M.D. *Cassinia* 11: 1–6.

Stresemann, E. (1975). *Ornithology from Aristotle to the Present.* Harvard University Press, Cambridge. (First published as *Die Entwicklung Der Ornithologie von Aristoteles bis zur Gegenwart*, F. W. Peters, Berlin, 1951. Translated by H. J. and C. Epstein, edited by G. W. Cottrell.)

Streshinsky, S. (1998). *Audubon: Life and Art in the American Wilderness.* University of Georgia, Athens.

Sultana, J. and J. J. Borg (2015). *History of Ornithology in Malta.* Birdlife Malta and Actrading, Marsa.

Sushkin, P. (1904). New birds from Palaearctic Asia. *Bulletin of the British Ornithologists' Club* 14(103): 42–6.

Swarth, H. S. (1926). James Hepburn, a little known Californian ornithologist. *Condor* 28(6): 249–54.

Swinhoe, R. (1861). *Account of the North China Campaign of 1860; Containing Personal Experiences of Chinese Character, and of the Moral and Social Condition of the Country; Together With a Description of the Interior of Pekin.* Smith, Elder and Co., London.

Swinhoe, R. (1862). Ornithological ramble in Foochow, in December 1861. *Ibis* 4(3): 253–65.

Swinhoe, R. (1863). The ornithology of Formosa, or Taiwan. *Ibis* 5(2): 198–219; 5(3): 250–311; 5(4): 377–435.

Swinhoe, R. (1873a). Notes on Chinese ornithology. *Ibis* 15(4): 361–72.

Swinhoe, R. (1873b). Birds found in Shanghai. *Ibis* 15(4): 423–7.

Sykes, J. (1926). *The Amalgamation Movement in English Banking, 1825–1924.* P. S. King and Son, London.

Taczanowski, L. (1883). Description des espèces nouvelles de la collection péruvienne de M. le Dr. Raimondi de Lima. *Proceedings of the Zoological Society of London* 1883(1): 70–2.

Taczanowski, L. (1893–93). *Fauna Ornithologique de la Siberie Orientale.* Memoires de l'Academie des Sciences de St Petersbourg (seventh series) No. 39. Eggers, St Petersburg.

Takahashi, Y. (1965). Biography of Robert Swinhoe 1836–1877. *Quarterly Journal of the Taiwan Museum* 18(3, 4): 335–40.

'TGB' (1907). Obituary notices of Fellow deceased: William Thomas Blanford. *Proceedings of the Royal Society* (series B) 79: 27–31.

Thompson, E. (1980). *The Sound of Many Waters: A History of Musquash.* Print'n'Press, St Stephen.

Thompson, J. and E. T. Jones (2004). *Civil War and Revolution on the Rio Grande Frontier: A Narrative and Photographic History.* Texas State Historical Association and the University of Texas, Austin.

Thompson, S. B. (1935). *Confederate Purchasing Operations Abroad.* University of North Carolina Press, Chapel Hill.

Thorsen, L. E., K. A. Rader and A. Dodd (eds) (2013). *Animals on Display: The Creaturely in Museums, Zoos and Natural History.* Penn State University Press, University Park.

Tigerstedt, Ö. (1940). *Kauppahuone Hackman, Erään vanhan Wiipurin kauppiassuvun vaiheet 1790–1879, Vol. 1.* Kustannusosakeyhtiö Otava (Otava Publishing Company), Helsinki.

Tigerstedt, Ö. (1952). *Kauppahuone Hackman, Erään vanhan Wiipurin kauppiassuvun vaiheet 1790–1879, Vol. 2.* Kustannusosakeyhtiö Otava (Otava Publishing Company), Helsinki.

Todd, A. (2001). *Abandoned: The Story of the Greely Arctic Expedition 1881–84.* Second edition. University of Alaska Press, Fairbanks.

Todes, D. P. (1989). *Darwin Without Malthus: The Struggle for Existence in Russian Evolutionary Thought.* Oxford University Press, Oxford.

Tree, I. (1991). *The Ruling Passion of John Gould: A Biography of the Bird Man.* Barrie and Jenkins, London.

Tree, I. (2004). *The Bird Man: The Extraordinary Story of John Gould.* Ebury Press, London.

Trevor-Battye, A. (1903). *Lord Lilford on Birds.* Hutchinson and Co., London.

Tristram, H. B. (1859–60). On the ornithology of northern Africa. *Ibis* 1(2): 153–62; 1(3): 277–301; 1(4): 415–35; 2(1): 68–83; 2(2): 149–65.

Tristram, H. B. (1860a). On the eggs of the Nutcracker and Parrot-billed Crossbill. *Ibis* 2(2): 168–70.

Tristram, H. B. (1860b). *The Great Sahara: Wanderings South of the Atlas Mountains.* John Murray, London.

Tristram, H. B. (1867). *The Natural History of the Bible: Being a Review of the Physical Geography, Geology, and Meteorology of the Holy Land: With a Description of Every Animal and Plant Mentioned in Holy Scripture.* Society for Promoting Christian Knowledge, London.

Tristram, H. B. (1882). Ornithological notes of a journey through Syria, Mesopotamia, and southern Armenia in 1881. *Ibis* 24(3): 402–19.

Tristram, H. B. (1884). *The Survey of Western Palestine: The Fauna and Flora of Palestine.* Committee of the Palestine Exploration Fund, London.

Tristram, H. B. (1894). Letter on *Coracias weigalli. Ibis* 36(2): 320.

Turner, M. (2013). Thomas Robson 1812–1883: the forgotten bird man. *Northumbrian Naturalist (Transactions of the Northumbrian Natural History Society)* 75: 42–52.

Twigg, T. (1830). *Twigg's Corrected List of the Country Bankers of England and Wales; With the Christian and Surnames of All Such as Take Out Licenses for Issuing Promissory Notes Payable on Demand.* T. Twigg, London.

Tyler, R. C. (1973). *Santiago Vidaurri and the Southern Confederacy.* Texas State Historical Association, Austin.

United States Naval War Records Office (1894–1917). *The War of the Rebellion: A Compilation of the Official Records of the Union and Confederate Navies.* Government Printing Office, Washington, DC.

Valledor de Lozoya, A. (2013). *Ostrero Canario: historia y biologia de la primera especie de la fauna española extinguida por el hombre* [Canary Islands Oystercatcher: history and biology of the first species of the Spanish fauna to be extinguished by humans]. Organismo Autónomo Parques Nacionales, Spain (in Spanish).

Vance Gillespie, Z. (2010). Vance, William. *Handbook of Texas Online*, Texas State Historical Association, www.tsha.utexas.edu/handbook/online/articles/view/VV/fva16, accessed 8 April 2017.

Vaughan, R. (1992). *In Search of Arctic Birds.* A. and C. Black, London.

Vaurie, C. (1972). *Tibet and Its Birds.* Witherby, London.

Venzke, J. F. (1990). The 1869/70 German Arctic Expedition. *Arctic* 43(1): 83–5.

Vibart, H. M. (rev. A. May) (2004). Irby, Leonard Howard Loyd (1836–1905). *Oxford Dictionary of National Biography*, Oxford University Press (online edition, 2006), www.oxforddnb.com/view/article/34109, accessed 8 April 2017.

Waddell, L. A. (1894). List of Sikhim birds and notes thereon. In H. H. Risley (ed.), *The Gazetteer of Sikhim*, pp. 198–234. Bengal Secretariat Press, Calcutta.

Waddell, L. A. (1899). *Among the Himalayas.* A. Constable and Co., Westminster.

Waddell, L. A. (1905). *Lhasa and Its Mysteries, with a Record of the Expedition of 1903–1904.* John Murray, London.

Wall, E. J. (1925). *The History of Three-Color Photography.* American Photographic Publishing Company, Boston.

Wallace, A. R. (1871). Review of Sharpe & Dresser's *A History of the Birds of Europe*, Part 1. *Nature* 3(78): 505.

Wallace, A. R. (1872). Review of Sharpe & Dresser's *A History of the Birds of Europe*, Parts 11–12. *Nature* 6(150): 390–1.

Wallace, A. R. (1875). Review of Dresser's *A History of the Birds of Europe, Including All the Species Inhabiting the Western Palæarctic Region*, Parts 35–36. *Nature* 3(78): 485.

Wallace, A. R. (1876). *The Geographical Distribution of Animals, With a Study of the Relations of Living and Extinct Faunas as Elucidating the Past Changes of the Earth's Surface.* Macmillan, London.

Wallace, A. R. (1880). *Island Life: Or, the Phenomenon and Causes of Insular Faunas and Floras, Including a Revision and Attempted Solution of the Problem of Ecological Climates.* Macmillan, London.

Wallace, A. R. (1900). *Studies Scientific and Social.* Two vols. Macmillan, London.

Waller, D. (2004). *The Pundits: British Exploration of Tibet and Central Asia.* Paperback edition. Kentucky University Press, Lexington. (First published 1990.)

Waller, P. J. (1981). *Democracy and Sectarianism: A Political and Social History of Liverpool 1868–1939.* Liverpool University Press, Liverpool.

Waller, P. J. (2004). Arthur Bower Forwood (1836–1898). *Oxford Dictionary of National*

Biography, Oxford University Press (online edition, 2006), www.oxforddnb.com/view/article/38693, accessed 8 April 2017.

Walters, M. P. (2003). *A Concise History of Ornithology*. Yale University Press, New Haven.

Walton (1906). On the birds of southern Tibet. *Ibis* 48(1): 57–84; 48(2): 225–56.

Warwick, A. R. (1972). *The Phoenix Suburb: A South London Social History*. Blue Boar Press, Richmond.

Watson, W. (1892). *The Adventures of a Blockade Runner; or, Trade in Time of War*. T. Fisher Unwin, London.

Wedderburn, W. (1913). *Allan Octavian Hume, C.B.: Father of the Indian National Congress*. T. Fisher Unwin, London.

Weintraub, J. (2015). Updating the life and death of A. L. Heermann. *Cassinia* 74–5: 63–5.

Wells, H. G. (1918). *Joan and Peter: The Story of an Education*. Macmillan, New York.

Wheelwright, H. W., *see* 'An Old Bushman'.

White, W. (1840). *History, Gazetteer, and Directory, of the East and North Ridings of Yorkshire [White's Directory]*. Robert Leader, Sheffield.

Whymper, S. L. (1906) Nesting of the Ibis-bill (*Ibidorhynchus struthersi*) and the Common Sandpiper (*Totanus hypoleucus*). *Journal of the Bombay Natural History Society* 17(2): 546–7.

Williams, C. W. (1982). *Texas' Last Frontier: Fort Stockton and the Trans-Pecos, 1861–1895*. Texas A. and M. University Press, College Station.

Wills, J. (1893). Notes on some Malagasy birds rarely seen in the interior. *Antananarivo Annual* 5(17): 119–20.

Wilson, A. (1808–14). *American Ornithology; or, the Natural History of the Birds of the United States: Illustrated with Plates Engraved and Colored from Original Drawings taken from Nature*. Nine vols. Bradford and Inskeep, Philadelphia.

Wilson, C. W. (ed.) (1881–83). *Picturesque Palestine, Sinai and Egypt*. Two vols. D. Appleton and Co., New York.

Wilson, J. B. and H. A. Wilson (1990). *The Story of Norwood*. Norwood Society, Norwood.

Wise, S. R. (1988). *Lifeline of the Confederacy: Blockade Running During the Civil War*. University of South Carolina Press, Columbia.

Witherby, H. F. (1905a). Letter – on the use of trinomials. *Ibis* 47(1): 140–1.

Witherby, H. F. (1905b). On a new subspecies of Tree-creeper from Algeria. *Bulletin of the British Ornithologists' Club* 15(112): 35–6.

Witherby, H. F. (1905c). On a new subspecies of Tawny Owl. *Bulletin of the British Ornithologists' Club* 15(112): 36–7.

Witherby, H. F. (1910). (Note on the scientific name of the Grey-headed Wagtail.) *British Birds* 9(3): 299.

Witherby, H. F. (1914). Rüppell's Warbler in Sussex. *British Birds* 8(4): 93–6.

Witherby, H. F. (1923). On the Red Grouse from Ireland and the Outer Hebrides. *British Birds* 17(5): 107.

Witherby, H. F. (1926). On some new British birds. *British Birds* 20(2): 11–12.

Witherby, H. F. (1940). Obituary – the Rev. F. C. R. Jourdain. *British Birds* 33(11): 287–93.

Wollaston, A. F. R. (1921). *Life of Alfred Newton, M.A., F.R.S.: Professor of Zoology and Comparative Anatomy in the University of Cambridge, 1866–1907*. John Murray, London.

Wolley, J. (1857). On the nest and eggs of the Waxwing (*Bombycilla garrula*, Temm.). *Proceedings of the Zoological Society of London* 25: 55–7.

Wolley, J. (1859). On the breeding of the Smew. *Ibis* 1(1): 69–76.

Wright, C. A. (1864a). List of the birds observed in the islands of Malta and Gozo. *Ibis* 6(1): 42–73, 6(2): 137–57.

Wright, C. A. (1864b). Appendix to list of birds observed in the islands of Malta and Gozo. *Ibis* 6(3, 4): 291–2.

Wright, C. A. (1865). Second appendix to a list of birds observed in the islands of Malta and Gozo. *Ibis* 7(4): 459–66.

Wright, C. A. (1869). Third appendix to a list of birds observed in the islands of Malta and Gozo. *Ibis* 11(3): 245

Wright, C. A. (1870). Fourth appendix to a list of birds observed in the islands of Malta and Gozo. *Ibis* 12(4): 488.

Wszolek, Z. K., D. P. Williams and R. A. Kyle (1990). Benedykt Dybowski – physician, explorer, scientist, political prisoner. *Mayo Clinic Proceedings* 65(10): 1381.

Wynn, G. (1981). *Timber Colony: A Historical Geography of Nineteenth Century New Brunswick.* University of Toronto Press, Toronto.

Yalden, D. W. (1999). *The History of British Mammals.* T. and A. D. Poyser, London.

Yalden, D. W. and U. Albarella (2009). *The History of British Birds.* Oxford University Press, Oxford.

Yanni, C. (2005). *Nature's Museums: Victorian Science and the Architecture of Display.* Princeton Architectural Press, New York.

Yapp, M. (2000). The legend of the great game. *Proceedings of the British Academy: 2000 Lectures and Memoirs* 111: 179–98.

Yarrell, W. (1843). *A History of British Birds.* Three vols. Van Voorst, London.

Young, G. M. (1953). *Portrait of an Age: Victorian England.* Second edition. Oxford University Press, Oxford.

Yuteng, Z. (1993). Yingguo bowuxejia Shiwenhou zai Taiwan de zilanshi diaocha jingguo ji xiangguan shiliao. *Taiwan shi yanjiu* 1: 132–51.

Zarudny, N. A. (1888). [Ornithological fauna of the Orenburg Oblast.] *Memoirs of the Academy of Sciences* Supplement 1: 1–333 (in Russian).

Zarudny, N. A. (1896). [Ornithological fauna of the Caspian region (Northern Persia, Transcaspian Province, Khiva khanate and Bukhara).] In *Materialy k poznaniyu flory i fauny Rossiĭskoĭ imperii II*, pp. 1–555. Moscow (in Russian).

Zoological Club (1933). *The Zoological Club 1866–1932.* Privately published, London.

Zuckerman, S. (ed.) (1976). *The Zoological Society of London 1826–1976 and Beyond.* Symposia of the Zoological Society of London, No. 40. Zoological Society of London and Academic Press, London.

Bibliography of Henry Dresser

Dresser, H. E. (ed.) (1896–97). Society for the Protection of Birds. Educational Series leaflets. Published by the Society for the Protection of Birds at the *Knowledge* office, London:

No. 1. Owls by M. Sharpe (1896)
No. 2. Woodpeckers by Sir H. Maxwell (1896)
No. 3. Starling by O. V. Aplin (1897)
No. 4. Swallows by T. Southwell (1897)
No. 5. Kingfisher by Sir E. Grey (1897)
No. 6. Osprey by J. A. Harvie-Brown (1897)
No. 7. Dippers by W. L. Mellersh (1897)
No. 8. Nightjar by M. Sharpe (1897)
No. 9. Titmice by S. Buxton (1897)
No. 10. Kestrel by J. Kelsall (1897)
No. 11. Plovers by J. A. Pease (1897)
No. 12. Terns, or Sea Swallows by T. Southwell (1897)
No. 13. Wagtails by W. Warde Fowler (1897)
No. 14. Chough by J. A. Harvie-Brown (1897)
No. 15. The Jay by J. Cordeaux (1897)
No. 16. Skuas by T. E. Buckley (1897)
No. 17. Flycatchers by J. R. B. Masefield (1897)
No. 18. Nightingale by J. H. Allchin (1897)
No. 19. Gulls by H. Saunders (1897)
No. 20. Leaf-Warblers by H. E. Dresser (1897)
No. 21. Pipits by W. H. Hudson (1897)
No. 22. Skylark by F. A Fulcher (1897)
No. 23. Grebes by J. R. B. Masefield (1897)
No. 24. Common Buzzard by H. A. Macpherson (1898)

Priced at 1d each, 3d for a dozen, 1s 6 d for 100; six assorted for 2d, twelve assorted for 4d; fifty for 1s.

Reports of the 'Close-Time' Committee, British Association for the Advancement of Science

British Association (1869). Applications for reports and researchers and involving grants of money. *Report of the British Association for the Advancement of Science* 1868: xlvii–xlviii.

Close-Time Committee (1870). Report on the practicability of establishing a "close-time" for British animals. *Report of the British Association for the Advancement of Science* 1869: 91–6.

— (1871). Report on the practicability of establishing a "close-time" for indigenous animals. *Report of the British Association for the Advancement of Science* 1870: 13–14.

— (1872). Report on the practicability of establishing "a close-time" for indigenous animals. *Report of the British Association for the Advancement of Science* 1871: 197.

— (1873). Report of the committee … appointed for the purpose of continuing the investigation on the desirability of establishing a "close time" for the preservation of indigenous animals. *Report of the British Association for the Advancement of Science* 1872: 320–1.

— (1874). Report of the committee … appointed for the purpose of continuing the investigation on the desirability of establishing a "close time" for the preservation of indigenous animals. *Report of the British Association for the Advancement of Science* 1873: 346–8.

— (1875). Report of the committee … appointed for the purpose of inquiring into the possibility of establishing a "close time" for the preservation of indigenous animals. *Report of the British Association for the Advancement of Science* 1874: 264–6.

— (1876). Report of the committee … appointed for the purpose of inquiring into the possibility of establishing a "close time" for the preservation of indigenous animals, and for watching Bills introduced into Parliament affecting this subject. *Report of the British Association for the Advancement of Science* 1875: 184–5.

— (1877). Report of the committee … appointed for the purpose of inquiring into the possibility of establishing a "close time" for the preservation of indigenous animals, and for watching Bills introduced into Parliament affecting this subject. *Report of the British Association for the Advancement of Science* 1876: 63–5.

— (1878). Report of the committee … appointed for the purpose of inquiring into the possibility of establishing a "close time" for the preservation of indigenous animals. *Report of the British Association for the Advancement of Science* 1877: 207–8.

— (1879). Report of the committee … appointed for the purpose of inquiring into the possibility of establishing a "close time" for indigenous animals. *Report of the British Association for the Advancement of Science* 1878: 146–9.

— (1879). Report of the committee … appointed for the purpose of inquiring into the possibility of establishing a close time the protection of indigenous animals. *Report of the British Association for the Advancement of Science* 1879: 165.

— (1880). Report of the committee … appointed for the purpose of inquiring into the possibility of establishing a close time for indigenous animals. *Report of the British Association for the Advancement of Science* 1880: 257.

BOU Lists of British Birds

Henry Dresser was a member of the Committees that produced the lists:

British Ornithologists' Union (Committee of) (1883). *A List of British Birds, Compiled by a Committee of the British Ornithologists' Union*. Van Voorst, London.

British Ornithologists' Union (Committee of) (1915). *A List of British Birds, Compiled by a Committee of the British Ornithologists' Union*. Second and revised edition. BOU, London.

Journal articles and books

*Contains descriptions of new species or subspecies of birds

Dresser, H. E. (1863). Occurrence of a white Redwing in Norfolk. *Zoologist* (first series) 21: 8484.
— (1865). Notes on the birds of Southern Texas. *Ibis* 7(3): 312–30; 7(4): 466–95.
— (1866). Notes on the birds of Southern Texas (concluded). *Ibis* 8(1): 23–46.
— (1866). Notes on the breeding of the Booted Eagle (*Aquila pennata*). *Proceedings of the Zoological Society of London* 1866(3): 377–80.
— (1867). List of birds noticed in East Finmark, with a few short remarks respecting some of them. By Ch. Sommerfeldt, parish priest of Naesseby. Translated and communicated by H. E. Dresser. *Zoologist* (second series) 2(April): 692–700; 2(June): 761–78.
— (1867). Notes on the breeding of the Booted Eagle (*Aquila pennata*) (communicated by the author and reprinted from the *Proceedings of the Zoological Society of London*). *Zoologist* (second series) 2(July): 803–7.
— (1869). Cuckow's eggs [letter]. *Nature* 1(8): 218.
— (1870). Exhibition of some eggs of the Little Gull (*Larus minutus*). *Proceedings of the Zoological Society of London* 1869(3): 530–1.
Sharpe, R. B. and H. E. Dresser (1870). On some new or little-known points in the economy of the Common Swallow (*Hirundo rustica*). *Proceedings of the Zoological Society of London* 1870(2): 244–9.
— and — (eds) (1870). *Aves. Zoological Record* 7.
— and — (1871). Notes on *Lanius excubitor* and its allies. *Proceedings of the Zoological Society of London* 1870(3): 590–600.
Dresser, H. E. (1871). Exhibition of rare European birds' eggs. *Proceedings of the Zoological Society of London* 1871(1): 102–4.
*Sharpe, R. B. and H. E. Dresser (1871). *A History of the Birds of Europe, Including All the Species Inhabiting the Western Palaearctic Region*. Parts 1–9. Privately published, London. (Note that Sharpe assisted only with the first thirteen parts of this work; his name appeared as lead author on the paper covers of parts 1–17.)
Dresser, H. E. (1871). Exhibition of a specimen of *Coccyzus americanus*. *Proceedings of the Zoological Society of London* 1871(2): 299.
*Sharpe, R. B. and H. E. Dresser (1871). On a new species of Long-tailed Titmouse from Southern Europe. *Proceedings of the Zoological Society of London* 1871(2): 312–13.
*— and — (1871). On two undescribed species of European birds (Picidae and Paridae). *Annals and Magazine of Natural History* (fourth series) 8: 436–7.
Dresser, H. E (1872). Exhibition of, and remarks upon, some skins and eggs of various species of *Reguloides* and *Phylloscopus*. *Proceedings of the Zoological Society of London* 1872(1): 25–6.
— (1872). Exhibition of some eggs of the Marbled Duck (*Querquedula marmorata*). *Proceedings of the Zoological Society of London* 1872(2): 605.
Sharpe, R. B. and H. E. Dresser (1872). *A History of the Birds of Europe, Including All the Species Inhabiting the Western Palaearctic Region*. Parts 10–13. Privately published, London. (Note that Sharpe assisted only with the first thirteen parts of this work, although his name appeared as lead author on the paper covers of parts 1–17.)
Dresser , H. E. (1872). *A History of the Birds of Europe, Including All the Species Inhabiting the Western Palaearctic Region*. Parts 14–15. Privately published, London. (Note

that Sharpe assisted only with the first thirteen parts of this work, although his name appeared as lead author on the paper covers of parts 1–17.)

Dresser, H. E. (1873). Exhibition of, and remarks upon, the skins of various eagles (*Aquila*). *Proceedings of the Zoological Society of London* 1872(3): 863–5.

★— (1873). *A History of the Birds of Europe, Including All the Species Inhabiting the Western Palaearctic Region.* Parts 16–24. Privately published, London. (Note that Sharpe assisted only with the first thirteen parts of this work; his name appeared as lead author on the paper covers of parts 1–17.)

— (1873). Exhibition of birds from the Ural. *Proceedings of the Zoological Society of London* 1873(2): 473.

— (1873). On certain species of *Aquila. Proceedings of the Zoological Society of London* 1873(2): 514–17.

— (1874). Bird-catchers [letter to the Editor]. *The Times*, 12 March, p. 10, col. e.

★— (1874). *A History of the Birds of Europe, Including All the Species Inhabiting the Western Palaearctic Region.* Parts 25–34. Privately published, London.

— (1874). Notes on the Small Spotted Eagle of Northern Germany, *Aquila maculata* (Gm.). *Annals and Magazine of Natural History* (fourth series) 13: 373–5.

★— (1874). On a new species of Marsh-Warbler. *Ibis* 16(4): 420–2.

— and W. T. Blanford (1874). Notes on the specimens in the Berlin Museum collected by Hemprich and Ehrenberg. *Ibis* 16(4): 335–43.

★Blanford, W. T. and H. E. Dresser (1874). Monograph of the genus *Saxicola*, Bechstein. *Proceedings of the Zoological Society of London* 1874(2): 213–41.

★Dresser, H. E. (1875). *A History of the Birds of Europe, Including All the Species Inhabiting the Western Palaearctic Region.* Parts 35–46. Privately published, London.

— (1875). On the nest and eggs of *Hypolais rama* (Sykes). *Proceedings of the Zoological Society of London* 1874(4): 655–6.

— (1875). Notes on Severtzoff's 'Fauna of Turkestan' (Turkestanskie Jevotnie). *Ibis* 17(1): 96–112; 17(2): 236–50; 17(3): 332–2.

— (1875). Notes on the nest and egg of *Hypolais caligata* and on the egg of *Charadrius asiaticus,* Pall., together with remarks on the latter species and *Charadrius veredus,* Gould. *Proceedings of the Zoological Society of London* 1875(1): 97–8.

— (1875). Notes on *Falco labradorus,* Aud., *Falco sacer,* Forster, and *Falco spadicus,* Forster. *Proceedings of the Zoological Society of London* 1875(2): 114–17.

— (1875). Letter: On *Carduelis caniceps. Ibis* 17(3): 387.

— (1875). Letter: On *Syvia rama* and an occurrence of the Eastern Golden Plover in Britain. *Ibis* 17(4): 513–14.

— (1875). Letter: On the journal 'Sylvan'. *Ibis* 17(4): 515.

★— (1876). *A History of the Birds of Europe, Including All the Species Inhabiting the Western Palaearctic Region.* Parts 47–56. Privately published, London.

— (1876). Notes on Severtzoff's *Fauna of Turkestan* (Turkestanskie Jevotnie). *Ibis* 18(1): 77–94; 18(2): 171–91; 18(3): 319–30; 18(4): 410–22.

— (1876). *Falco labradorus* (Labrador Falcon). *Ornithological Misellany* 1: 185–91.

— (1876). Remarks on a hybrid between the Black Grouse and the Hazel Grouse. *Proceedings of the Zoological Society of London* 1876(2): 345–7.

★— (1876). On a new species of Broad-billed Sandpiper. *Proceedings of the Zoological Society of London* 1876(3): 674–5.

★— (1876). On a new species of *Tetraogallus. Proceedings of the Zoological Society of London* 1876(3): 675–7.

— (1876). *Reprint of Eversmann's Addenda ad celeberrimi Pallasii Zoographiam Rosso-asiaticam.* Privately published, London.

— (1877). *A History of the Birds of Europe, Including All the Species Inhabiting the Western Palaearctic Region.* Parts 57–64. Privately published, London.

— (1878). *A History of the Birds of Europe, Including All the Species Inhabiting the Western Palaearctic Region.* Parts 65–72. Privately published, London.

— (1879). *A History of the Birds of Europe, Including All the Species Inhabiting the Western Palaearctic Region.* Parts 73–6. Privately published, London.

— (1879). Translation from the Swedish of W. Meves, 1854, Oef. Ak. Förh. No. 8, On the change of colour in birds, through and irrespective of moulting. *Zoologist* (third series) 3(27): 81–9.

— (1880). *A History of the Birds of Europe, Including All the Species Inhabiting the Western Palaearctic Region.* Parts 77–82. Privately published, London.

— (1881). *A List of European Birds Including All Species Found in the Western Palaearctic Region.* Privately published, London.

— (1881). Exhibition of examples of *Saxicola deserti* and *Picus pubescens* – the former shot in Clackmannanshire, the latter in France. *Proceedings of the Zoological Society of London* 1881(2): 453.

— ('1881'= 1882). *A History of the Birds of Europe, Including All the Species Inhabiting the Western Palaearctic Region.* Part 83–4 (final part). Privately published, London.

— (1883). Exhibition of, and remarks upon, the identity of *Melittophagus boehmi* and *Merops dresseri. Proceedings of the Zoological Society of London* 1882(4): 634.

— (1883). Exhibition of, and remarks upon, a specimen of *Merops philippinus*, stated to have been obtained near the Snook, Seaton Carew. *Proceedings of the Zoological Society of London* 1883(1): 1.

— (1884). Letter: On *Otocorys brandti* and *Otocorys longirostris. Ibis* 26(1): 116–18.

— (1884). Exhibition of, and remarks upon, some Ringed Pheasants from Corea. *Proceedings of the Zoological Society of London* 1883(4): 466.

— (1884). *A Monograph of the Meropidae, or Family of the Bee-eaters.* Parts 1–3. Privately published, London.

— (1885). Exhibition of, and remarks upon, specimens of *Sylvia nisoria* and *Hypolais icterina*, killed in Norfolk. *Proceedings of the Zoological Society of London* 1884(4): 477–8.

— (1885). The species of British-killed Spotted Eagles determined. *Zoologist* (third series) 9(102): 230–1.

— (1885). Bartram's Sandpiper, Little Bustard, and Hoopoe in Cornwall. *Zoologist* (third series) 9(102): 232.

— (1885). Letter: On *Sylvia nisoria* and *S. melanocephala. Ibis* 27(4): 453–4.

— (1885). *A Monograph of the Meropidae, or Family of the Bee-eaters.* Part 4. Privately published, London.

— (1886). On the Wren of St Kilda. *Ibis* 28(1): 43–5.

— (1886). Exhibition of, and remarks upon, specimens of the American Killdeer Plover (*Aegialitis vocifera*) and the Desert Wheater (*Saxicola deserti*) killed in Great Britain. *Proceedings of the Zoological Society of London* 1885(4): 835–6.

— (1886). *A Monograph of the Meropidae, or Family of the Bee-eaters.* Part 5. Privately published, London.

⋆— (1888). Exhibition of, and remarks upon, some specimens of a Titmouse obtained by Dr. Guillemard in Cyprus. *Proceedings of the Zoological Society of London* 1887(4): 563.

— (1888). Letter: On *Syrrhaptes paradoxus* in Jersey. *Ibis* 30(3): 376.

⋆— (1888). Exhibition, and remarks upon, an example of a new species of Shrike (*Lanius raddei*) from the Transcaspian District. *Proceedings of the Zoological Society of London* 1888(3): 291.

— (1889). Exhibition of, and remarks upon, some eggs of the Adriatic Black-headed Gull and the Slender-billed Gull obtained in Andalusia. *Proceedings of the Zoological Society of London* 1889(3): 316.

— (1889). Notes on birds collected by Dr. G. Radde in the Transcaspian Region. *Ibis* 31(1): 85–92.

— (1890). Notes on birds collected by Dr. G. Radde in the Transcaspian Region. *Ibis* 32(3): 342–4.

— (1890). Three weeks on the Guadalquivir. *The Naturalist* 174: 17–38.

★— (1890). Notes on the Racquet-tailed Rollers. *Annals and Magazine of Natural History* (sixth series) 6: 350–1.

— (1890). Letter: On different forms of African Rollers. *Ibis* 32(3): 384–6.

— (1891). Notes on *Eurystomus orientalis. Ibis* 33(1): 99–102.

— (1891). On a collection of Birds from Erzeroom. *Ibis* 33(3): 364–70.

— (1891). Notes on some of the rarer Western Palaearctic birds. *Ibis* 33(3): 360–4.

— (1892). Friedrich Wilhelm Meves. *Ibis* 34(1): 191–2.

— (1892). Remarks on *Lanius lahtora* and its allies. *Ibis* 34(2): 288–93.

— (1892). Remarks on *Lanius excubitor* and its allies. *Ibis* 34(3): 374–80.

★— (1892). Remarks on the Palaearctic White-breasted Dippers. *Ibis* 34(3): 380–7.

— (1893). Letter: On omission in the synonymy of the Meropidae in the 'Catalogue of birds in the British Museum'. *Ibis* 35(1): 151–2.

— (1893). On *Acredula caudata* and its allied forms. *Ibis* 35(2): 240–3.

— (1893). Notes on the synonymy of some Palaearctic birds. *Ibis* 35(3): 375–80.

★Salvadori, T. and H. E Dresser (1893). Exhibition and description of a new species of *Acredula* from Macedonia. *Bulletin of the British Ornithologists' Club* 1(4): 15. (Abstract dated 31 December 1892, but presumably not widely available until early 1893. Note, the original description was published under Dresser's name alone in error by the publishers: see *Bulletin of the British Ornithologists' Club* 1(5): 23(n).)

★Dresser, H. E. (1893). Exhibition and description of *Cryptolopha xanthopygia* obtained from the Island of Palawan by Mr J. Whitehead. *Bulletin of the British Ornithologists' Club* 1(6): 31.

— (1893). *A Monograph of the Coraciidae, or Family of the Rollers.* Privately published, Farnborough.

— (1894). Remarks on some specimens of Central-Asiatic Shrikes. *Ibis* 36(3): 384–5.

— (1895). Notes on several rare Palaearctic Birds. *Proceedings of the Zoological Society of London* 1895(2): 311–12.

— (1895). *Supplement to A History of the Birds of Europe, Including All the Species Inhabiting the Western Palaearctic Region.* Parts 1–6. Privately published, London.

— (1896). *Supplement to A History of the Birds of Europe, Including All the Species Inhabiting the Western Palaearctic Region.* Parts 7–9. Privately published, London.

— (1897). Pallas's Willow Warbler shot at Cley-next-the-sea, Norfolk. *Proceedings of the Zoological Society of London* 1896(4): 856.

— (1897). Recent additions to the British Avifauna. *Zoologist* (fourth series) 1(667): 5–7.

— (1897). Notes on Pallas's Willow Warbler and some other rare European Warblers. *Transactions of the Norfolk and Norwich Naturalists' Society* 6: 280–90.

— (1897). Obituary notice of Heinrich Gätke. *Zoologist* (fourth series) 1(669): 139–40.

— (1897). Osprey in Dorset. *Zoologist* (fourth series) 1(677): 508.

— (1898). Rare partridges in Leadenhall Market. *Zoologist* (fourth series) 2(683): 215.

— and E. Delmar Morgan (1899). On new species of birds obtained in Kan-su by M. Berezovsky. *Ibis* 41(2): 270–6.

Dresser, H. E. (1901). Exhibition of specimens of 'three-colour' printing. *Bulletin of the British Ornithologists' Club* 11(79): 59.

— (1901). On variation in plumage of *Emberiza citrinella* (exhibition, on behalf of

E. S. Montagu, of a red-throated variety obtained near Cambridge). *Bulletin of the British Ornithologists' Club* 11(80): 66.

— (1901). Notice of reproductions of an illustration of the Labrador Falcon by Joseph Wolf. *Zoologist* (fourth series) 5(724): 400.

— (1901). On some rare or unfigured Palaearctic birds' eggs. *Ibis* 43(3): 445–9.

— (1901). On variation in plumage of *Emberiza citrinella* (further exhibition, on behalf of E. S. Montagu, of a red-throated variety obtained near Cambridge). *Bulletin of the British Ornithologists' Club* 11(81): 69–70.

— (1901). Local variation in Grouse [letter]. *The Field* 98(2552, 23 November): 829.

— (1902). On some rare Palaearctic birds' eggs. *Ibis* 44(2): 177–80.

— (1902). Letter: On *Emberiza erythrogenys*, Brehm. *Ibis* 44(2): 352.

★— (1902). On a new pheasant from Japan. *Ibis* 44(4): 656–7.

— (1902). Exhibition of eggs of rare species of birds, collected by Mr Zarudny in Transcaspia and E. Persia. *Bulletin of the British Ornithologists' Club* 12(90): 83.

— (1902). *A Manual of Palaearctic Birds*. Part 1. Privately published, London.

— (1902). Exhibition of eggs of *Falco altaicus* and *Parus cypriotes*. *Bulletin of the British Ornithologists' Club* 13(91): 18.

— and H. Trueman Wood (1902). The reproduction of colours by photography. *Nature* 67(1728): 127–9.

Dresser, H. E. (1903). Exhibition of eggs of rare Palaearctic birds [collected in Transcaspia by Zarudny and sent by him]. *Bulletin of the British Ornithologists' Club* 13(95): 50.

— (1903). *A Manual of Palaearctic Birds*. Part 2. Privately published, London.

— (1903). On some rare or unfigured Palaearctic birds' eggs. *Ibis* 45(1): 88–9.

— (1903). Notes on the synonymy of some Palaearctic birds. *Ibis* 45(1): 89–91.

— (1903). Birdsnesting in lower Hungary, Bosnia, &c. *The Field* 101(2615, 7 February): 222.

— (1903). On some rare and unfigured Palaearctic birds' eggs. *Ibis* 45(3): 404–7.

— (1904). Breeding of the Knot on the Taimyr Peninsula. *Bulletin of the British Ornithologists' Club* 14(102): 32. (Meeting date 16 December 1903, abstract dated 29 December, but probably not widely available until early 1904.)

— (1904). On some rare or unfigured Palaearctic birds' eggs. *Ibis* 46(1): 106–12.

— (1904). On some rare and unfigured Palaearctic birds' eggs. *Ibis* 46(2): 280–3.

— (1904). On some rare or unfigured Palaearctic birds' eggs. *Ibis* 46(4): 485–9.

— (1904). Exhibition of nest and eggs of Rose-finches. *Bulletin of the British Ornithologists' Club* 15(110): 26.

— (1904). On the late Dr. Walter's Ornithological researches in the Taimyr Peninsular. *Ibis* 46(2): 228–35.

— (1905). *Eggs of the Birds of Europe, Including All the Species Inhabiting the Western Palaearctic Area*. Parts 1–2. Privately published, London.

— (1905). Exhibition of some birds' eggs from Persia. *Bulletin of the British Ornithologists' Club* 15(112): 38.

— (1905). Exhibition of types of three new species from Tibet. *Bulletin of the British Ornithologists' Club* 15(112): 38.

— (1905). Exhibition of the parent bird, nest and eggs of *Cossypha gutturalis*. *Bulletin of the British Ornithologists' Club* 15(116): 76.

★— (1905). Descriptions of three new species of birds obtained during the recent expedition to Lhassa. *Proceedings of the Zoological Society of London* 1905(1): 54–5.

— (1905). An oological journey to Russia. *Ibis* 47(2): 149–58.

— (1905). On some rare or unfigured Palaearctic birds' eggs. *Ibis* 47(4): 525–7.

— (1905). Exhibition of some Tibetan eggs. *Bulletin of the British Ornithologists' Club* 16(120): 38.

— (1905). On Mr. Buturlin's discovery of the breeding-place of Ross's Rosy Gull in NE Siberia. *Bulletin of the British Ornithologists' Club* 16(120): 41.

— (1906). Exhibition of eggs of Ross's Rosy Gull. *Bulletin of the British Ornithologists' Club* 16(125): 97.

— (1906). *Eggs of the Birds of Europe, Including All the Species Inhabiting the Western Palaearctic Area.* Parts 3–6. Privately published, London.

— (1906). On some Palaearctic birds' eggs from Tibet. *Ibis* 48(2): 337–47.

— (1906). Note on the egg of Ross's Rosy Gull. *Ibis* 48(3): 610–11.

— (1906). Letters on the taking of the eggs of the Great Skua in Iceland. *Ibis* 48(3): 611–12; 48(4): 737–8.

— (1906). Obituary of Canon Henry Baker Tristram, D.D., F.R.S., etc. etc. *Zoologist* (fourth series) 10(778): 155–6.

— (1907). On behalf of Mr. S. A. Buturlin exhibited and made remarks on examples of nine species of Siberian birds. *Bulletin of the British Ornithologists' Club* 19(130): 43.

— (1907). Exhibition of young in down of *Rhodostethia rosea, Tringa maculata,* and *Limosa novae zealandiae* from N. E. Siberia. *Bulletin of the British Ornithologists' Club* 19 (135): 109.

— (1907). On some rare Palaearctic birds' eggs. *Ibis* 49(2): 322–4.

— (1907). On the breeding-habits of the Rosy Gull and the Pectoral Sandpiper by S. A. Buturlin (communicated by H. E. Dresser). *Ibis* 49(4): 570–3.

— (1907). *Eggs of the Birds of Europe, Including All the Species Inhabiting the Western Palaearctic Area.* Parts 7–10. Privately published, London.

— (1907). Obituary of Professor A. Newton. *Zoologist* (fourth series) 11(793): 272–3.

— (1908). Exhibition of, and remarks on, various species of birds. *Bulletin of the British Ornithologists' Club* 21(140): 52–3.

— (1908). Exhibition of eggs of *Hypolais icterina, H. caligata, Motacilla ocularis, Carpodacus erythrinus, Emberiza leucocephala,* and *Pratincola maura. Bulletin of the British Ornithologists' Club* 21(143): 98–9.

— (1908). *Eggs of the Birds of Europe, Including All the Species Inhabiting the Western Palaearctic Area.* Parts 11–16. Privately published, London.

— (1908). Further notes on rare Palaearctic birds' eggs. *Ibis* 50(3): 486–90.

— (1908). On the Russian Arctic Expedition of 1900–1903. *Ibis* 50(3): 510–17; 50(4): 593–9.

— (1908). Baron Toll's Russian Arctic Expedition. *The Field* 112(2911, 10 October): 630–1.

— (1908). Exhibition of some rare eggs, viz. *Lampronetta fischeri,* Brandt, and *Phylloscopus viridianus. Bulletin of the British Ornithologists' Club* 23(146): 39.

— (1908). An old record of the Little Bunting in Essex. *British Birds* 1(12): 385.

— (1909). *Eggs of the Birds of Europe, Including All the Species Inhabiting the Western Palaearctic Area.* Parts 17–20. Privately published, London.

— (1909). Exhibition of two examples of a rare wader (*Pseudoscolopax taczanowskii,* Verr.) from Western Siberia [together with egg]. *Bulletin of the British Ornithologists' Club* 23(149): 60–1.

— (1909). On the occurrence of *Pseudoscolopax taczanowskii* in Western Siberia. *Ibis* 51(3): 418–21.

— (1909). On the males of the White-spotted Bluethroat. *Ibis* 51(3): 561.

— (1910). Exhibition of eggs of the Slender-billed Curlew (*Numenius tenuirostris*), the Siberian form of the Common Curlew (*N. lineatus*), and Swinhoe's Snipe (*Gallinago megala*). *Bulletin of the British Ornithologists' Club* 25(156): 38–9.

— (1910). *Eggs of the Birds of Europe, Including All the Species Inhabiting the Western Palaearctic Area.* Parts 21–4. Privately published, London.

— (1910). Proceedings of the Fifth International Congress of Ornithologists. *Ibis* 52(4): 710–13.

Notices about Dresser's collection

Anon. (1899). Notice of the acquisition of Mr H. E. Dresser's collection of birds by the Manchester Museum. *Ibis* 41(4): 663–4.

Anon. (1899). Sale of H. E. Dresser's ornithological collection. *Zoologist* (fourth series) 3(698): 384.

Anon. (1899) Notice of the acquisition of H. E. Dresser's collection by Manchester Museum. *Science Gossip* (new series) 6(64): 119.

Anon. (1912). The Dresser collection of birds' eggs. *Ibis* 54(1): 215–16.

Anon. (1931). The oological collection at Cambridge. *Ibis* 73(2): 355–7.

Brewer, T. M. (1877). A run through the museums of Europe. *Popular Science Monthly* 11(25): 472–81.

Biographical notices

Anon. (1905). Portraits of European ornithologists. *Condor* 7(3): 67.

Evans, A. H. (ed.) (1909). Henry Eeles Dresser. In Biographical notices of the original members of the British Ornithologists' Union, of the principal contributors to the first series of 'The Ibis' and of the officials, pp. 219–20 (and photograph). *Ibis* 50, Supplement 1 (Jubilee Supplement): 71–232.

Anon. (1910). International Ornithological Congress. *Bird Notes and News* 4 (2): 13.

Obituaries of Dresser

Anon. (1915). 'Mr. H. E. Dresser'. *The Times*, 6 December, p. 6, col. d.

Anon. (1915). Obituary – Mr. Henry Eeles Dresser. *Bird Notes and News* 6(8): 121–1.

Anon. (1915). Obituary – H. E. Dresser. *Nature* 96(2406): 403.

Anon. (1916). Obituary – H. E. Dresser. *Ibis* 58(2): 340–2.

Anon. (1916). Obituary notice – Henry Dresser. *Emu* 15: 267.

Anon. (1917). Obituary: Dresser Henry Eeles 1838–1915. *Aquila* 23: 597–9.

Buturlin, S. A. (1916). [Henry E. Dresser – obituary.] *Nasha Okhota* [*Our Hunting*] 1916(3): 3–4 (in Russian).

Buturlin, S. A. (1916). [Henry Eeles Dresser – obituary.] *Messager Ornithologique* [*Ornitologiskii Vyestnik*] 7: 71–4 (in Russian).

Fortune, R. (1916). In memoriam: Henry Eeles Dresser. *The Naturalist* 708, 485 of new series: 25–6.

Harting, J. E. (1915). Obituary – Henry Dresser. *Field*, 11 December, p. 978.

Rothschild, W. (1916). Henry Eeles Dresser [obituary]. *British Birds* 9(8): 194–6.

Stone, W. (1916). Notes and news – obituary of Henry Eeles Dresser. *Auk* 33(2): 232.

Index of birds

Common names follow the BOU British List (Harrop *et al.*, 2013) for species recorded in Britain and the IOC World Birds List, Master IOC list v5.4 for other species. Where these differ the IOC name is in parentheses. See appendices 1–3. Page numbers in *italic* refer to illustrations; 'n.' after a page number indicates the number of a note on that page.

General index

Page numbers in *italic* refer to illustrations.

EU authorised representative for GPSR:
Easy Access System Europe, Mustamäe tee 50,
10621 Tallinn, Estonia
gpsr.requests@easproject.com

www.ingramcontent.com/pod-product-compliance
Lightning Source LLC
Chambersburg PA
CBHW040843100426
42812CB00014B/2597